Sensor Technology and Devices

Related Titles

The Fiber-Optic Gyroscope, Hervé C. Lefèvre

Industrial Microwave Sensors, Ebbe G. Nyfors and Pertti V. Vainikainen

Introduction to Sensor Systems, Shahen A. Hovanessian

Optical Fiber Sensors, Volume I: Principles and Components, John Dakin and Brian Culshaw

Optical Fiber Sensors, Volume II: Systems and Applications, John Dakin and Brian Culshaw

Sensor Technology and Devices, Ljubisa Ristic

For further information on these and other Artech House titles, contact:

Artech House
685 Canton Street
Norwood, MA 01602
617-769-9750
Fax: 617-762-9230
Telex: 951-659
email: artech@world.std.com

Artech House
6 Buckingham Gate
London SW1E6JP England
+44 (0)71 973-8077
Fax: +44 (0) 71-630-0166
Telex: 951-659
email: artech@world.std.com

Sensor Technology and Devices

edited by

Ljubisa Ristic

Artech House
Boston • London

Library of Congress Cataloging-in-Publication Data
Sensor Technology and Devices/Ljubisa Ristic, editor
Includes bibliographical references and index.
ISBN 0-89006-532-2
1. Interface circuits. 2. Detectors—Design and construction. 3. Semiconductors. I. Ristic Ljubisa
TK7868I58S46   1994   94-2361
621'.2–dc20   CIP

A catalogue record for this book is available from the British Library

© 1994 ARTECH HOUSE, INC.
685 Canton Street
Norwood, MA 02062

International Standard Book Number: 0-89006-532-2
Library of Congress Catalog Card Number: TK7868I58S46   1994

10 9 8 7 6 5 4 3 2 1

Those haunted by imagination
might be blessed by a rare moment
of getting to the other side,
seeing unseen,
reaching unknown.
But it is you, reality,
we dare to change.

Contents

x

Preface

In the last several years I have taught a sequence of graduate-level courses in solid-state sensors at the University of Alberta. During these endowments, I struggled to find a comprehensive text book on sensor technology. My graduate students, who had to deal with a number of technical papers, wondered why there was no unified text book on this subject. The most probable reason is the novelty of the field. Interest in microsensors has grown dramatically only in the last decade because of the wide range of applications for them. At the same time, sensor technology has become one of the most rapidly changing fields today. Another reason stems from the very nature of sensor devices; solid-state sensors are so diversified that this area can truly be described as a "multidisciplinary field." Thus, it is difficult for any single person to follow up on all the activity in this field. These are the reasons that I have asked some of my colleagues, experts on sensors, to join me in writing on relevant and specific topics that could be combined into an integral book. The result is this new book on sensor technology and devices.

All of the fourteen authors are intimately familiar with the field, and they bring an enormous amount of personal experience to this book. Without them, writing this book would have been impossible, and I am truly indebted to them. I would also like to acknowledge many other people who have contributed directly or indirectly to this book. My special thanks go to previous graduate students T. Smy, M. T. Doan, M. Paranjape, M. Parameswaran, K. Chau, A. C. Dhaded, T. Q. Troung, T. R. Sych, A. Nathan, and D. R. Briglio, whose presence in the Sensor Group at the University of Alberta made life so enjoyable. Their excitement about new ideas and their healthy skepticism strongly motivated me to endure as a researcher and pedagogue. I also express my appreciation to my colleagues, Drs. W. Allegretto, H. P. Baltes, M. Brett, I. Filanovsky, D. J. Harrison, and W. B. Joerg, professors and cofounders of the Sensor and Material Group at the University of Alberta, for many inspiring discussions. I benefited greatly from their wisdom and sincerity. My special thanks go also to Dr. R. Popovic for many open discussions on magnetic field sensors as well as sensors in general. Finally, I thank R. C. Wells for suggestions on the wafer-bonding material.

I am also grateful to Motorola for creating a stimulating environment for development of sensor technology and sensor products. My current work at Motorola is one of my most enjoyable professional experiences. Motorola's constant support of me and my colleagues made this book possible.

The book is intended for practicing engineers, scientists, and advanced graduate students who seek a broader understanding on important subjects in the sensor field. Many of the presented topics are considered crucial in today's sensor technology. The topics in the book are arranged in logical order in the form of five sections. Besides the introduction to a sensor world, four other important areas are covered: modeling, sensor technology, devices and applications, and specific circuits for sensors. Each chapter is self-contained, and readers interested in a particular subject will be able to find the needed information easily.

The first chapter provides introductory information on sensors and sensing systems. The second chapter describes basic models and existing problems in the simulation of sensors and actuators. Chapters 3 through 6 cover the key aspects of modern sensor technology; bulk and surface micromachining, direct wafer bonding, and sensor packaging. Chapters 7 through 11 deal with the device theory of different sensors and their applications. Examples include magnetic field sensors, thermal sensors, photosensors, CCDs, and sensors for automotive applications. Finally, Chapters 12 and 13 describe specific circuits for sensor applications.

It is my sincere wish that the combination of the variety of topics and depth offered in this book will make it a valuable reference as well as a useful teaching text.

Lj. Ristic
Phoenix, Arizona
September 1993

Chapter 1
Sensing the Real World

Lj. Ristic and R. Roop
Motorola

Sensors are devices that provide an interface between electronic equipment and the physical world. They help electronics to "see," "hear," "smell," "taste," and "touch." In their interface with the real world, sensors typically convert nonelectrical physical or chemical quantities into electrical signals. This chapter provides a brief introduction to the world of sensors. The discussion includes a classification of sensors and a description of the basic parameters used for sensor characterization. Also, the basic aspects of sensing systems are discussed. Finally, the chapter concludes with comments on the growth of the sensor industry in the years to come.

1.1 INTRODUCTION

Microsensors have become an essential element of process control and analytical measurement systems, finding countless applications in, for example, industrial monitoring, factory automation, the automotive industry, transportation, telecommunications, computers and robotics, environmental monitoring, health care, and agriculture; in other words, in almost all spheres of our life. The main driving force behind this progress comes from the evolution in the signal processing. With the development of microprocessors and application-specific integrated circuits (IC), signal processing has become cheap, accurate, and reliable—and it increased the intelligence of electronic equipment. In the early 1980s a comparison in performance/price ratio between microprocessors and sensors showed that sensors were behind. This stimulated research in the sensor area, and soon the race was on to develop sensor technology and new devices. New products and companies have emerged from this effort, stimulating further advances of microsensors. Application of sensors brings new dimensions to products in the form of convenience, energy savings, and safety [1]. Today, we are witnessing an explosion of sensor applications. Sensors can be found in many products, such as microwave and gas ovens, refrigerators, dishwashers, dryers, carpet cleaners, air conditioners, tape recorders, TV and stereo sets, compact and videodisc players. And this is just a beginning.

1.2 SENSOR CLASSIFICATION

Sensing the real world requires dealing with physical and chemical quantities that are diverse in nature. From the measurement point of view, all physical and chemical quantities (measurands) can be divided into six signal domains [2].

1. The thermal signal domain: the most common signals are temperature, heat, and heat flow.
2. The mechanical signal domain: the most common signals are force, pressure, velocity, acceleration, and position.
3. The chemical signal domain: the signals are the internal quantities of the matter such as concentration of a certain material, composition, or reaction rate.
4. The magnetic signal domain: the most common signals are magnetic field intensity, flux density, and magnetization.
5. The radiant signal domain: the signals are quantities of the electromagnetic waves such as intensity, wavelength, polarization, and phase.
6. The electrical signal domain: the most common signals are voltage, current, and charge.

As mentioned, sensors convert nonelectrical physical or chemical quantities into electrical signals. It should be also noted that the principle of operation of a particular sensor is dependent on the type of physical quantity it is designed to sense. Therefore, it is no surprise that a general classification of sensors follows the classification of physical quantities. Accordingly, sensors are classified as thermal, mechanical, chemical, magnetic, and radiant.

There is also a classification of sensors based on whether they use an auxiliary energy source or not. Sensors that generate an electrical output signal without an auxiliary energy source are called *self-generating* or *passive*. An example of this type of sensor is a thermocouple. Sensors that generate an electrical output signal with the help of an auxiliary energy source are called *modulating* or *active*. Figure 1.1 shows symbolic presentations of self-generating and modulating sensors. Here, s_1 represents the input signal, s_2 is the output signal, and a_1 is the auxiliary energy source. In modulating sensors, the auxiliary energy serves as a main source for the output signal, and the measured physical quantity modulates it. This class of sensors includes magnetotransistors and phototransistors. Modulating sensors are the best choice for the measurement of weak signals.

In addition to the preceding classifications, there are many others based on some common features. A good example is automotive sensors, where the common

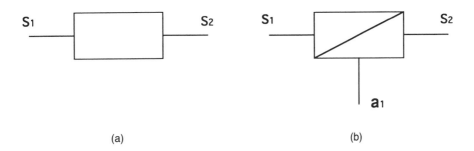

Figure 1.1 Symbolic presentation of self-generating and modulating sensor: (a) self-generating sensor;
(b) modulating sensor, where s_1 is the input signal, s_2 is the output signal, a_1 is the auxiliary
energy source.

feature is the application in automobiles for engine and vehicle control. A curious
reader can find more information about the classification of sensors in a recently
published book on silicon sensors [3].

1.3 SENSOR PARAMETERS

Performance of sensors, like other electronic devices, is described by parameters.
The following section briefly describes the most common sensor parameters. This
section is included to help in understanding the devices described in the forthcoming
chapters.

- Absolute sensitivity is the ratio of the change of the output signal to the change
  of the measurand (physical or chemical quantity).
- Relative sensitivity is the ratio of a change of the output signal to a change in
  the measurand normalized by the value of the output signal when the measurand
  is 0.
- Cross sensitivity is the change of the output signal caused by more than one
  measurand.
- Direction dependent sensitivity is a dependence of sensitivity on the angle between the measurand and the sensor.
- Resolution is the smallest detectable change in the measurand that can cause a
  change of the output signal.
- Accuracy is the ratio of the maximum error of the output signal to the full-scale
  output signal expressed in a percentage.
- Linearity error is the maximum deviation of the calibration curve of the output
  signal from the best fitted straight line that describes the output signal.
- Hysteresis is a lack of the sensor's capability to show the same output signal at
  a given value of measurand regardless of the direction of the change in the
  measurand.

- Offset is the output signal of the sensor when the measurand is 0.
- Noise is the random output signal not related to the measurand.
- Cutoff frequency is the frequency at which the output signal of the sensor drops to 70.7% of its maximum.
- Dynamic range is the span between the two values of the measurand (maximum and minimum) that can be measured by sensor.
- Operating temperature range is the range of temperature over which the output signal of the sensor remains within the specified error.

It should be pointed out that in addition to these common parameters, other parameters are often used to describe other unique properties of sensors.

1.4 A SEAMLESS SENSOR SYSTEM

Sensing systems are generally used for process control and measurement instrumentation. A simple block diagram of a sensing system is shown in Figure 1.2 [4,5]. As can be seen, the term *transducer* is used for both the input and the output blocks of the sensing system. The role of the input transducer is to get information from the real world about a physical or chemical quantity; in other words, to "sense the world." This is the reason why input transducers are commonly called *sensors*. Often the electrical signals generated by sensors are weak and have to be amplified or processed in some way. This is done by the signal processing part of the sensing system. Finally, the role of the output transducer is to convert an electrical signal into a form acceptable for our senses or to initiate some "action," for example, opening or closing a valve. For this reason, output transducers are often called *actuators*. A simple block diagram of the sensing system, as just described, helps to grasp the basic concept of sensing, but it really does not tell the whole story.

Much has been written about the phenomenal development of microelectronics and the strong influence of microprocessors and other integrated circuits on sensing systems. Figure 1.3 shows a typical sensing system composed of the many devices of modern microelectronics [6]. Following the signal path in Figure 1.3, one can see that the electrical signals created by sensors are amplified, converted to digital form, and transferred to a microprocessor. The microprocessor also controls a variety of actuators through the interface circuits, where the signals are converted back to analog form and used to drive the actuators. The entire sensing system thus can form

Figure 1.2 Simple block diagram of the sensing system.

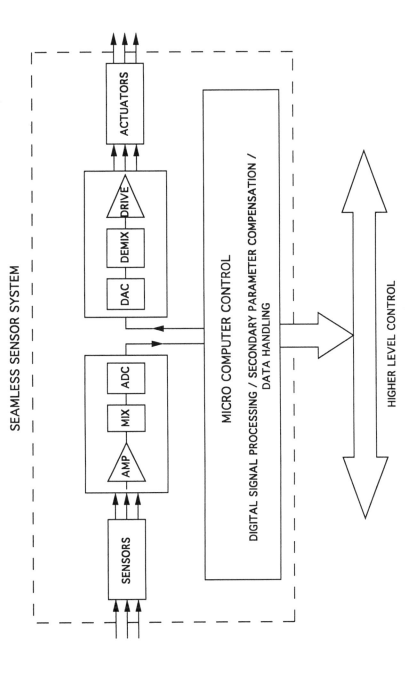

Figure 1.3 Seamless sensor system on the chip.

a closed control loop. Also, the microprocessor may communicate with a higher level control computer, making the sensing system, shown in Figure 1.3, part of a larger system. Currently, the type of sensing system shown in Figure 1.3 is spatially distributed and made of separate functional blocks. Point-to-point wiring is typically used for the electrical connection between the blocks. Many experts expect in the future that such sensing systems will be integrated into a single chip, forming a "smart" sensor or "seamless" sensor system, where boundaries between the functional blocks will not be apparent.

Is this a dream? Over the past 10 years new technologies have been developed, such as wafer bonding and bulk and surface micromachining, allowing production of many practical sensors and actuators. Many of the process steps from these technologies are compatible with standard IC processing [7–9], which makes the idea of seamless sensor systems one step closer to reality. It should be pointed out that challenging processing issues are still to be solved for a full realization of this dream, but the benefits of an integrated system, such as self-testing, digital interface, PROM-based digital compensation, and increased accuracy, as well as high reliability, are too compelling to stop short of not solving the problems. By the year 2000 the seamless sensor system will have become a reality.

1.5 SENSOR INDUSTRY GROWTH

The sensor industry has grown rapidly in recent years to reach $5 billion in 1990, Figure 1.4 [10,11]. The market share is rather evenly divided between the United States, Japan, and Europe, each having about a third of the world sensor market, Figure 1.4(b). The rest of the world has about 14% of the world sensor market. The trend is expected to continue through the year 2000. The compound annual growth by the year 2000 is projected to be 8%. Much of this growth is due to the rapid development of silicon sensor technology and its capability to deliver reliable, small, and cost-effective devices for widespread applications.

It should be emphasized that a main reason for the rapid growth of silicon sensors is legislation affecting the automotive industry and requirements to build better and safer vehicles, which means, among other things, vehicles with more electronics. Since the 1960s, the percentage of a car's cost from the electronic system has been gradually increasing as shown in Figure 1.5 [12]. By the year 2000 automotive electronics will average around 25% of the cost of a vehicle. A significant portion of automotive electronics will be sensing systems. For example, federal regulations in the United States require safety and environmental features such as air bags and monitoring of exhaust emissions, as well as improved mileage. This, in turn, requires efficient accelerometers and gas and pressure sensors. The automotive market is, no doubt, the biggest segment of the sensor market. The revenues for the automotive sensor market in 1990 were $2.3 billion, 46% of the total sensor market.

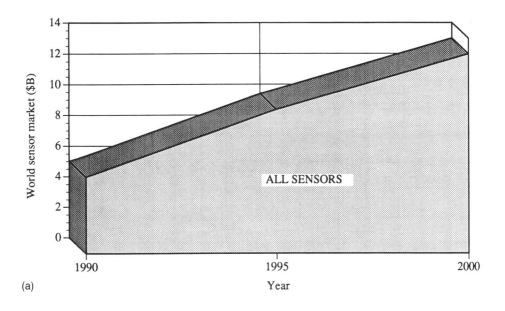

(a)

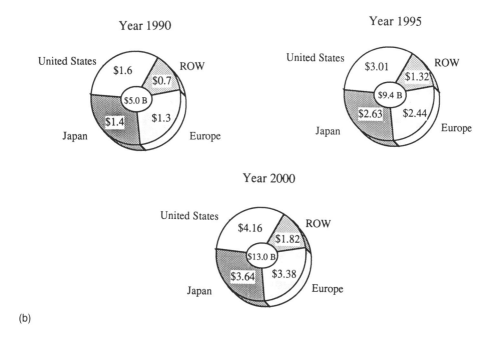

(b)

Figure 1.4 World sensor market: (a) the total sensor market is expected to grow to $13 billion by the year 2000; (b) world sensor market by regions, where ROW is the rest of the world.

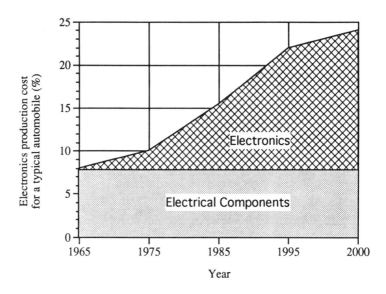

Figure 1.5 The cost of electronics in the car.

This amount is expected to grow to $6.2 billion by the year 2000, Figure 1.6 [11,13]. The market is also rather equally divided among the United States, Japan, and Europe; and it is predicted to stay that way through the year 2000, Figure 1.6(b). It should be also noted that the same type of sensors presently used will still represent the major portion of applications in the year 2000. An example is shown in Figure 1.7 for the U.S. automotive sensor market [11,13]. The five major sensors (speed, temperature, accelerometer, position, and pressure) will still dominate the market.

Finally, it should be pointed out, that the second largest application of sensors is in the industrial sector. Figure 1.8 shows an example for the U.S. market where industrial applications are about 30% of the total market [11,13]. The other important applications are in the biomedical and computer industries. This share of the market is expected to stay steady through the year 2000.

1.6 SUMMARY

The focal point of this chapter was an introduction to the world of sensors. The basic classification of sensors was presented as well as the description of common parameters used for sensor characterization. The discussion also included a simple description of sensing system and an elaboration on seamless sensor systems. Finally, the comments on the current and future sensor market are presented.

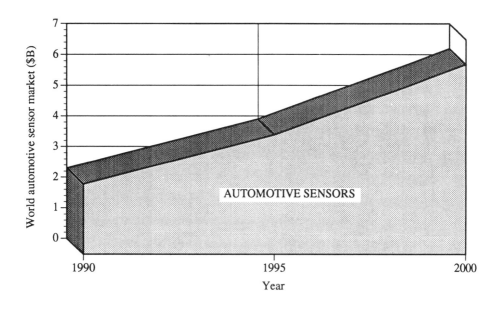

(a)

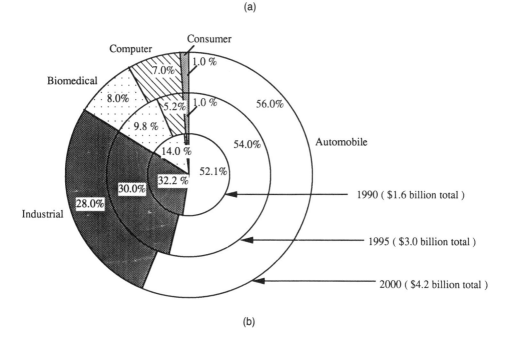

(b)

Figure 1.6 World automotive sensor market: (a) the total automotive sensor market is expected to grow to $6.2 billion by the year 2000; (b) world automotive sensor market by regions.

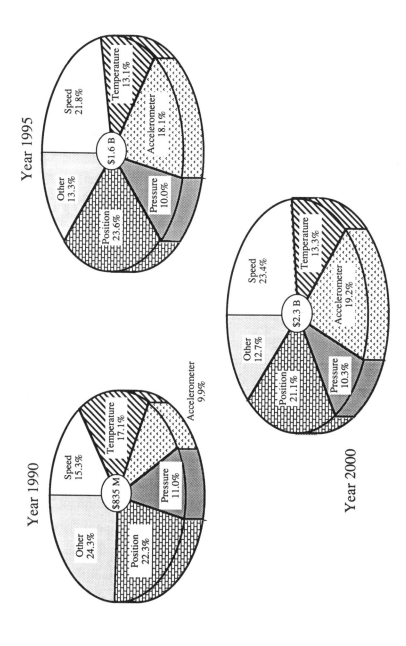

Figure 1.7 The U.S. automotive sensor market by application.

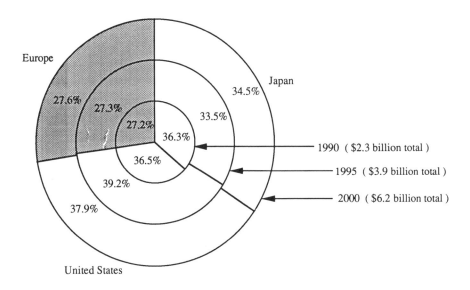

Figure 1.8 The U.S. sensor market by application.

REFERENCES

[1] T. Kobayashy, "Solid-State Sensors and Their Applications in Consumer Electronics and Home Appliances," *Tech Digest*, Transducers '85, Int. Conf. on Solid-State Sensors and Actuators, 1985, p. 8.

[2] K. S. Lion, "Transducers: Problems and Prospects," *IEEE Trans. Industr. Electron. Contr. Instrum.*, Vol. IECI-16, 1969, p. 2.

[3] S. Middelhoek and S. A. Audet, *Silicon Sensors*, Boston, Academic Press, 1989.

[4] S. Middelhoek and D. J. W. Noorlag, "The Present and Future of Silicon Microtransducers," *Proc. des J. d'Electronique et de Microtechnique*, Lausanne, 1980, p. 57.

[5] S. Middelhoek and D. J. W. Noorlag, "Silicon Micro-Transducers," *J. Phys. E. Sci. Instrum.*, Vol. 14, 1981, p. 1343.

[6] K. D. Wise and N. Najafi, "The Coming Opportunities in Microsensor Systems," *Tech. Digest*, Transducers '91, Int. Conf. on Solid-State Sensors and Actuators, 1991, p. 2.

[7] Lj. Ristic, "CMOS Technology: A Base for Micromachining," *Microelectronics J.*, Vol. 20, 1989, p. 153.

[8] M. Parameswaran, "Microelectronic and Micromechanical Sensors and Actuators in CMOS Technology—A Novel Scheme towards Regularization and Integration," Ph.D. dissertation, University of Alberta, 1990.

[9] M. W. Putty, S. Chang, R. T. Howe, A. L. Robinson, and K. D. Wise, "Process Integration for Active Polysilicon Resonant Microstructures," *Sensors and Actuators*, Vol. A20, 1989, p. 143.

[10] Battelle Institut, "Sensors: Miniaturization and Integration," unpublished multiclient study, Frankfurt, 1989.

[11] Motorola, Marketing Department, private communication, 1993.

[12] M. Baumler and L. J. Olsson, "Prometheus: Sensors for Automobile of the Future," in *The Sensor Industry in Europe*, ed. G. Tschulena, Proc. of a Battelle Europe Conf. 1989, p. 77.

[13]. "Automotive Sensor Demand Forecast," *BIS Market Report*, March 1992.

Chapter 2

Modeling and Simulation
of Microsensors and Actuators

W. Allegretto
University of Alberta

2.1 INTRODUCTION

Sensors or "input transducers" are employed to convert a physical or chemical input into an electronic signal that can be readily processed, stored, or transmitted. Microsensors are miniaturized sensors, often fabricated using standard silicon technologies and with the integration of the associated relevant circuitry on the same chip [1–5]. It is a common practice to speak of microsensors in terms of the input that will be converted; that is, magnetic, thermal, radiant, mechanical, chemical, and so forth. By design any of these inputs should perturb the electrical transport within the relevant microsensor, and the magnitude of this perturbation is ultimately employed to gauge the size of the input.

The ever increasing demand for microsensors implies a constant need for the development of new or improved devices. Traditionally, this involves several design and fabrication iteration cycles until the desired specifications are met. The main purpose of simulation is simply to reduce the number of trial and error steps in this development. By providing insight into the functioning of sensors, modeling can lead to the optimal design of present devices or to the feasibility of projected structures, [1,2,6,7].

The behavior of semiconductor devices is governed, under a series of assumptions, by a system of nonlinear partial differential equations and associated boundary and initial value conditions [8–11]. In very special situations, one or all of these equations can be solved analytically. This approach is very valuable as a first heuristic tool, but the extreme nature of the conditions needed for its applicability limits its usefulness. For the microsensor structures commonly encountered it is essential to solve numerically at least some of the equations in order to obtain a meaningful description of sensor behavior.

Section 2.2, recalls the equations that describe microsensor operation. Section 2.3 introduces various discretization procedures, and Section 2.4 presents selected illustrative results dealing with magnetic, thermal-flow and mechanical sensors. This chapter does not discuss chemical or optical sensors. The interested reader may consult [1–5,12,13] for a discussion of these sensors and further references.

The examples are based on methods and software developed over the past few years at the University of Alberta and found reliable and reasonably easy to employ. For several of the devices considered, it is also possible to use off-the-shelf software packages, either directly or with some modifications. For mechanical or thermal effects and design, for example, such codes as SENSIM [14], ANSYS [15], MSC-NASTRAN [16], OYSTER [17], and many others are available and can be used to advantage [6,7,18]. With some thought, it is even possible to employ circuit design tools such as SPICE for modeling [19]. The number and versatility of such packages will undoubtedly continue to increase. Consequently, although for magnetic sensor simulation we presently are only aware of the recently announced commercial package GENSIM [20], we conjecture that many other such packages will soon be available.

Because simulation of devices involves the numerical solution of a system of parabolic equations, it would seem reasonable to always deal with this problem by employing general purpose partial differential equation solvers, or by adapting solvers designed for other purposes, with at most some very minor modifications. The difficulty with this approach is that standard discretization procedures that work well in other cases may fail in suitably resolving the equations involved here, a fact that has been known for some time [21]. Furthermore, the dependent variables in the basic equations may differ by many orders of magnitude, not only from each other but also in different device subregions. Consequently some form of careful scaling needs to be first performed, [8,21].

Finally, we observe that we shall consider only microsensors in silicon technology, and most of the physical and process parameters (such as mobility, generation-recombination, thermal conductivity) can be adequately described by the same equations as in standard device simulations. A thorough description of these may be found in [3,8,22]. We will not present theoretical considerations on the equations involved or numerical convergence results. The interested reader may find these and related arguments discussed in [9,10,21–27] and the numerous references therein. It is interesting to note that existence and uniqueness questions still remain even in the case of the simplified system that governs thermistor behavior, [28,29] in realistic device situations.

2.2 MODELING EQUATIONS

2.2.1 The Basic Semiconductor Equations

The nonlinear system of partial differential equations that is commonly used to describe the electrical processes arising in semiconductor devices is derived in [8] and explicitly given by

$$-\text{div} \,(\varepsilon \,\text{grad}\, \psi) = q(p - n + N_D - N_A) \tag{2.1}$$

$$-q\,\frac{\partial n}{\partial t} + \text{div}(\mathbf{J}_n) = q(R - G) \tag{2.2}$$

$$q\,\frac{\partial p}{\partial t} + \text{div}(\mathbf{J}_p) = -q(R - G) \tag{2.3}$$

$$\rho c\,\frac{\partial T}{\partial t} - \text{div}[K(T)\,\text{grad}\, T] = H \tag{2.4}$$

where

ψ = electrostatic potential,
T = lattice temperature,
n, p = carrier concentrations,
$\mathbf{J}_n, \mathbf{J}_p$ = current densities,
ε = material permittivity,
q = elementary charge,
N_D, N_A = ionized donor and acceptor distributions,
$K(T)$ = thermal conductivity,
ρ = mass density,
c = specific heat,
H = heat sources in the system,
R, G = recombination and generation rates.

The carrier concentrations n, p are assumed to satisfy Maxwell-Boltzmann statistics; that is,

$$n = n_{ie} \exp[(\psi - \varphi_n)/V_T] \tag{2.5}$$

$$p = n_{ie} \exp[(\varphi_p - \psi)/V_T] \tag{2.6}$$

where V_T is the thermal voltage (kT/q), n_{ie} is the effective intrinsic concentration, and φ_n, φ_p are the Fermi potentials.

These equations, usually termed the *drift-diffusion model*, are a result of numerous approximations [8] but are felt to be applicable to microsensors in view of the physical dimensions and operating conditions usually encountered in real applications. To our knowledge, simulations based on the "hydrodynamic model" [30–36], or on Monte Carlo simulations [37,38] are being presently contemplated but they have so far not been implemented for sensors. As mentioned earlier, expressions for the coefficients R, G, H, K may also be found in [8,22] and elsewhere. These in general are functions of the unknowns n, p, T. For the sake of completeness, we briefly recall that the net generation-recombination term in the commonly used Shockley-Read-Hall model is given by

$$R - G = \frac{np - n_i^2}{\alpha n + \beta p + \gamma}$$

where α, β, γ denote terms that depend on the carrier lifetimes. The thermal conductivity K takes the form

$$K(T) = \frac{1}{a + bT + cT^2}$$

for constants a, b, c [8]. Finally the thermal generation term H in the simplest form is represented by the Gaur-Navon model as $H = (\mathbf{J}_n + \mathbf{J}_p) \cdot (\nabla \psi)$. More sophisticated models, as well as models for mobilities, concentrations and other process dependent quantities may also be found in [8].

Finally, we observe that the form of $\mathbf{J}_n$, $\mathbf{J}_p$ varies with the application. As given in [8] and elsewhere, in case of zero magnetic field, we have:

$$\mathbf{J}_n = zD_n[\text{grad}(n) - n\,\text{grad}(\psi/V_T)] + q\mu_n n\alpha_{sn}\,\text{grad}(T) \tag{2.7}$$

$$\mathbf{J}_p = -qD_p[\text{grad}(p) + p\,\text{grad}(\psi/V_T)] - q\mu_p p\alpha_{sp}\,\text{grad}(T) \tag{2.8}$$

where α_{sn}, α_{sp} denotes Seebeck coefficients, μ_n, μ_p denotes drift mobilities, and D_n, D_p denotes concentration diffusion constants. In the case of a small magnetic field $\mathbf{B}$ [1,2,11,39,40], $\mathbf{J}_m$, $\mathbf{J}_p$ are to be replaced in (2.2), (2.3) by $\mathbf{J}_{nB}$, $\mathbf{J}_{pB}$ given by

$$\mathbf{J}_{nB} + \mu_n^* \mathbf{J}_{nB} \times \mathbf{B} = \mathbf{J}_n \tag{2.9}$$

$$\mathbf{J}_{pB} - \mu_p^* \mathbf{J}_{pB} \times \mathbf{B} = \mathbf{J}_p \tag{2.10}$$

with $\mathbf{J}_n$, $\mathbf{J}_p$ as given in the zero field case by (2.7) and (2.8), and μ_n^*, μ_p^* denoting the Hall mobilities [1,11,41].

2.2.2 Boundary Conditions

It is important to observe that in many sensor applications, equations (2.1)–(2.4) need to be solved on different regions. Indeed, (2.2) and (2.3) will clearly apply only to parts of the device where currents are present, whereas (2.1) and (2.4) will hold over the entire device. It is easy to see an example where this is the case: in a temperature-flow sensor the quantity being sensed may determine a boundary condition on the outside of the device, that is, on the outside of oxide layers where no current flows. A similar observation applies to the potential equation (2.1) in the case of magnetic field sensors.

Equations (2.1)–(2.4) are solved on the various regions, as required, subject to a mixture of Dirichlet and Neumann boundary conditions. The precise nature of these conditions depends on the problem under consideration, but the potential and carrier densities at (ideal) ohmic contacts are explicitly given [8,22] by assuming thermal equilibrium, zero space charge, and taking into account the built-in potential. Specifically,

$$\left.\begin{array}{l} n = [(N^2/4) + n_{ie}^2]^{1/2} + N/2 \\ p = [(N^2/4) + n_{ie}^2]^{1/2} - N/2 \\ \psi = V_a + V_T \sinh^{-1}[N/2n_{ie}] \end{array}\right\} \qquad (2.11)$$

Similarily, at temperature sinks (if any), the temperature is assumed fixed at some given value.

It is invariably the case that the simulated part of the device involves artificial "open" boundaries, which are introduced to restrict the computation to the physically significant region. At these "open" boundaries Neumann conditions are usually imposed [8]. For the potential equation (1), the standard condition is: grad $(\psi) \cdot \mathbf{S} = 0$, where $\mathbf{S}$ denotes the outward normal. Since it is felt that this condition is not satisfied at the edges of the active regions in the case of magnetic field sensors due to the effects of the external field, this boundary condition is kept but equation (2.1) is solved over a considerably larger region [42]. For the current equations, (2.2) and (2.3), the "open" boundary condition is $\mathbf{J}_n \cdot \mathbf{S} = \mathbf{J}_p \cdot \mathbf{S} = 0$. Finally, with the temperature equation (4) we associate the condition

$$k(T)\frac{\partial T}{\partial \mathbf{S}} + h(T - T_\infty) = 0 \qquad (2.12)$$

where T_∞ denotes ambient temperature and h represents the heat transfer coefficient. Condition (2.12), with h independent of T, allows the treatment of both convection and radiation if T is near T_∞ [43,44]. In the general case, h must also be taken to be a function of T, T_∞.

2.2.3 Simplification of the Problem

So far the equations and conditions have been stated in generality. For many situations, however, it suffices to deal with only a simplified set of equations and conditions. Consider, for example, the case of a Hall device in steady state with uniform doping ($n = n_0$) and constant temperature. In this case $\mathbf{J}_{pB} = \mathbf{0}$ and $\mathbf{J}_{nB}$ takes the "drift" form: $\mathbf{J}_{nB} = A \ \text{grad}(\psi)$, with

$$A = \frac{-qn_0\mu_n}{[1 + (\mu_n^*|\mathbf{B}|)^2]} \begin{pmatrix} 1 & -|\mathbf{B}|\mu_n^* \\ |\mathbf{B}|\mu_n^* & 1 \end{pmatrix}$$

System (2.1)–(2.4) may thus be replaced by the much simpler linear equation:

$$\text{div}(A \ \text{grad} \ \psi) = 0. \tag{2.13}$$

A similar situation arises in the case of a temperature-flow sensor involving polysilicon resistors. Such material has been the subject of numerous investigations in the past 15 years [45–53]. The potential barriers at the polysilicon grain boundaries were analyzed under a variety of simplifying assumptions—all grains the same, thermionic theory, depletion layer approximation yields the formula

$$\mathbf{J}_n = \sigma \ \text{grad}(\psi) \tag{2.14}$$

with $\sigma^{-1} \sim T^\alpha \exp [E_a/kT]$ with $\alpha < 1$ and E_a is activation energy. Lu et al. [50] suggest $\alpha = 1/2$ at medium doping levels. System (2.1)–(2.4) may be now replaced by

$$\text{div}(\sigma \ \text{grad} \ \psi) = 0 \tag{2.15}$$

and equation (2.4). This system of two equations is still nonlinear, but it is much easier to handle than the original one.

2.2.4 Combination of Mechanical and Electrical Effects

So far we have dealt with only the equations that describe electrical behavior. For many sensor applications, however, the mechanical performance of the structure is also of paramount importance, and we now consider the associated equations. It is common practice to employ the equations of classic linear elasticity [54,55]. Such theory is restricted to small deflections in particular, and its practical application to the complex sandwich structures encountered in micromachined sensors [46] is not yet fully validated. Specifically, such theory cannot predict the nonlinear effects that may appear [2,56].

We recall [54,55] that the relevant equations and boundary conditions governing the small displacement ω of a thin plate may be found by minimizing the strain energy V, using the equation

$$V = \frac{1}{2} \int\int \left[D_x \left(\frac{\partial^2 \omega}{\partial x^2}\right)^2 + 2D_1 \frac{\partial^2 \omega}{\partial x^2}\frac{\partial^2 \omega}{\partial y^2} + D_y \left(\frac{\partial^2 \omega}{\partial y^2}\right)^2 + 4D_{xy} \left(\frac{\partial^2 \omega}{\partial x \partial y}\right)^2 \right.$$

$$\left. + N_x \left(\frac{\partial \omega}{\partial x}\right)^2 + N_y \left(\frac{\partial \omega}{\partial y}\right)^2 + 2N_{xy} \frac{\partial \omega}{\partial x}\frac{\partial \omega}{\partial y} \right] dx\, dy - \int\int fu\, dx\, dy$$

$$- \int M_n \frac{\partial u}{\partial n}\, ds + \int \left(Q_n - \frac{\partial M_{nt}}{\partial s}\right) u\, ds \tag{2.16}$$

The double integrals are taken over the entire plate, whereas the single integrals involve part of the boundary of the plate. Here D denotes bending rigidity; f, the effective load; M_n, the bending moments; $Q_n - (\partial M_{nt}/\partial s)$, the transverse forces distributed on the boundary; and N the forces acting in the middle plane. In a realistic structure, the simplest ways to obtain the physical constants is by averaging over the thickness of the plate [55]. With a clamped edge we associate the boundary condition $\omega = \partial \omega/\partial n = 0$. If an edge is free or if bending moments-transverse forces are present, then the relevant boundary conditions are recovered naturally from (2.16) with little effort. Once ω is known, the relevant components of the stress σ may be calculated as follows [54,55]:

$$\sigma_{xx} = \frac{-zE}{1 - \nu^2} \left[\frac{\partial^2 \omega}{\partial x^2} + \nu \frac{\partial^2 \omega}{\partial y^2} \right] + \sigma_1$$

$$\sigma_{yy} = \frac{-zE}{1 - \nu^2} \left[\nu \frac{\partial^2 \omega}{\partial x^2} + \frac{\partial^2 \omega}{\partial y^2} \right] + \sigma_2 \tag{2.17}$$

$$\sigma_{xy} = -zG \left[\frac{\partial^2 \omega}{\partial x \partial y} \right] + \sigma_3$$

where E, G denote Young's modulus and shear modulus, respectively; ν is Poisson's ratio; and σ_1, σ_2, σ_3 denote the built-in process-dependent stresses due to packaging, thermal effects, and so on.

We remark that the natural boundary conditions associated with (2.16) are sometimes modified by the introduction of an "edge factor" to allow for deformation of the supporting rim [55]. An approximate value of this factor is found by comparison with actual measurements [55].

Once the value of σ is known, the fractional change in resistivity in a piezo-electric layer may be found by simple multiplication [55]. For a capacitative pressure transducer, the relative change in capacitance involves a simple integration [55].

Specifically, the plane stress approximation yields for the fractional change of resistance the formula:

$$\frac{\Delta R}{R} = \pi_l \sigma_l + \pi_t \sigma_t + \pi_s \sigma_s \qquad (2.18)$$

where ΔR denotes the resistance change, R is the unstressed resistance, σ_l is the normal stress parallel to the current path, σ_t is the transverse stress perpendicular to the current, and σ_s is the shear stress. In the same way $\pi_{l,t,s}$ denotes the shear piezoresistive coefficients. For capacitative sensors, the capacitance between a movable diaphragm and the fixed plate is given by

$$C = \int\int \frac{\varepsilon_0}{[s - \omega(x, y)]} \, dxdy + C_p \qquad (2.19)$$

where ε_0 is the dielectric constant of the cavity, C_p is the parasitic capacitance, and s is the zero pressure distance between diaphragm and fixed plate. The double integral is taken over the entire diaphragm. Finally, because stress affects, in particular, the energy gap and carrier mobility [57–59], the change in the output characteristics in bipolar transistors can be calculated as a function of σ. We remark that if the deflections involved are large, then equation (2.16) needs to be replaced by more complicated expression [54, p. 415].

Because modern sensors and microstructures involve the growth, deposition, and etching of numerous thin films at different process steps leading to complex structures subject to various thermal and mechanical loads [60], the simulation procedure we briefly described has led in recent years to considerably more involved approaches, [6,7,61]. These in particular involve the use of multiple grids and the "numerical growth" of each film layer at the various specified temperatures with suitable numerical conditions at the interfaces [7]. One is thus able to also predict the consequences on the behavior of the devices of the packaging stresses.

2.3 DISCRETIZATION PROCEDURES

It is clear from Section 2.2 that microsensor modeling usually involves the numerical resolution of a system of nonlinear partial differential equations. The simplest approach is to use a (nonlinear) Gauss-Seidel iteration technique: In equation (2.1) all variables are held constant except for the potential ψ, which is updated. The new value of ψ and the old p are then used in (2.2) to update n. Finally equation (2.3) is employed to update p, and the entire process is repeated. These procedures, also known as *Gummel's algorithm*, work well if the device currents and R,G are small; that is, in cases where the system (2.1)–(2.3) is weakly coupled. As is well known

[62], the classic approach in more general cases involves the application of accelerated modified Newtonian schemes, and the consequent reduction to a series of linear problems. We briefly recall that the basic procedure can be viewed abstractly as follows: if we need to solve the equation $f(u) = 0$, then we make an initial guess $u = u_0$ and then employ the truncated expression $0 = f(u_0 + \delta u) \cong f(u_0) + f'(u_0)(\delta u)$ to compute the correction δu. We then replace u_0 by $u_0 + \delta u$ and repeat until δu is deemed to be sufficiently small. In isothermal semiconductor problems, we observe that u_0 could actually be the triple (ψ_0, n_0, p_0) and, in this case, f' represents a 3×3 matrix with partial differential equations as coefficients. It is also possible to mix these procedures. For example, for a given (n, p) the potential equation may be written in nonlinear form and treated by a Newtonian scheme. Once the new value of ψ is found, it is substituted into the current equations, which are then solved by Gauss-Seidel iteration. The new values of (n, p) are then substituted in the potential equation and the process repeated.

2.3.1 Accelerated Nonlinear Procedures

It has been found very important to accelerate the preceding procedure in various ways. One of these procedures [63] involves replacing u_0 by $u_0 + t\delta u$, where the parameter t is selected by a descent method, that is, so that an associated "energy" functional J is minimized; hence, we select $t = t_0$, where t_0 is the minimum of $J(u_0 + t\delta u)$. For nonvariational problems for which no functional J exists, other acceleration algorithms can be based on solutions of differential equations. It may also be advantageous to replace $f'(u_0)$ by related simpler expressions. We illustrate explicitly these abstract remarks by considering the simple procedure we have employed successfully for a system such as (2.1)–(2.3) in the isothermal steady state case and in the presence of a magnetic field. We apply Gummel's algorithm basically as follows. Given a guess ψ_0, n_0, p_0 we apply the Newtonian scheme to (2.1) and obtain

$$-\text{div}(\epsilon \text{ grad } \psi) + \left(\frac{q}{V_T}\right)(p_0 + n_0)\psi = \left(\frac{q}{V_T}\right)(p_0 + n_0)\psi_0 + \rho_0 \qquad (2.20)$$

where

$$p_0 = n_i \exp\left[\frac{(\varphi_p^0 - \psi^0)}{V_T}\right] \qquad n_0 = n_i \exp\left[\frac{(\psi^0 - \varphi_n^0)}{V_T}\right]$$

$$\rho_0 = q(p_0 - n_0 + N) \qquad N = N_D - N_A$$

With the quasi-Fermi potentials kept constant and applying the acceleration algorithm mentioned previously, we solve (2.1). We next employ a simple iterative process to find the solution of the continuity equations for the given ψ. At each iteration step, the generation recombination terms on the righthand side of the continuity equations are evaluated at the last updates of n, p. It is essential in this procedure to begin with reasonably good starting guesses. To do this, the applied bias is incremented in steps from the zero bias case and the problem solved at each step. The starting guesses for any given step are made by linearly extrapolating from the values calculated in the preceding steps. We remark that, at this level of abstraction, the procedure is the same as what is used in the $\mathbf{B} = 0$ case. What makes the magnetic sensor simulation different is the discretization of the continuity equations, as we indicate later.

As another example, consider system (2.4), (2.15), which arises in the thermistor problem. We have found a simple iterative procedure quite satisfactory for several applications. Specifically, to ensure stability, we first discretize $\partial T / \partial t$ in (2.4) by means of a backward Euler procedure and then linearize (2.4) to obtain

$$-\mathrm{div}(k(T_{i+1,j}) \operatorname{grad} (T_{i+1,j+1})] + \rho c \frac{T_{i+1,j+1}}{\Delta t} = \sigma(T_{i+1,j})|\operatorname{grad} \psi_{i+1,j}|^2 + \frac{\rho c}{\Delta t} T_i \quad (2.21)$$

where Δt denotes time step, T_i denotes T at time $i(\Delta t)$, $T_{i,j}$ denotes the jth iteration for T at time $i(\Delta t)$, and $\psi_{i,j}$ denotes the jth iteration for ψ at time $i(\Delta t)$. Once $T_{i+1,j+1}$ is obtained from (2.21), ψ is updated from

$$-\mathrm{div}[\sigma(T_{i+1,j+1}) \operatorname{grad}(\psi_{i+1,j+1})] = 0 \quad (2.22)$$

The initial guesses are provided by the values calculated at the previous time step. The procedure outlined in equations (2.21) and (2.22) is repeated until self-consistent values of ψ_{i+1}, T_{i+1} are found.

As mentioned earlier and indicated by these examples, we eventually need to resolve a series of linear problems, and we now briefly discuss some of the procedures employed. We shall present results for two dimensions, but much of what we state also holds for three dimensions with obvious modifications. The computational demands of a three-dimensional problem are of course much greater.

2.3.2 Methods of Solution and Grid Generation

Several methods are available for the solution of the linear problems just constructed: finite differences, finite elements, finite boxes, boundary methods, and others [8, 64–67]. Each has advantages and disadvantages, and the choice is to some extent made on individual preference and needs.

The first step—and arguably the most important—involves the discretization of the simulation domain in space or time. To avoid convergence problems and for accuracy, the generated mesh should be able to accommodate "arbitrary" shapes reasonably well and to take into account the variation of the physical and material parameters. To minimize computing effort, clearly the generation of many irrelevant grid nodes should be avoided, particularly in view of the relatively large physical size of the active regions that arise in microsensor simulation. In a classic finite difference approach, one decomposes the region into rectangles or cubes. This method is less flexible in dealing with complicated regions and mesh refinements. Furthermore, it tends to introduce nodes in irrelevant regions, but it is relatively simple to implement even in three dimensions, and has other advantages as well. For example, the nodes adjacent to any given node are known a priori, making adjacency information irrelevant and preconditioning in iterative methods fairly easy to apply. The interested reader is referred to [8] for more discussion on this approach.

Passing to other methods and focusing on the two-dimensional case first, we recall that a variety of mesh generation techniques has been considered and implemented. Many of these are discussed in the survey articles of Thacker [68] and Simpson [69] (see also [70]). We recall that the different approaches involved include curvilinear coordinate transformations, local grid refinement, vertex triangulation, boundary contraction, and modified quadtree representation. The curvilinear transformation procedure involves construction of a one-to-one mapping of the complicated device region to a simple one. Such a mapping is then used to transform the mesh on the simple region to a mesh on the more complicated one [71,72]. The main disadvantage is of course the construction of a suitable coordinate transformation that is similar to the creation of an irregular mesh [68]. The local mesh refinement procedure involves first the creation of a coarse approximate mesh, which is then subdivided to produce a finer one. Each element of the original mesh is tested by a refinement criterion, and if required, the element is then subdivided. This process is carried out until the refinement criterion is no longer applicable to any of the elements [73]. In the vertex triangulation approach, the mesh nodes are placed first in the device region and a suitable mesh is then built to join them [74]. The boundary contraction method generates stripes of elements along the boundary. This shrinks the region, and the process stops when the entire original region is covered [75]. The modified quadtree representation uses a recursive subdivision process to a region or subregion by assigning the subregions into a complete quadrant or a cut quadrant [76]. For these methods—indeed for any method—a large amount of input data is needed to generate a first reasonable mesh in regions with complicated boundaries. Any refinement procedure that may follow must be completely automatic to avoid error.

In general triangular mesh elements can approximate complicated planar boundaries better than rectangular elements, however, most planar mesh generation techniques use only triangular elements or a mixture of triangular and rectangular elements. Finally, there is a natural correspondence between the number of nodal points in an element and the degree of the local polynomial approximation to the solution. In general, to minimize computational effort, the minimum degree possible is used; for example, piecewise linear approximation to second-order equations.

As may be expected, grid generation is much more complicated in three dimensions. The simplest approaches employ cubes or prismatic elements, but more general three-dimensional generators involving tetrahedrons are now known [77–81].

To control the errors introduced [80,82–84], we also need to avoid extreme angles and be able to interactively refine any part of the mesh. Some refinement is usually present in the original mesh, based both on the problem at hand and on previous experience. Such a priori refinements are possibly combined with a posteriori steps based on error estimates [85–88].

Once the grid is generated, the equations are reduced to a matrix problem. The precise nature of the matrices involved depends on the method followed.

We illustrate some of these procedures by considering first in more detail an adaptation of the "box" method [89], which we have found easy and effective. For the sake of clarity, we illustrate this method by applying it to the case of a steady-state magnetic field sensor (like (2.1)–(2.3)). To avoid extreme angles or large numbers of nodes and be able to interactively refine any part of the mesh, we have developed a grid generator [90], based on a hybrid scheme that combines procedures in [91,92] with a simpler generator. The procedure is roughly as follows: Given a simulation region, we first decompose it into convex subregions whose area depends on the overall device shape as well as on the local changes and size of the physical parameters. Each subregion is then automatically decomposed into triangular elements that are close to either equilateral or right triangles. Element and node matching at the interfaces between regions completes the grid generation procedure. In these subregions, triangles with obtuse angles may be generated. Examples of such grids are shown in Figure 2.1. Note that we select right triangles in regions of slow variable change or change in only one direction. Such triangulation requires few nodes, whereas division into equilateral triangles requires relatively more nodes and is employed in regions where conditions and variables change rapidly along curves. It is important to observe for what follows that the perpendicular bisectors of the sides of such triangles must meet within the triangle or on its boundary. The vast majority of microsensor device configurations exhibit some form of physical symmetry. It has been found very useful to incorporate several mirroring procedures in our generator. In this way, only a part of the grid is actually generated, and the rest of the mesh is obtained by reflection.

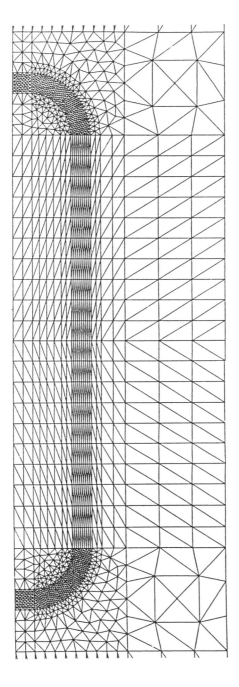

Figure 2.1(a) Grid generated for magnetic sensor simulation [1].

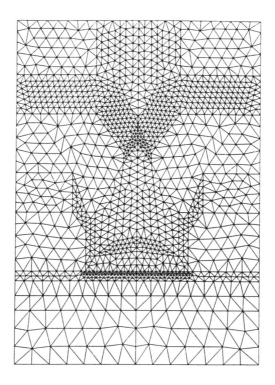

Figure 2.1(b) Grid generated for thermal analysis of contacts and vias [107].

2.3.3 Equation Discretization

Once the grid is generated, and after suitable variable scaling [8], we select any node i and construct the associated cell Ω_i by constructing the perpendicular bisectors of the element edges that meet at node i, Figure 2.2. If we denote the boundary of Ω_i by $\partial\Omega_i$, then Gauss's divergence theorem gives:

$$\int_{\Omega_i} \operatorname{div}(\mathbf{F}) = \int_{\partial\Omega_i} \mathbf{F} \cdot \mathbf{n} d(\partial\Omega_i) \tag{2.23}$$

for a vector field $\mathbf{F}$. We assume ψ is spatially linear within an element and denote ψ at node i by ψ_i. If we apply (2.23) to the linearized equation (2.20) we obtain for any given element the contribution:

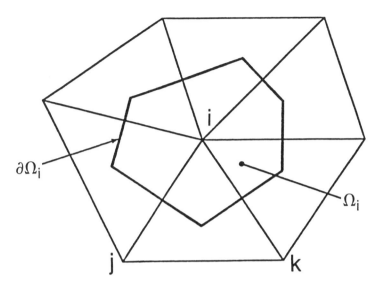

Figure 2.2 Cell associated with node i [1].

$$V_t \, \varepsilon \left[\frac{d_{ij}}{l_{ij}} (\psi_i - \psi_j) + \frac{d_{ik}}{l_{ik}} (\psi_i - \psi_k) \right] + \int_A \left(\frac{q}{V_T} \right) (p_0 + n_0) \psi dA$$

$$= \int_A \left[\frac{q}{V_T} (p_0 + n_0) \psi_0 + \rho_0 \right] dA \quad (2.24)$$

to the discretized equation for node i. Here ε is assumed constant in the cell; d_{ij}, d_{ik}, l_{ij}, l_{ik} denote the distances shown in Figure 2.3 and; A represents the part of the cell

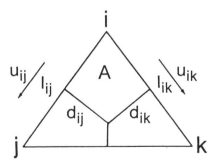

Figure 2.3 Lengths associated with discretization within an element [1].

Ω_i located within the element being considered. The integrals in (2.24) are usually evaluated by assuming that the quantities involved are linear in the element.

An analogous technique is applied to the continuity equations, keeping in mind the extension to triangular grids of the Scharfetter-Gummel approach [93,94]. Here, the asymmetries introduced by the magnetic field are present. If we denote $\mathbf{J}_n$ by (2.7) and $\mathbf{J}_{nB}$ by (2.9), then, disregarding the terms involving grad (T), we obtain

$$\mathbf{J}_n \cdot \mathbf{u}_{ij} \exp\left[\frac{-\psi}{V_T}\right] = qD_n \, \mathrm{grad}\left[n \exp\left(\frac{-\psi}{V_T}\right)\right] \cdot \mathbf{u}_{ij} \tag{2.25}$$

where $\mathbf{u}_{ij}$ denotes the unit vector along the side connecting node i to node j, Figure 2.2. Integrating (2.25) by treating $\mathbf{J}_n \cdot \mathbf{u}_{ij}$ and μ_n as constants but ψ as linear, and repeating for the side joining node i to node k yields

$$\mathbf{J}_n \cdot \mathbf{u}_{ij} = \frac{qD_n}{l_{ij}}\left[n_j B\left(\frac{\psi_{ji}}{V_T}\right) - n_i B\left(\frac{\psi_{ij}}{V_T}\right)\right]$$

$$\mathbf{J}_n \cdot \mathbf{u}_{ik} = \frac{qD_n}{l_{ik}}\left[n_k B\left(\frac{\psi_{ki}}{V_T}\right) - n_i B\left(\frac{\psi_{ik}}{V_T}\right)\right] \tag{2.26}$$

As is customary in the zero field cases, B here denotes the Bernoulli function. Equations (2.9) and (2.26) and an elementary calculation can be employed to relate $\mathbf{J}_{nB}$ in the element to the nodal values n_i, n_j, n_k. Assembling contributions to the equation for n_i on an element by element basis, we arrive at a nonsymmetric matrix equation. Some analytical thought [95] shows that in realistic situations this matrix is definite, and consequently a wide variety of matrix solvers are still applicable. The procedure employed for the hole continuity equation is identical.

Once the matrices are assembled, the resulting equations may be solved either by iterative or direct methods [96–99]. We have often used a family of direct solvers given in SPARSPAK [100,101]. Although these solvers were found to give very good results, the storage requirements that we encountered in the simulation of various magnetic field sensitive bipolar devices indicate that an iterative scheme, such as a generalized preconditioned conjugate gradient or residual method, would be preferable. For three-dimensional simulations, an iterative scheme appears essential. The introduction of parallel procedures is now also being seen [33,102].

Terminal currents are evaluated once the solution to the variables has been achieved. A common feature of most magnetic sensors is the desirability of the accurate calculation of a current imbalance between a pair of electrodes. This imbalance is used as a measure of the magnitude and direction of the magnetic field. The technique we have employed for this current calculation is analogous to the approach in [103] and based on the weak definition of the equations. We observe that the total current through an electrode α is given by

$$I_\alpha = \int_\Omega (\mathbf{J}_{nB} + \mathbf{J}_{pB}) \cdot \mathrm{grad}(\xi)d\Omega \qquad (2.27)$$

where ξ is a smooth function which takes on the value 1 at electrode α and 0 at all other electrodes. Here Ω represents the device domain. The validity of equation (2.27) is immediate from the divergence theorem and equations (2.2) and (2.3). Conservation of total electron or hole current can also be based on equation (2.27). We have found this procedure extremely simple to implement and to yield good results.

It is a simple matter to adapt these procedures to cases of magnetic field sensors governed by linear equations or to the thermal-flow problems mentioned in Section 2.2. For more details we refer the interested reader to [104–109].

The preceding piecewise linear discretization ideas fail a priori for the case of sensors that lead to fourth-order differential equations, such as those governing the mechanical behavior of microstructures. If one deals with square diaphragms, constant loads, and no holes then a 13-point finite difference approximation is fairly straightforward to implement [55]. If however one is interested in the rapid simulation of structures that could vary considerably in shape and boundary conditions, without performing involved software modifications, then a finite element approach may be advantageous. We have employed a conforming method using piecewise cubics—compare this with ANSYS [15]—that guarantees convergence [83], albeit possibly giving somewhat worse results in some cases when compared with nonconforming methods.

Our specific choice was motivated in considerable part by the desire to employ without change the grid generator we described earlier. Once this grid was generated, following the Clough-Tocher macro-triangle approach [110], we divided each triangle into three daughter subtriangles and then constructed C^1 basis functions by piecing together a cubic from each subtriangle. With a reasonable amount of effort, the connection matrix [83] can be explicitly calculated and equation (2.16) gives rise to a matrix equation in a straightforward manner. This part of the procedure is analogous to what was used in earlier codes such as NASTRAN [16]. Because the equations of linear elasticity were employed, none of the aforementioned nonlinear procedures needed to be invoked.

2.3.4 Error Checking

Finally, we briefly describe our error checking procedures. New codes are always first run on simple problems whose answers are explicitly known either analytically or from earlier verified codes. Such procedures as current balance checks are always performed when applicable. These steps are taken to eliminate routine coding mistakes. Other error reducing procedures involve the grid node distribution. Our grid generator attempts to place nodes such as to reasonably distribute the errors, although we have been usually unable to obtain precise practical estimates on how this should be done. In view of the numerous assumptions involved in obtaining the various mathematical models [8], the obvious question arises of whether the results obtained reflect reasonably accurately the physical device behavior. If the basic assumptions introduce serious errors, then no fiddling with the code will suffice. In summary, the ultimate validity of the simulations lies in the comparison of the numerical with the experimental results [14,41,105].

2.4 RESULTS AND DISCUSSION

To illustrate the material presented in the other sections, we choose a selection of specific simulation results.

2.4.1 Hall Devices

We first consider the distribution of carrier transport for a variety of Hall devices under various configurations of discontinuous magnetic induction [105]. Such devices have applications in reading magnetic tapes and disks and magnetic ink patterns in banknotes and credit cards [111,112]. There have been many analytical and numerical investigations of Hall plates in the case of uniform magnetic fields, and we refer to [65,105] for further references. In the presence of nonuniform magnetic inductions, we recall the results of Hlasnik and Kokavec [113], Fluitman [114], Brunner [115], De Mey [116], and Baird [117] using different approaches and for a variety of device configurations. We observe that some care must be taken in the calculations, due to the discontinuities in **B**. We have dealt with this problem by always considering the equations to hold in weak form, furthermore the "open boundaries" condition was always taken to be the natural boundary condition associated with the equation; that is, $A \cdot \text{grad } \Psi = 0$, with A as given in (2.13). Note that this incorporated automatically the magnetic field **B** in the boundary condition. In Figure 2.4 we present the distributions of carrier transport for a variety of Hall devices under various configurations of discontinuous magnetic induction. These were obtained [105] using the procedures outlined in the earlier sections. A comparison of experimental and numerical output responses is illustrated in Figure 2.5. This clearly indicates the agreement between the experimental measurements and our analysis.

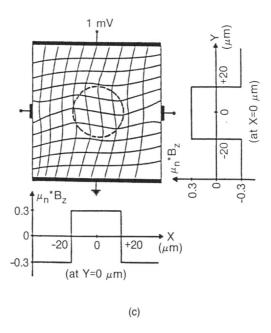

(c)

Figure 2.4 continued.

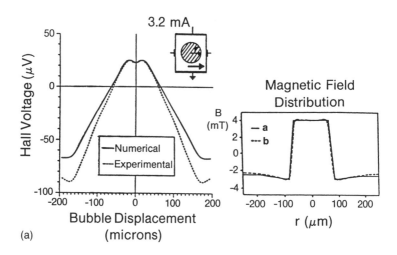

(a)

Figure 2.5 Comparison of experimental and numerical results for (a) bubble, (b) strip. The bubble distribution used in simulations is denoted by *a* and the measured distribution by *b* [1].

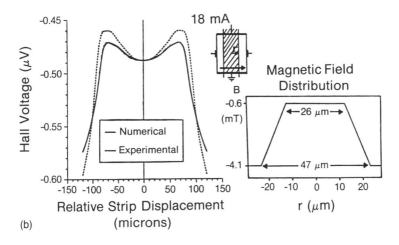

(b)

Figure 2.5 continued.

2.4.2 Magnetotransistor

We next consider the problem of the simulation of the Hall voltage in the base region of a vertical magnetic transistor [41]. The simulation domain is shown in Figure 2.6, with Figure 2.7 illustrating the actual device structure. We observe that the "oxide" layer in Figure 2.6 is introduced to define the boundary of the computational domain for the potential equations, as mentioned earlier. This is a considerably more computationally demanding problem than that of the Hall plates.

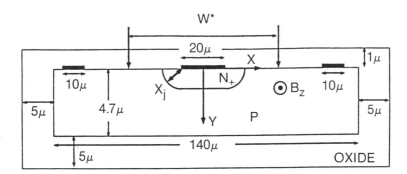

Figure 2.6 Two-dimensional simulation geometry of the magnetotransistor base region surrounded by an artificial oxide region [1].

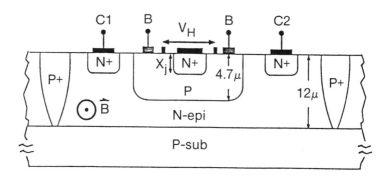

Figure 2.7 Cross-sectional view of dual-collector magnetotransistor fabricated in bipolar technology [1].

The coefficients in the relevant equations are determined by assuming the previously mentioned models with the following additions. First, we assume that the model of Caughey-Thomas [118] holds for the drift mobilities $\mu_{n,p}$, and we choose the Hall mobilities $\mu_{n,p}^* = r_{n,p} \cdot \mu_{n,p}$ with the Hall coefficient factor $r_{n,p} = 1.2$. The carrier lifetimes in the Shockley-Read-Hall model [8,22] are selected in accordance with formulas suggested by Fossum [22]. We assume a Gaussian distribution for the acceptor concentration in the base region, with a maximum value of 5×10^{17} cm^{-3} and a junction depth of 4.7 μm. In the emitter region, the peak donor concentration is set at 6×10^{19} cm^{-3} with a fitted exponential distribution based on magnetotransistor structures fabricated in standard bipolar technology. The junction depth for both lateral and vertical emitter diffusion is set at 2.5 μm. The full details of the parameter selection, and hence of the equation coefficients, may be found in [41], to which we refer the interested reader.

Once the equation coefficients are determined, the problem is discretized and a numerical solution found in accordance with the procedures described earlier. In particular, we found it useful in obtaining or accelerating convergence to begin the simulation at small values of applied bias. This (easier) problem is solved first and the bias is increased in small steps ($\approx .1$V). The process is repeated at each step until the desired operating conditions are reached. At any given bias we begin with initial guesses that come from linear extrapolation from the values found for the previous bias voltages.

Figures 2.8 and 2.9 indicate simulation results for $V_{BE} = .85$V and $|\mathbf{B}| = 2$T, parallel to the chip surface. We observe the symmetry of the equipotential lines in the emitter vicinity, Figure 2.8, and note the carrier deflection, Figure 2.9, caused by the magnetic field. These numerical predictions are in reasonable agreement with experimental results obtained by Hall probe measurements [41]. In view of the difficulties found in obtaining practical estimates on the overall simulation error, such agreement is of paramount importance. Both approaches indicate that while Hall

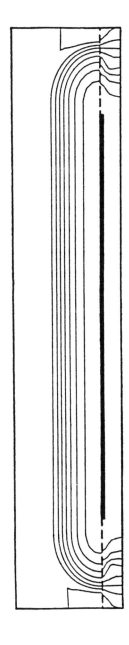

Figure 2.8 Equipotential lines near the emitter for $V_{BE} = .85V$ and $B = 2T$ [1].

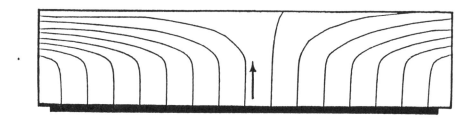

Figure 2.9 Electron flow lines in the same region for V_{BE} = .85V and B = 2T [1].

fields are present in the emitter-base junction vicinity, they are too weak to substantiate the validity of the emitter injection modulation model.

2.4.3 Micromachined Flow Sensor

The period of the 1970s and early 1980s saw the recognition of silicon as a mechanical material for solid state sensors and actuators [119,120]. Of specific interest to us here are CMOS microbridges employed as temperature or fluid flow sensors. These descendants of the classic hot wire anemometer have been realized in a variety of ways [46,108,121–124]. Basically, they consist of a sandwich microbridge structure, made up of polysilicon heater and sensor resistors between oxide layers, Figure 2.10. The sensor is exposed to a gas flow, c, or a different outside ambient tem-

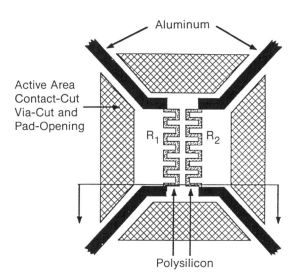

Figure 2.10 Layout of the suspended microbridge used for temperature-flow sensing [108].

perature, T_∞, or both. Variations in c or T_∞ are monitored by checking the variation of resistance in the sensor resistor. A determination of the sensitivity of such devices as functions of T_∞, c, and the geometry and material composition of the bridges is essential in determining the suitability of the sensor.

For the two-dimensional simulation, we replace the structure of interest by an equivalent structure, Figure 2.11. The equations involved in this simulation are (2.4) and (2.15). In general, no further reduction can be made to this system; however the special geometry of the equivalent structure leads to the consideration of the special simplification $\Psi = \Psi(z)$, $T = T(x, y)$ and the reduction of the system (2.4), (2.15) to the single equation

$$\rho c \frac{\partial T}{\partial n} - \text{div}(K(T) \, \text{grad} \, T) = \sigma(T)\left[\frac{V}{L}\right] \qquad (2.4')$$

where V denotes the applied bias and L the resistor length. Observe that once the temperature is known, we can calculate the sensor resistance R_s merely by evaluating $R_s = L/\{\iint \sigma[T(x, y)]dxdy\}$ where the double integral is taken only over the resistance cross section.

We set the polysilicon film doping level at 10^{19} cm^{-3}. This is considerably higher than that usually encountered in the literature, but it is the value provided to us for the standard 3μ CMOS process of the Canadian Microelectronics Center. The activation energy is chosen to be 0.1 eV and the applied voltage is 1.0V in most

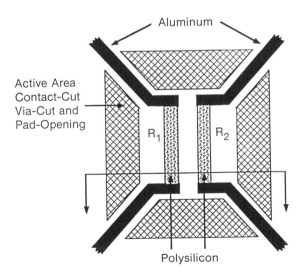

Figure 2.11 Computational device domain used in the analysis [108].

cases. Because the resistor length is 200 μm, we obtain a resulting electric field strength of 5,000 Vm^{-1}. In every case, we assume the ambient fluid to be dry air at 300K, and the fluid velocity c is measured in terms of the average Reynolds number Re_L ($= c\rho L/\mu$). We first investigated the steady-state effects on the sensor resistor R_s of varying the distance between sensor and heater resistors, Figure 2.12. We next discretized the time derivative in (2.4') by means of a backward Euler procedure; and in Figure 2.13, we present the time response of the sensor element.

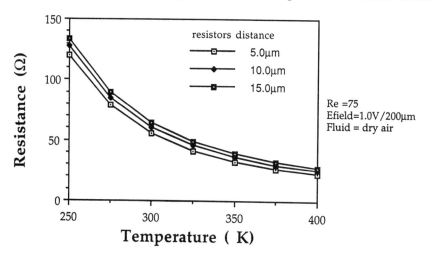

Figure 2.12 Effects on sensor resistance of varying the distance between heater and sensor resistors [108].

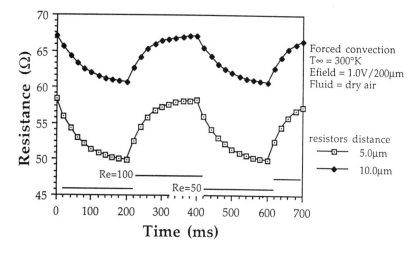

Figure 2.13 Time response of the sensor element [108].

Re_L was changed every 200 ms between 50 and 100 and R_s was calculated every 20 ms. For resistor separations of 5.0 μm and 10.0 μm the rise-fall time constants have been calculated by the simulations to be 55 ms/65 ms and 65 ms/75 ms, respectively. The simulation indicates considerably slower response times (four to five times) for these structures as compared with available results in the literature for bare wire sensors [123].

Most of the heat-transfer problems previously considered in the classic engineering literature dealt with the simulation of structures much larger than the microstructures considered here. It is therefore not obvious that various known procedures for determining the heat transfer coefficients are still applicable, and experimental verification is ongoing. Possibly, simulation in combination with experimental results may be used to obtain realistic values for these coefficients in practical situations.

2.4.4 Dynamic Microstructures

As a final example, we consider the problem of simulating the flexing of a variety of dynamic microstructures, such as diaphragms and microbridges under different loading [122,125–127].

Consider, for example, a 10-μm thick n-type epilayer that is grown on a p-type $\langle 100 \rangle$ silicon wafer. Electrochemically controlled anisotropic etching may be employed to create a diaphragm. Assume first that all four sides are clamped. In accordance with [128,129], we set Poisson's ratio at .1025 and choose Young's modulus $E = .162 \times 10^{12}$ N/m^2 and the shear modulus $G = .073 \times 10^{12}$ N/m^2. When we impose a load of 100 mm Hg, we observe a symmetric deflection with a maximum of ≈ 1.2 μm, see Figure 2.14. If we consider a square diaphragm with an opening at the center of the structure and the same uniform load, we find a maximum deflection of about the same size at the middle of the opening edges, Figure 2.15. For illustration, the magnitude of the stress σ due to the bending can be evaluated by computing the maximum of $\{|\lambda_1|, |\lambda_2|\}$ where λ_1, λ_2 are the eigenvalues of the matrix

$$\begin{pmatrix} \sigma_{xx} & \sigma_{xy} \\ \sigma_{xy} & \sigma_{yy} \end{pmatrix}$$

The stress magnitude due to the deflection is shown in Figure 2.16. The high stresses at hole domain corners are anticipated, because it is well known that the governing theory fails in such regions and plastic effects or even cracks may appear.

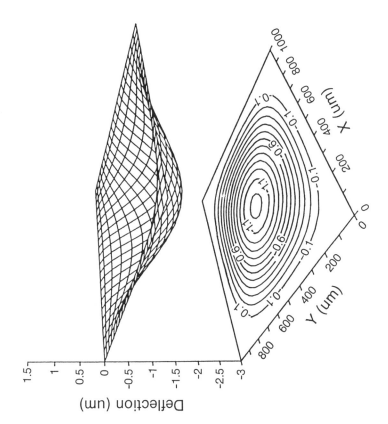

UNIFORM LOAD : 100 mm Hg
MAX DEFLECTION : 1.233 μm

Figure 2.14 Diaphragm deflection under a uniform load [122].

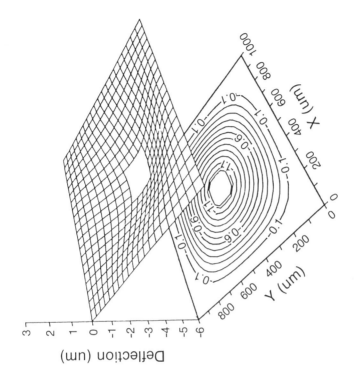

UNIFORM LOAD : 100 mm Hg
MAX DEFLECTION: 1.198 μm

Figure 2.15 Deflection of diaphragm with a central hole under a uniform load condition [122].

42

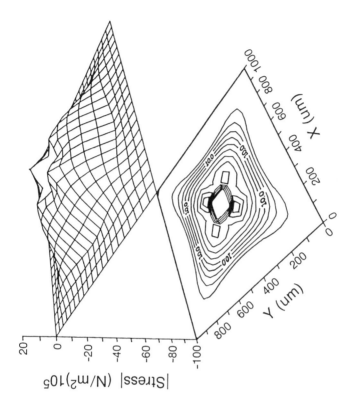

UNIFORM LOAD : 100 mm Hg

MAX |STRESS| : 2.735 x 10⁶ N/m²

Figure 2.16 Stress magnitude due to deflection of diaphragm with a central hole under a uniform load condition [122].

2.5 SUMMARY

We conclude by observing that in this chapter we have presented the equations and boundary conditions that describe the relevant physical effects and their interactions in a variety of microsensors. The equations form the basis of numerical discretization and solution schemes that we have also indicated and that furnish a highly desirable design and analysis tool. Several illustrative examples were also given. Optimization of device structure, sensitivity of possible new devices, extraction of pertinent electrical and mechanical parameters can be investigated rapidly and relatively inexpensively in this manner.

REFERENCES

[1] H. P. Baltes, W. Allegretto, and A. Nathan, "Microsensor Modeling," in *Simulation of Semiconductor Devices and Processes*, ed. G. Baccarani, and M. Rudan, Bologna, Italy, Tecnoprint, 1988, pp. 563–577.

[2] A. Nathan and H. Baltes, "Sensor Modeling," in *Sensors*, Vol. 1, ed. T. Grandke and W. H. Ko, Weinheim, Germany, VCH, 1989, pp. 45–77.

[3] S. Middelhoek and S. A. Audet, *Silicon Sensors*, New York, Academic Press, 1989.

[4] R. S. Muller, R. T. Howe, S. D. Senturia, R. L. Smith and R. M. White, eds., *Microsensors*, New York, IEEE Press, 1991.

[5] S. Middelhoek and J. Van der Spiegel, eds., *State of the Art of Sensor Research and Development*, Lausanne, Switzerland, Elsevier, 1987.

[6] B. Puers, E. Peeters, and W. Sansen, "CAD Tools in Mechanical Sensor Design," *Sensors and Actuators*, Vol. 17, 1989, pp. 423–429.

[7] F. Pourahmadi, P. Barth, and K. Petersen, "Modeling of Thermal and Mechanical Stresses in Silicon Microstructures," *Sensors and Actuators*, Vols. A21–23, 1990, pp. 850–855.

[8] S. Selberherr, *Analysis and Simulation of Semiconductor Devices*, Vienna, Springer-Verlag, 1984.

[9] P. A. Markowich, C. A. Ringhofer, and C. Schmeiser, *Semiconductor Equations*, New York, Springer-Verlag, 1990.

[10] M. S. Mock, *Analysis of Mathematical Models of Semiconductor Devices*, Dublin, Boole Press, 1983.

[11] A. Nathan, H. Baltes, and W. Allegretto, "Review of Physical Models for Numerical Simulation of Semiconductor Microsensors, *IEEE Trans. Computer-Aided Design*, Vol. 9, 1990, pp. 1198–1208.

[12] J. L. Gray, "Two-Dimensional Modeling of Silicon Solar Cells," Ph.D. dissertation, Purdue University, 1982.

[13] J. Geist and H. P. Baltes, "High Accuracy Modeling of Photodiode Front-Region Quantum Efficiency," *Appl. Opt.*, Vol. 28, 1989, pp. 3929–3939.

[14] K. W. Lee and K. D. Wise, "SENSIM: A Simulation Program for Solid State Pressure Sensors," *IEEE Trans. Electron Devices*, Vol. ED-29, 1982, pp. 39–41.

[15] *User's Manual*, ANSYS Engineering Analysis System, Swanson Analysis Systems Inc., Houston, 1989.

[16] A. K. Aziz, *The Mathematical Foundations of the Finite Element Method with Applications to Partial Differential Equations*, New York, Academic Press, 1972.

[17] G. M. Koppelman, "OYSTER, A Three-Dimensional Structural Simulator for Microelectro Mechanical Design," *Sensors and Actuators*, Vol. 20, 1989, pp. 179–185.

[18] S. B. Crary, "Thermal Management of Integrated Microsensors," *Sensors and Actuators*, Vol. 12, 1987, pp. 303–312.

[19] N. Swart, A. Nathan, M. Shams, and M. Parameswaran, "Numerical Optimization of Flow-Rate Microsensors Using Circuit Simulation Tools," to be published.

[20] *GENSIM*, Integrated Systems Lab, ETH, Zurich, Switzerland.

[21] P. A. Markowich, *The Stationary Semiconductor Equations*, New York, Springer-Verlag, 1986.

[22] G. Baccarani, M. Rudan, R. Guerrieri, and P. Ciampolini, "Physical Models for Numerical Device Simulation," in *Process and Device Modeling*, ed. W.L. Engl, Amsterdam, North Holland, 1986, pp. 107–158.

[23] F. Brezzi and P. A. Markowich, "A Convection-Diffusion Problem with Small Diffusion Coefficient Arising in Semiconductor Physics," *Boll. U.M.I.*, Vol. 7, No. 26, 1988, pp. 903–930.

[24] H. Gajewski and R. Gröger, "On the Basic Equations for Carrier Transport in Semiconductors," *J. Math. Anal. and Appl.*, Vol. 113, 1986, pp. 12–35.

[25] T. I. Seidman and G. M. Troianiello, "Time-Dependent Solutions of a Nonlinear System Arising in Semiconductor Theory," *Nonlinear Analysis T.M.A.*, Vol. 9, 1985, 1137–1157.

[26] J. W. Jerome, Consistency of Semiconductor Modeling: An Existence-Stability Analysis for the Stationary Van Roosbroeck System, *SIAM J. Appl. Math.*, Vol. 45, 1985, pp. 565–590.

[27] R. E. Bank, ed., *Computational Aspects of VLSI Design with an Emphasis on Semiconductor Device Simulation*, Lectures in Applied Mathematics, Vol. 25, Providence, RI, American Mathematical Society, 1990.

[28] H. Xie and W. Allegretto, "$C^\alpha(\overline{\Omega})$ Solutions of a Class of Nonlinear Degenerate Elliptic Systems Arising in the Thermistor Problem," *SIAM J. Math. Anal*, to appear.

[29] W. Allegretto and H. Xie, "Existence of Solutions for the Time-Dependent Thermistor Equations," to be published.

[30] K. Bløtekjaer, "Transport Equations for Electrons in Two-Valley Semiconductors," *IEEE Trans. Electron Dev.*, Vol. ED-17, No. 1, 1970, pp. 38–47.

[31] M. Rudan and F. Odeh, "Multi-Dimensional Discretization Scheme for the Hydrodynamic Model of Semiconductor Devices," *COMPEL*, Vol. 5, 1986, pp. 149–183.

[32] M. Rudan, F. Odeh, and J. White, "Numerical Solution of the Hydrodynamic Model for a One-Dimensional Semiconductor Device," *COMPEL*, Vol. 6, 1987, pp. 151–170.

[33] C. L. Gardner, P. J. Lanzkron, and D. J. Rose, "A Parallel Block Iterative Method for the Hydrodynamic Device Model," *IEEE Trans. Computer-Aided Design*, Vol. 10, 1991, pp. 1187–1192.

[34] W. Quade, M. Rudan, and E. Scholl, "Hydrodynamic Simulation of Impact-Ionization Effects in P-N Junctions," *IEEE Trans. Computer-Aided Design*, Vol. 10, 1991, pp. 1287–1294.

[35] C. L. Gardner, J. W. Jerome, and D. J. Rose, Numerical Methods for the Hydrodynamic Device Model: Subsonic Flow, *IEEE Trans. Computer-Aided Design*, Vol. 8, 1989, pp. 501–507.

[36] J. W. Roberts and S. G. Chamberlain, "Energy-Momentum Transport Model Suitable for Small Geometry Silicon Device Simulation," *COMPEL*, Vol. 9, 1990, pp. 1–22.

[37] F. Venturi, E. Sangiorgi, R. Brunetti, W. Quade, C. Jacoboni, and B. Riccó, "Monte Carlo Simulations of High Energy Electrons and Holes in Si-n-MOSFET's," *IEEE Trans. Computer-Aided Design*, Vol. 10, 1991, pp. 1276–1286.

[38] P. Lugli, "Monte Carlo Simulation of Semiconductor Devices and Processes," in *Simulation of Semiconductor Devices and Processes*, ed. G. Baccarani and M. Rudan, Bologna, Tecnoprint, 1988, pp. 3–18.

[39] O. Madelung, *Introduction to Solid State Theory*, Berlin, Springer-Verlag, 1978.

[40] H. P. Baltes and R. S. Popovic, "Integrated Semiconductor Magnetic Field Sensors," *Proc. IEEE*, Vol. 74, 1986, pp. 1107–1132.

[41] A. Nathan, K. Maenaka, W. Allegretto, H. P. Baltes, and T. Nakamura, "The Hall Effect in Magnetotransistors," *IEEE Trans. Electron Devices*, Vol. ED-36, 1989, pp. 108–117.

[42] W. Allegretto, A. Nathan, and H.P. Baltes, "Two-Dimensional Numerical Analysis of Silicon Bipolar Magnetotransistors, *Proc. NASECODE V. Conf.*, 1987, pp. 87–92.

[43] F. Kreith and M. S. Bohn, *Principles of Heat Transfer*, 4th ed., New York, Harper and Row, 1986.

[44] M. N. Ozisik, *Basic Heat Transfer*, Malabar, FL, Krieger, 1987.

[45] T. Kamins, *Polycrystallino Silicon for Integrated Circuit Applications*, Boston, Kluwer-Academic Publishers, 1988.

[46] M. Parameswaran, "Microelectronic and Micromechanical Sensors and Actuators in CMOS Technology—A Novel Scheme Towards Regularization and Integration," Ph.D. thesis, University of Alberta, Edmonton, Canada, 1990.

[47] J. Y. W. Seto, "The Electrical Properties of Polycrystalline Silicon Films," *J. Appl. Phys.*, Vol. 46, 1975, pp. 5247–5254.

[48] G. Baccarani, B. Ricco, and G. Spadini, "Transport Properties of Polycrystalline Silicon Films," *J. Appl. Phys.*, Vol. 49, 1978, pp. 5565–5570.

[49] G. Korsh and R. S. Muller, "Conduction Properties of Lightly Doped Polycrystalline Silicon," *Solid-State Electronics*, Vol. 21, 1978, pp. 1045–1051.

[50] N. C. Lu, L. Gerzberg, C. Y. Lu, and J. Meindl, "Modeling and Optimization of Monolithic Polycrystalline Silicon Resistors," *IEEE Trans. on Electron Devices*, Vol. ED-28, 1981, pp. 313–839.

[51] M. M. Mandurah, K. C. Saraswat, and T. I. Kamins, "A Model for Conduction in Polycrystalline Silicon—Part 1: Theory," *IEEE Trans. on Electron Devices*, Vol. ED-28, 1981, pp. 1163–1171.

[52] T. Yoshihara, A. Yasuoka and H. Abe, "Conduction Properties of Chemically Deposited Polycrystalline Silicon," *J. Electrochemical Society*, Vol. 127, 1980, pp. 1603–1607.

[53] O. Zucker, W. Langheinrich, and J. Meyer, "Investigation of Polysilicon for Its Application to Multi-Functional Sensors," *Sensors and Actuators*, Vols. 25–27, 1991, pp. 647–651.

[54] S. Timoshenko and S. Woinowsky-Krieger, *Theory of Plates and Shells*, New York, McGraw-Hill, 1959.

[55] K. W. Lee, "Modeling and Simulation of Solid-State Pressure Sensors," Ph.D. thesis, University of Michigan, Ann Arbor, 1982.

[56] K. Suzuki, T. Ishihara, M. Hirata, and H. Tanigawa, "Nonlinear Analysis of a CMOS Integrated Silicon Pressure Sensor," *IEEE Trans. Electron Devices*, Vol. ED-34, 1987, pp. 1360–1367.

[57] J. J. Wortman, J. R. Hauser, and R. M. Burger, "Effect of Mechanical Stress on p-n Junction Device Characteristic," *J. Appl. Phys.*, Vol. 35, 1964, pp. 2122–2131.

[58] B. Puers, L. Reynaert, W. Snoeys, and W. M. C. Sansen, "A New Uniaxial Accelerometer in Silicon Based on the Piezojunction Effect," *IEEE Trans. Electron Devices*, Vol. ED-35, 1988, pp. 764–770.

[59] W. Rindner and I. Braun, "Resistance of Elastically Deformed Shallow p-n Junctions," *J. Appl. Phys.*, Vol. 34, 1963, pp. 1958–1970.

[60] "Small Machines, Large Opportunities," Report of the NSF Workshop on Microelectromechanical Systems Research, Princeton, NJ, 1988.

[61] F. L. Deibel, "Calculating Residual Manufacturing Stresses in Braze Joints Using ANSYS," *IEEE Trans. Electron Devices*, Vol. ED-34, 1987, pp. 1214–1217.

[62] J. M. Ortega and W. C. Rheinboldt, *Iterative Solution of Nonlinear Equations in Several Variables*, New York, Academic Press, 1970.

[63] R. E. Bank and D. J. Rose, "Parameter Selection for Newton-like Methods Applicable to Nonlinear Partial Differential Equations," *SIAM J. Numer. Anal*, Vol. 17, 1980, pp. 806–822.

[64] W. L. Engl, H. K. Dirks, and B. Meinerzhagen, "Device Modeling," *Proc. IEEE*, Vol. 71, 1983, pp. 10–33.

[65] G. De Mey, "Potential Calculations in Hall Plates," *Advances in Electronics and Electron Physics*, Vol. 61, 1983, pp. 1–62.

[66] M. Guvenc, "Finite-Element Analysis of Magnetotransistor Action," *IEEE Trans. Electron Devices*, Vol. ED-35, 1988, pp. 1851–1860.

[67] C. C. Lee, A. L. Palisoc and J. M. W. Baynham, "Thermal Analysis of Solid State Devices Using the Boundary Element Method," *IEEE Trans. on Electron Devices*, Vol. ED-35, 1988, pp. 1151–1153.

[68] W. C. Thacker, "A Brief Review of Techniques for Generating Irregular Computational Grids," *Int. J. for Num. Math. in Eng.*, Vol. 15, 1980, pp. 1335–1341.

[69] R. B. Simpson, "A Survey of Two Dimensional Finite Element Mesh Generation," *Proc. 9th Manitoba Conference in Numerical Math. and Computing*, 1979, pp. 49–124.

[70] J. F. Thompson, Z. U. A. Warsi, and C. Wayne Mastin, *Numerical Grid Generation*, New York, North Holland, 1985.

[71] W. J. Gordon and C. A. Hall, "Construction of Curvilinear Coordinate System and Applications to Mesh Generation," *Int. J. for Num. Meth. in Eng.*, Vol. 7, 1973, pp. 461–477.

[72] Z. M. V. Kovacs and M. Rudan, "Boundary Fitted Coordinate Generation for Device Analysis on Composite and Complicated Geometries," *IEEE Trans. Computer-Aided Design*, Vol. 10, 1991, pp. 1242–1250.

[73] I. Babuska and W. C. Rheinboldt, "Error Estimates for Adaptive Finite Element Computations," *SIAM J. Numer. Anal.*, Vol. 15, 1978, pp. 736–754.

[74] J. C. Cavendish, "Automatic Triangulation of Arbitrary Planar Domains for the Finite Element Method," *Int. J. for Num. Meth. in Eng.*, Vol. 8, 1974, pp. 679–696.

[75] A. Bykat, "Automatic Generation of Triangular Grid: I—Subdivision of General Polygon into Convex Subregions, II—Triangulation of Convex Polygons," *Int. J. for Num. Meth. in Eng.*, Vol. 10, 1976, pp. 1329–1342.

[76] M. A. Yerry and M. S. Shephard, "A Modified Quadtree Approach to Finite Element Mesh Generation," *IEEE Computer Graphics and Applications*, Vol. 3, 1983, pp. 39–46.

[77] P. Conti and W. Fichtner, "Automatic Grid Generation for Third Device Simulation," in *Simulation of Semiconductor Devices and Processes*, ed. G. Baccarani and M. Rudan, Bologna, Tecnoprint, 1988, pp. 497–505.

[78] P. Ciampolini, A. Pierantoni, and G. Baccarani, "Efficient 3-D Simulation of Complex Structures," *IEEE Trans. Computer-Aided Design*, Vol. 10, 1991, pp. 1141–1149.

[79] G. Heiser, C. Pommerell, J. Weis, and W. Fichtner, "Three-Dimensional Numerical Semiconductor Device Simulation: Algorithms, Architectures, Results," *IEEE Trans. Computer-Aided Design*, Vol. 10, 1991, pp. 1218–1230.

[80] W. M. Coughran, Jr., M. R. Pinto, and R. K. Smith, "Adaptive Grid Generation for VLSI Device Simulation," *IEEE Trans. Computer-Aided Design*, Vol. 10. 1991, pp. 1259–1275.

[81] P. Conti, N. Hitschfeld, and W. Fichtner, "Ω—An Octree-Based Mixed Element Grid Allocator for the Simulation of Complex 3-D Device Structures," *IEEE Trans. Computer-Aided Design*, Vol. 10, 1991, pp. 1231–1241.

[82] J. F. Burgler, W. M. Coughran, Jr., and W. Fichtner, "An Adaptive Grid Refinement Strategy for the Drift-Diffusion Equations," *IEEE Trans. Computer-Aided Design*, Vol. 10, 1991, pp. 1251–1258.

[83] G. Strang and G. Fix, *An Analysis of the Finite Element Method*, Englewood Cliffs, NJ, Prenctice-Hall, 1973.

[84] I. Babuska and A. K. Aziz, "On the Angle Condition in the Finite Element Method," *SIAM J. Numer. Anal.*, Vol. 13, 1976, pp. 214–216.

[85] P. Ciampolini, A. Forghieri, A. Pierantoni, M. Rudan, and G. Baccarani, "A Flexible and Efficient Adaptive Refinement Scheme Using a Local Solution Procedure," in *Simulation of Semiconductor Devices and Processes*, ed. G. Baccarani and M. Rudan, Bologna, Tecnoprint, 1988, pp. 507–518.

[86] W. Schilders, "A Novel Approach to Adaptive Meshing for the Semiconductor Problem," in

Simulation of Semiconductor Devices and Processes, ed. G. Baccarani and M. Rudan, Bologna, Tecnoprint, 1988, pp. 519–528.

[87] K. Deljouie-Rakhshandeh, "A Self-Adaptive Approach for Numerical Device Simulation," in *Simulation of Semiconductor Devices and Processes*, ed. G. Baccarani and M. Rudan, Bologna, Tecnoprint, 1988, pp. 529–548.

[88] R. Ismail and G. Amaratunga, "Adaptive Meshing Schemes for Simulating Dopant Diffusion," *IEEE Trans. Computer-Aided Design*, Vol. 9, 1990, pp. 276–289.

[89] M. Rudan, R. Guerrieri, P. Ciampolini, and G. Baccarani, "Discretization Strategies and Software Implementation for a General Purpose 2D-Device Simulator, in "Problems and New Solutions for *Device and Process Modeling*, ed. J.J.H. Miller, Dublin, Boole, 198, pp. 110–121.

[90] K. Chau, W. Allegretto, A. Nathan, and W. Joerg, "A Hybrid Mesh Generation Procedure for Numerical Simulation of Microsensors," to be published.

[91] B. Delaunay, "Sur la sphère vide," *Bull. Acad. Sci. USSR (VII), Classe Sci., Mat. Nat.*, 1934, pp. 793–800.

[92] B. Joe, "Delaunay Triangulation Meshes in Convex Polygons," *SIAM J. Sci. Stat. Comput.*, Vol. 7, 1986, pp. 514–539.

[93] P. E. Cottrell and E. M. Buturla, "Two-Dimensional Static and Transient Simulation of Mobile Carrier Transport in a Semiconductor," in *Proc. NASECODE I*, ed. B. T. Browne and J. J. H. Miller, Dublin, Boole Press, 1979, pp. 31–64.

[94] D. L. Scharfetter and H. K. Gummel, "Large-Signal Analysis of a Silicon Read Diode Oscillator," *IEEE Trans. Electron Devices*, Vol. ED-16, 1969, pp. 64–77.

[95] W. Allegretto, A. Nathan, and H. P. Baltes, "Numerical Analysis of Magnetic-Field-Sensitive Bipolar Devices," to be published.

[96] I. S. Duff, A. M. Erisman, and J. K. Reid, *Direct Methods for Sparse Matrices*, Oxford, Clarendon Press, 1986.

[97] L. A. Hageman and D. M. Young, *Applied Iterative Methods*, New York, Academic Press, 1981.

[98] K. Wu, G. R. Chin, and R. W. Dutton, "A STRIDE Towards Practical 3-D Device Simulation-Numerical and Visualization Considerations," *IEEE Trans. Computer-Aided Design*, Vol. 10, 1991, pp. 1132–1140.

[99] Z. Zhao, Q. Zhang, G. Tan, and J. Xu, New Preconditions for CGS Iteration in Solving Large Sparse Nonsymmetric Linear Equations in Semiconductor Device Simulation, *IEEE Trans. Computer-Aided Design*, Vol. 10, 1991, pp. 1432–1440.

[100] A. George and J. W. Liu, "The Design of a User Interface for a Sparse Matrix Package," *ACM Trans. Math. Software*, Vol. 5, 1979, pp. 139–162.

[101] A. George, J. W. Liu, and E. Ng, *User Guide for SPARSPAK*, Waterloo Sparse Linear Equations Package, 1980.

[102] D. M. Webber, E. Tomacruz, R. Guerrieri, T. Toyabe, and A. Sangiovanni-Vincentelli, "A Massively Parallel Algorithm for Three-Dimensional Device Simulation," *IEEE Trans. Computer-Aided Design*, Vol. 10, 1991, pp. 1201–1209.

[103] C. L. Wilson and J. L. Blue, "Accurate Current Calculation in Two Dimensional MOSFET Models," *IEEE Trans. Electron Devices*, Vol. ED-32, 1985, pp. 2060–2068.

[104] W. Allegretto, Y. S. Mun, A. Nathan, and H. P. Baltes, "Optimization of Semiconductor Magnetic Field Sensors Using Finite Element Analysis," in *Proc. NASECONDE IV Conf.*, 1985, pp. 129–133.

[105] A. Nathan, W. Allegretto, H. P. Baltes, and Y. Sugiyama, "Carrier Transport in Semiconductor Magnetic Domain Detectors," *IEEE Trans. Electron Devices*, Vol. ED-34, 1987, pp. 2077–2085.

[106] A. Nathan, W. Allegretto, H. P. Baltes, Y. Sugiyama, and M. Tacano, "Position Sensitive Geometric Factor for Semiconductor Hall Devices Detecting Magnetic Domains," *Transducers '87, Digest of Technical Papers*, 1987, pp. 536–537.

[107] W. Allegretto, A. Nathan, K. Chau, and H. P. Baltes, "Two Dimensional Numerical Simulations

in Electrothermal Behaviour in Very Large Scale Integrated Contacts and Vias," *Canad. J. Physics*, Vol. 67, 1989, pp. 212–217.

[108] K. Chau, W. Allegretto, and Lj. Ristic, "Thermal Modeling of CMOS Temperature/Flow Microsensors," *Canad. J. Physics*, Vol. 69, 1991, p. 212.

[109] G. Ghione, P. Golzio, and C. Naldi, "Thermal Analysis of Power GaAs MESFETS," in *Proc. NASECODE V Conf.*, 1987, pp. 195–200.

[110] R. W. Clough and J. L. Tocher, "Finite Element Stiffness Matrices for Analysis of Plates in Bending," *Proceedings Conf. on Matrix Methods in Structural Mechanics*, Wright-Patterson Air Force Base, Dayton, Ohio, 1966.

[111] A. W. Vinal, "Considerations for Applying Solid State Sensors to High Density Magnetic Disc Recordings," *IEEE Trans. Magnetics*, Vol. MAG-20, 1984, pp. 681–686.

[112] Y. Sugiyama, "Fundamental Research on Hall Effects in Inhomogeneous Magnetic Fields," *Res. Electrotech. Lab.*, No. 838, Electrotech. Lab., Tokyo, 1983.

[113] I. Hlasnik and J. Kokavec, "Hall Generator in Inhomogeneous Field and Dipole Notion of the Hall Effect," *Solid-State Electron*, Vol. 9, 1966, pp. 585–594.

[114] J. H. J. Fluitman, "On the Calculation of the Response of (Planar) Hall Effect Devices to Inhomogeneous Magnetic Fields," *Sensors and Actuators*, Vol. 2, 1981–82, pp. 155–170.

[115] J. Brunner, "Der Halleffect in inhomogenen Magnet Field," *Solid-State Electron*, Vol. 2, 1960, pp. 172–175.

[116] G. De Mey, "Hall Effect in a Nonhomogeneous Magnetic Field," *Solid-State Electron*, Vol. 20, 1977, pp. 139–142.

[117] A. W. Baird, "The Hall Cross in a Perpendicular Inhomogeneous Magnetic Field," *IEEE Trans. Magnetics*, Vol. MAG-15, 1979, pp. 1138–1141.

[118] D. M. Caughey and R. F. Thomas, "Carrier Mobilities in Silicon Empirically Related to Doping and Field," *Proc. IEEE*, Vol. 52, 1967, pp. 2192–2193.

[119] K. E. Petersen, "Micromechanical Light Modulator Array Fabrication on Silicon," *Appl. Phys. Lett.*, Vol. 31, 1977, pp. 521–523.

[120] K. E. Petersen, "Silicon as a Mechanical Material," *Proc. IEEE*, Vol. 70, 1982, pp. 420–457.

[121] M. A. Huff, S. D. Senturia, and R. T. Howe, "A Thermally Isolated Microstructure Suitable for Gas Sensing Applications," *Tech. Digest*, IEEE Solid State Sensors and Actuators Workshop, Hilton Head, SC, June 1988, pp. 47–50.

[122] K. Chau, W. Allegretto, and Lj. Ristic, "Simulation of Silicon Microstructures," *Sensors and Materials*, Vol. 2, No. 5, 1991, p. 253.

[123] C. H. Mastrangelo and R. S. Muller, "A Constant-Temperature Gas Flowmeter with a Silicon Micromachined Package," *Tech. Digest*, IEEE Solid State Sensors and Actuators Workshop, Hilton Head, SC, June 1988, pp. 43–46.

[124] D. Moser, R. Lenggenhager, and H. Baltes, "Silicon Gas Flow Sensors Using Industrial CMOS and Bipolar IC Technology," *Sensors and Actuators*, Vols. 25–27, 1991, pp. 577–581.

[125] T. Tschan, N. De Rooij, A. Bezinge, S. Ansermet, and J. Berthoud, "Characterization and Modelling of Silicon Piezoresistive Accelerometers Fabricated by a Bipolar-Compatible Process," *Sensors and Actuators*, Vols. 25–27, 1991, pp. 605–609.

[126] R. Frank, "Pressure Sensors Merge Micromachining and Microelectronics," *Sensors and Actuators*, Vol. 28, 1991, pp. 93–103.

[127] H. J. M. Geijselaers and H. Tijdeman, "The Dynamic Mechanical Characteristics of a Resonating Micro-bridge Mass-Flow Sensor," *Sensors and Actuators*, Vol. 29, 1991, pp. 37–41.

[128] W. Riethmuller and W. Benecke, "Thermally Excited Silicon Microstructures," *IEEE Trans. Electron Devices*, Vol. ED-35, 1988, pp. 758–763.

[129] D. Gardner and P. A. Flinn, "Mechanical Stress as a Function of Temperature in Aluminimum Films," *IEEE Trans. Electron Devices*, Vol. ED-35, 1988, pp. 2160–2169.

Chapter 3
Bulk Micromachining Technology

Lj. Ristic, H. Hughes, and F. Shemansky
Motorola

3.1 INTRODUCTION

The significant progress made in recent years with microsensors and microactuators is due primarily to advances in micromachining technology. Micromachining technology allows the fabrication of mechanical structures with very small dimensions (in the micrometer range). Generally, micromachining technology can be classified as either bulk micromachining or surface micromachining. Bulk micromachining technology is based on single crystal silicon etching, and the micromechanical structures developed with this technology are made of either silicon crystal or deposited or grown layers on silicon. Surface micromachining, on the other hand, utilizes deposited or grown layers on top of the substrate to fabricate micromechanical devices. There are considerable differences between these two approaches; therefore, they are described in two separate chapters. Chapter 3 will focus on bulk micromachining, and Chapter 4 will cover surface micromachining.

Micromechanical structures produced using micromachining technology can be divided into three groups: static, dynamic, and kinematic. Static micromechanical devices include fixed three-dimensional structures such as nozzles [1–3], cavities [4], capillary columns [5,6], circular orifices, and miniature electrical connectors [7]. Examples of dynamic micromechanical structures are diaphragms and membranes [8–12], microbridges [13–16], cantilever beams [13,17–24], and resonators [17,18,25,26]. The common characteristic for dynamic devices is that they require controlled displacement to accomplish the desired function. The kinematic group includes micromotors, microgears, *pin* joints, springs, cranks, and sliders [27–34]. At the present time, micromechanical structures from the first two groups are used mainly as components of larger sensor systems. The kinematic (movable) devices are considered to be essential in the future for microrobotics and microsurgery.

In this chapter, the basic concepts of bulk micromachining technology are discussed. Included in the discussion are wet anisotropic etching and related topics such as crystallographic orientation–dependent etch rate, patterning, selectivity, and etchant variations. Processing issues of integrated sensor systems are then illustrated using CMOS technology as an example. Finally, some practical applications of bulk micromachining are outlined.

3.2 BASIC CONCEPT OF BULK MICROMACHINING

Micromechanical structures fabricated using bulk micromachining concepts are typically constructed either of silicon crystal or composite materials deposited or grown on the silicon substrate. Because silicon crystal is frequently used as a core material for bulk micromachined structures, we start this chapter with a short discussion of silicon as a mechanical material.

3.2.1 Silicon as a Mechanical Material

Single crystal silicon is composed of atoms arranged in a diamond structure with a cubic symmetry. Although silicon has a tendency to cleave along crystallographic planes because of the bulk defects in the crystal, it is a mechanical material with high strength. Important mechanical properties of silicon are shown in Table 3.1. It should be pointed out that silicon crystal shows a yield strength two times higher than steel and a Young's modulus very close to that of steel, yet silicon has a density very close to aluminum and it is only one-third the density of steel. It does not deform plastically—silicon crystal is entirely elastic; therefore, it does not show mechanical hysteresis. The hardness of silicon is close to quartz and higher than the hardness of iron, tungsten, and aluminum. Silicon's thermal expansion coefficient is one-fifth that of steel and its thermal conductivity is about 50% higher than the

Table 3.1
Important Mechanical Properties of Silicon Crystal

Property	Measure
Crystal structure	Diamond, 8 atoms/unit cell
Melting point	1415°C
Thermal expansion	$2.5 \times 10^{-6}/°C$
Density	2.3 g/cm^3
Young's Modulus	1.9×10^{12} dyn/cm^2
Yield strength	6.9×10^{10} dyn/cm^2
Knoop hardness	850 kg/m^2

thermal conductivity of steel. All of these attributes suggest that silicon is really an excellent mechanical material.

In addition, silicon is highly stress sensitive and demonstrates a significant piezoresistive effect. Piezoresistivity is a phenomena in which an external applied stress is manifested in a change in material resistivity. Since the discovery of the piezoresistivity effect in silicon in 1954 by C.S. Smith [35], silicon has been used for strain gauge applications because the piezoresistive effect in silicon is about 100 times greater than in metals. The piezoresistive effect is also used extensively in pressure sensors and accelerometers. The piezoresistive effect theory is rather detailed with a full explanation involving the band-gap model of semiconductors. We will outline here just major conclusions.

The parameter commonly used for quantifying the piezoresistive effect is the gauge factor. The gauge factor relates the relative change of the resistivity to the strain. It depends on crystallographic orientation of the material. For example, *p*-type silicon shows a maximum gauge factor for the ⟨111⟩ orientation whereas *n*-type silicon shows a maximum gauge factor for the ⟨100⟩ plane. The *n*-type silicon shows a negative gauge factor (the resistivity decreases with stress increase) whereas *p*-type silicon has a positive gauge factor. The gauge factor for *p*-type silicon is larger than the gauge factor for *n*-type silicon. As a consequence, piezoresistors are usually designed as *p*-type resistors. The gauge factor also depends on impurity concentration and temperature. It increases when the impurity concentration decreases: silicon with higher resistivity shows a stronger piezoresistive effect. However, lightly doped semiconductors show a large gauge factor temperature dependency. Therefore, in the design of sensors that use the piezoresistive effect, the compromise between a higher sensitivity and the possible need for temperature compensation has to be considered.

3.2.2 Wet Anisotropic Etching

Wet anisotropic etching is a process of preferential directional etching of material using liquid source etchants. Anisotropic silicon etching (crystal orientation-dependent etching) was reported in the late 1960s and early 1970s [36–40]. (Much of this early work has been admirably reviewed by Bean [41].) However, the full potential of this technique was not recognized until the late 1970s and early 1980s [17–25].

Numerous anisotropic etchants are used for silicon, such as hydrazine–water solution [39,42,43], EDP (ethylenediamine–pyrocatechol–water) [38,41,44,45], KOH (potassium hydroxide–water) [41,46,47], TMAH (tetramethylammonium hydroxide) [48,49], and CsOH (cesium hydroxide–water) [50]. They are composed of a primary component (hydrazine, ethylenediamine, potassium hydroxide, tetramethylammonium hydroxide, cesium hydroxide), a complexing agent (isopropyl alcohol, catechol), and a diluent (water). The most frequently used etchant is potassium hydroxide but other etchants are also used, depending on specific requirements. The unique features of various etchants are discussed later. Here, we start with the common features of anisotropic etching.

3.2.2.1 Crystallographic Orientation–Dependent Etch Rate

A basic feature of anisotropic etchants is that their etch rates are strongly dependent on crystallographic orientation. More specifically, ⟨111⟩ surfaces etch at a slower rate than all other crystallographic planes. This indicates that the dissolution rate will be a function of the crystal orientation of the silicon wafer. Qualitatively, anisotropic etching is a function of the areal density of atoms (number of atoms per square centimeter), the energy needed to remove an atom from the surface, and geometric screening effects (three-dimensional distribution of atoms in the lattice).

As a consequence of anisotropy, it is possible to develop unique structures not otherwise feasible. For example, consider a ⟨100⟩-oriented silicon wafer with an etched hole in a layer of silicon dioxide that covers the surface. When exposed to an anisotropic etchant this will create a truncated pyramidal shaped pit as shown in Figure 3.1(a). The pit is bounded by ⟨111⟩ crystallographic planes that exhibit a very low etch rate. The ⟨111⟩ planes have an inclination of 54.7°. The orientation dependence of the EDP etch rate on a ⟨100⟩-oriented silicon wafer is shown in Figure 3.1(b).

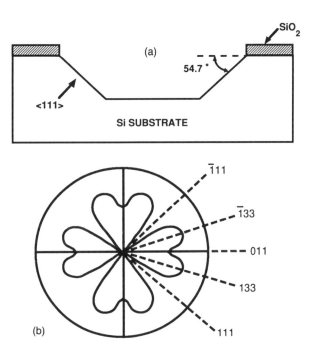

Figure 3.1 (a) Truncated pyramidal pit bounded by silicon ⟨111⟩ crystallographic planes; (b) EDP etch rate dependence on crystallographic orientation in ⟨100⟩ silicon.

The maxima of etch rates coincide with $\langle 133 \rangle$ planes, and the etch rate on $\langle 111 \rangle$ planes is almost 0 [24,51]. Exploiting this and the crystal symmetry, it is possible to fabricate shapes like the one in Figure 3.2(a) or V-grooves with uniform walls defined by the $\langle 111 \rangle$ planes. The truncated pyramidal pit in Figure 3.2(b) was obtained by etching a $\langle 100 \rangle$-oriented silicon wafer in EDP. Similar very deep, narrow structures with uniform vertical sidewalls are possible, if the orientation of the wafer surface is $\langle 110 \rangle$ [24,25,46,47]. It should be pointed out that, in many cases, it is difficult to determine which crystallographic planes will etch fastest because the etch rate will vary as a function of the etchant formulation and local etching conditions.

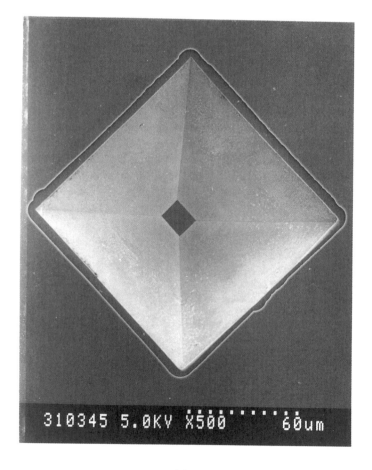

310345 5.0KV X500 60um

(a)

Figure 3.2 (a) SEM of etched truncated pyramidal pit; (b) V-groove anisotropically etched in silicon.

Figure 3.2 continued.

3.2.2.2 *Patterning*

The geometry of three-dimensional structures etched through the openings in a mask, in addition to the other factors, depends on the shape of the opening itself. To obtain the desired structure the edges of the opening must be oriented correctly. For example, to obtain the truncated pyramidal shaped pit shown in Figure 3.2(b), the edges of the opening have to be aligned in the $\langle 110 \rangle$ directions (Figure 3.3). The mask shown in Figure 3.3 has only concave corners, which are not usually undercut if the opening is oriented properly. If the edges of the opening are misaligned, con-

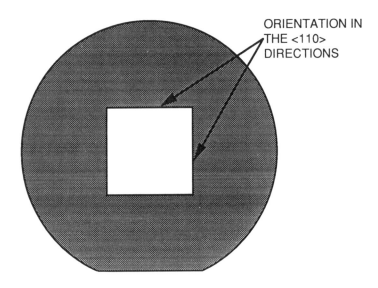

Figure 3.3 Proper alignment of the mask opening.

cave corners and edges will be undercut [7,47]. Figure 3.4 shows undercutting of edges not properly aligned. Table 3.2 indicates the relationship between the properly oriented elemental shapes, surface orientation, and the three-dimensional structures produced.

Very often the geometry of the opening includes convex corners, and irrespective of alignment conditions, these corners will inevitably be undercut [39,41,43,51]. The mechanism for this phenomenon is not completely clear. It has been found experimentally that the undercutting depends on the total etching time

Figure 3.4 Undercutting of improperly aligned edges. Improperly aligned edges are the corners on the right side.

Table 3.2
The Relationship Between Opening, Surface Orientation, and Structure

Opening (Window)	Surface Orientation	Structure
Square	$\langle 100 \rangle$	Pyramidal pit or truncated pyramidal pit
Rectangle	$\langle 100 \rangle$	Rectangular pit (trench)
Circle	$\langle 100 \rangle$	Pyramidal pit
Arbitrarily shaped close pattern	$\langle 100 \rangle$	Rectangular pit
Square or rectangle	$\langle 110 \rangle$	Hole with vertical side walls

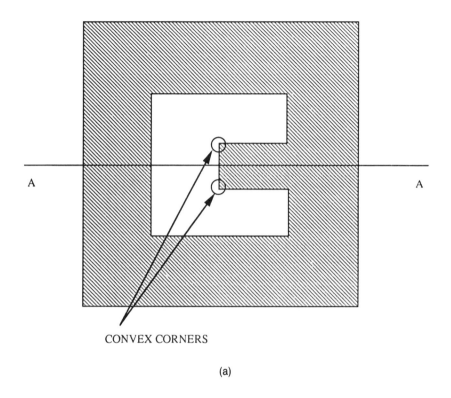

CONVEX CORNERS

(a)

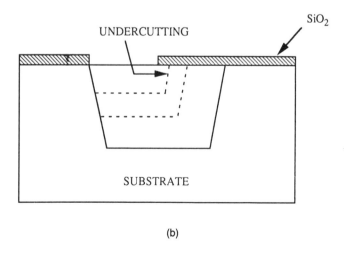

(b)

Figure 3.5 (a) Mask opening with convex corners; (b) sideview of undercutting convex corners.

and the amount of local surface area actively attacked by the etch. Figure 3.5(a) illustrates the pattern with two convex corners and the undercutting phenomena taking place during silicon etching, Figure 3.5(b). A schematic representation of undercutting on four convex corners is shown in Figure 3.6. Several different examples of undercutting are shown in Figures 3.7–3.9. Figure 3.7 shows a spiral structure with undercutting starting at the convex corners. Figure 3.8 shows an array of cantilevers with the front of undercutting starting at the convex corners and then progressing from the tip of the cantilever toward the supporting point. Finally, Figure 3.9, shows a full aspect of undercutting. The oxide on the silicon surface is patterned as a central square supported by four tethers in diagonal fashion. When anisotropic etching begins, undercutting of the four tethers takes place because they are not aligned to the $\langle 110 \rangle$ directions. The tethers are fully undercut, creating four convex corners. This in turn causes further undercutting of the oxide that is patterned as a central square. Figure 3.9 shows the structure with silicon still supporting the oxide in the center. If the etching continues, the central square will be fully undercut and suspended. Buckling of the tethers is caused by compressive stress in the oxide.

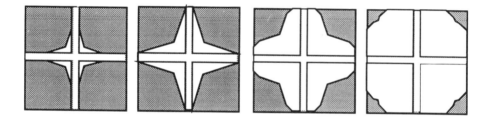

Figure 3.6 Schematic representation of undercutting propagating for the pattern with four convex corners.

Figure 3.7 Undercutting in a spiral structure.

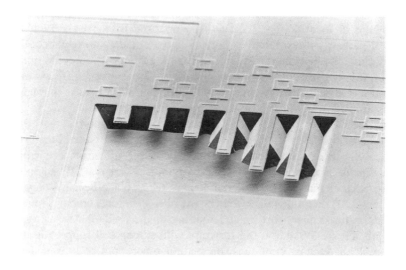

Figure 3.8 Undercutting progression in a cantilever array.

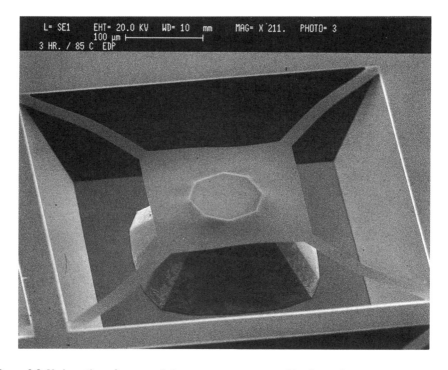

Figure 3.9 Undercutting of a suspended mass structure supported by four tethers.

3.2.2.3 Selectivity

Anisotropic etchants can be highly material selective, which indicates that they may be masked by different materials (silicon dioxide, silicon nitride, chromium, gold, and so on). For example, because the hydrazine–isopropyl alcohol–water solution does not attack silicon dioxide and aluminum films [39], either of these could be used as a masking material. EDP does not attack gold, chromium, silver, or tantalum; and the etch rate of silicon dioxide and silicon nitride is very low compared to silicon [25]. However, EDP may attack aluminum. Cesium hydroxide has a silicon $\langle 110 \rangle$ to silicon dioxide etch ratio of 5500 and also exhibits low tantalum etch rate [50], allowing either to be used as a masking material. A disadvantage of a potassium hydroxide solution is that it etches silicon dioxide much faster than the other anisotropic etches, which almost precludes the use of silicon dioxide alone as a masking material when using KOH [25]. Tetramethylammonium hydroxide shows higher selectivity to silicon dioxide and silicon nitride and, if properly prepared, can be selective to aluminum.

It should be pointed out that photoresist is not usually suitable as a masking material because of the long etch times (typically several hours or longer) with caustic etchants used to produce deep structures. The etch conditions can result in lift-off or degradation of the photoresist material.

3.2.3 Anisotropic Etchants

In this section, we describe briefly the most common etchants currently used for anisotropic etching. For historical reasons, we start with ethylenediamine-pyrocatechol, followed by potassium hydroxide (currently the most frequently used etchant for sensors), tetramethylammonium hydroxide, and finally cesium hydroxide.

3.2.3.1 Ethylenediamine–Pyrocatechol–Water (EDP)

Crishal and Harrington [37] were among the first to report anisotropic silicon etching using a hydrazine–pyrocatechol mixture. Hydrazine is very toxic, and in a later paper, Finne and Klein [38] reported the development of an ethylenediamine–pyrocatechol–water based mixture as an anisotropic silicon etchant. They were able to detail many of the essential features of the process as it is used today. The etch rate as a function of the water content of the etchant was studied over the entire possible range of concentrations from 0 to 100%. The etch rate shows a maximum at a composition corresponding to a mole ratio of water to ethylenediamine of 2. The etch rate falls to 0 at the extreme ends of the water concentration scale. The etch rate curve shape was found to be independent of the resistivity of the silicon in the range studied.

The pyrocatechol content of the etch was also studied. The rate was reduced when the pyrocatechol was omitted from the etchant composition, falling from about 35 μm/hr at the maximum 3.7 mole percent pyrocatechol to 18 μm/hr with no pyrocatechol. These observations indicate that water and amine, unlike pyrocatechol, are necessary ingredients in the etch. The authors suggested that the rate controlling step may be due to concentration of hydroxyl ions, OH^-, in the solution.

A chemical mechanism for the anisotropic etching of silicon was proposed by Finne and Klein [38]. It was noted that water is a necessary component in the etch. This is based on the observation that no silicon is etched in anhydrous amine or anhydrous amine–pyrocatechol mixtures. The overall reaction was represented as

$$2NH_2(CH_2)_2NH_2 + Si + 3C_6H_4(OH)_2 \rightarrow 2NH_2(CH_2)_2NH_3^+ + [Si(C_6H_4O_2)_3]^{--} + 2H_2$$
$$(3.1)$$

The proposed mechanism indicates that the reaction should not be limited by water depletion, and the etch rate should remain constant over long periods of time in the presence of sufficient amine and complexing agent. Silanol groups (Si-OH) are converted to hydrated silica prior to dissolution of the pyrocatecholato complex.

Many studies have been done on the chemistry of etching with EDP [7,37,38,43], but the toxic nature of this etchant has greatly reduced its use.

3.2.3.2 Potassium Hydroxide (KOH)

The use of alcoholic potassium hydroxide, KOH, as a silicon anisotropic etchant has been described in detail by J.B. Price [52]. In the Price work, silicon etching in ternary mixtures of KOH, H_2O, and isopropyl alcohol (IPA) were used to study the etch rate of silicon as a function of etchant composition, etch temperature, silicon orientation, and resistivity. Use of this anisotropic etchant was also reported by Mok and Salama in the manufacture of a V-groove Schottky-barrier FET for UHF applications [53].

Typically heterogeneous reactions, such as the etching of silicon in potassium hydroxide–isopropyl alcohol–water mixtures, are controlled by one of the following processes:

1. Diffusion of reactant molecules through the boundary layer to the solid surface,
2. Adsorption of reactant molecules on the solid surface,
3. Surface reaction,
4. Reaction product desorption,
5. Diffusion of by-products back across the boundary layer into the bulk of the solution.

The temperature dependence of diffusion controlled reactions (for example, items 1 and 4), typically yields an apparent activation energy, $E_a \approx 4.5$ kcal/mole (0.2 eV).

Price reported apparent activation energies of 12–16 kcal/mole. Diffusion controlled reactions may have a dependence on the stir rate although none was reported by Price. Therefore, it was concluded that diffusion was not the rate controlling step and surface processes were most likely rate limiting in this case. A relation was also found between the reaction rate and the available number of bonds for the various silicon crystal orientations. Therefore, it was assumed that the rate controlling step in this study was the oxidation of silicon surface atoms by reactant reduction; for example,

$$Si + H_2O + 2KOH \rightarrow K_2SiO_3 + 2H_2 \tag{3.2}$$

Seidel [54] also studied the mechanism of anisotropic silicon etching. Based upon the alkalinity of the solution, the development of hydrogen bubbles during etching, the inverse fourth-power law for the etch rate (for highly boron doped silicon), and the fourth-order dependence of the etch rate on the water concentration (for high KOH concentrations), he proposed the following mechanism:

$$Si + 2OH^- \rightarrow Si(OH)_2^{++} + 4e^- \tag{3.3}$$

$$4H_2O + 4e^- \rightarrow 4OH^- + 2H_2 \tag{3.4}$$

$$Si(OH)_2^{++} + 4OH^- \rightarrow SiO_2(OH)_2^{--} + 2H_2O \tag{3.5}$$

The overall reaction is summarized as

$$Si + 2OH^- + 2H_2O \rightarrow SiO_2(OH)_2^{--} + 2H_2 \tag{3.6}$$

Price used a two-phase mixture; phase one being isopropyl alcohol rich with a small concentration of water and KOH (see Figure 3.10); and phase two, a water and KOH-rich mixture with a small concentration of IPA. The variables $X_2(IPA)$, $X_2(H_2O)$, and $X_2(KOH)$ represent the composition of phase two in equilibrium with phase one of composition $X_1(IPA)$, $X_1(H_2O)$, and $X_1(KOH)$. $X(IPA)$, $X(H_2O)$, and $X(KOH)$ represent the composition of the entire system in weight percent. ($X_1^!(KOH)$ is the composition of the binary mixture of KOH and water prior to the addition of IPA.)

A typical manufacturing etch solution is made up of 5000 ml of 40% aqueous KOH with sufficient IPA added to ensure separation into two phases. Additional IPA is added to produce sufficient liquid on top of the solution (typically *ca.* 2 cm will suffice) to ensure two-phase operation. Etchants are generally used between 80°C and 90°C.

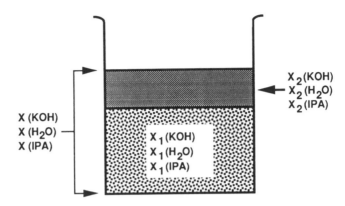

Figure 3.10 Two-phase etch system.

The effect of the etchant composition, saturated with isopropyl alcohol, on an etch rate of ⟨100⟩ crystallographic silicon–orientation is shown in Figure 3.11 [52]. The etch rate of ⟨100⟩ Si increases with KOH concentration quickly and reaches a maximum near 30 wt%–40 wt% at about 1 μm/min. The anisotropy ratio, ⟨100⟩/⟨111⟩, equals 34 at a concentration of 40 wt% KOH with IPA saturated in the etch-

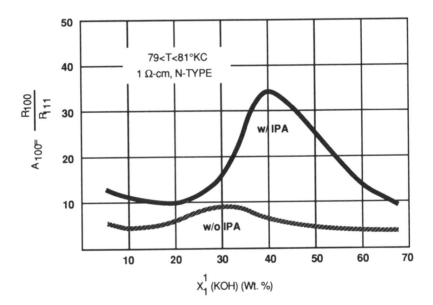

Figure 3.11 Silicon etch selectivity (A_{100}) as a function of KOH concentration with and without IPA at 80°C.

ant. Without IPA, the anisotropy ratio is only 8. Thus, the presence of IPA is vital to achieve a superior anisotropy ratio. Similar results are found using $\langle 110 \rangle$-oriented silicon as opposed to $\langle 111 \rangle$ material.

The resistivity dependence of the etch rate has also been determined for the various commonly used crystal orientations of silicon. The $\langle 111 \rangle$ etch rate, 0.02 μm/min, was found to be independent of resistivity for the dopants As, P, B, and Sb in 40 wt% KOH at 80°C. The $\langle 110 \rangle$ etch rate is 0.37 μm/min and independent of the resistivity. The $\langle 100 \rangle$ etch rate, 0.94 μm/min, is also independent of the doping concentration, N_x, for As, P, and Sb. For boron, however, the $\langle 110 \rangle$ etch rate decreases at very high doping levels from 0.94 μm/min at $N_B = 10^{18}$ atoms per cm³ to only 0.02 μm/min at $N_B = 10^{20}$ atoms per cm³.

3.2.3.3 Tetramethylammonium Hydroxide (TMAH)

The compatibility of anisotropic etching with materials used in IC fabrication is a necessary condition to make the system integration fully manufacturable. This is not always achievable with the currently available anisotropic etchants. Other difficulties include low etch rate ratio of silicon to oxide (with KOH), the toxic nature of etchants (hydrazine, ethylenediamine, pyrocatechol), and the high cost (cesium hydroxide).

Tabata et al. [48] reported on the use of quaternary ammonium hydroxide, tetraethylammonium hydroxide (TEAH) and tetramethylammonium hydroxide (TMAH),

$$\left[\begin{array}{c} CH_3 \\ | \\ CH_3 - N - CH_3 \\ | \\ CH_3 \end{array} \right]^+ OH^-$$

in electrochemical etching of $\langle 100 \rangle$ and $\langle 111 \rangle$ silicon. Solutions of various concentrations from 5 wt% to 40 wt% were used. The maximum etch rate of $\langle 100 \rangle$ silicon was observed with 5 wt% with TMAH, however, formation of hillocks were observed at <15 wt%. Maximum etch rate for TEAH was with 20 wt% with hillocks appearing at <20 wt%. TMAH in a concentration of 22 wt% at 90°C provided 1 μm/min for $\langle 100 \rangle$ with the lowest $\langle 100 \rangle/\langle 111 \rangle$ etch ratio. The latter etch rate is lower than observed for KOH or EDP but similar to that observed with NH₄OH. TMAH was readily controllable and the etch rate was constant over long etching times, as it does not decompose below 130°C. The etch ratio of $\langle 100 \rangle/\langle 111 \rangle$ was 35 at 5 wt% and 90°C and decreased with increasing concentration and lower temperature. The etch rate of SiO₂ was almost four orders of magnitude lower than for $\langle 100 \rangle$ silicon. LPCVD Si₃N₄ showed no appreciable etching.

Schnakenberg, Benecke, and Lange [49] reported a maximum etch rate of 39 μm/hr and a $\langle 100 \rangle/\langle 111 \rangle$ silicon etch rate ratio of 50 with 2 wt% TMAH at 80°C. The pH value was found to be the most significant factor affecting the etching characteristics. For pH < 13, hillock formation was observed. Also for pH < 13, the silicon surface needed to be free of oxide or a high density of hillocks, and strong gas evolution was observed. The selectivities of various dielectrics increased when the solution was doped with silicon at 13.5 g silicon per liter etchant (Table 3.3). These dopant levels were also sufficient to passivate aluminum via a native aluminum oxide. Hence aluminum can be used as a metallization layer in silicon-doped TMAH etchants.

Silicon $\langle 100 \rangle$ etch rates were found to be higher than those observed for EDP, hydrazine–water, AHW (ammonium hydroxide–water), and tetraethylammonium hydroxide and lower than for KOH-based etchants. The $\langle 100 \rangle/\langle 111 \rangle$ silicon etch rate ratios in doped TMAH are comparable to KOH, EDP, hydrazine, AHW, and better than those found for TEAH (tetraethylammonium hydroxide).

When the TMAH is doped with silicon, its pH value decreases. This corresponds to a decrease in hydroxyl ions in the bulk solution. Hydroxyl ions from the bulk solution probably do not contribute significantly to the silicon etching mechanism since an increase in $\langle 100 \rangle$ etch rate was observed for concentrated TMAH solutions doped with silicon. The pH value could decrease through the dissociation of the silicon etching reaction product, which acts as a weak acid:

$$Si(OH)_4 \leftrightarrow SiO_2(OH)_2^{--} + 2H^+ \tag{3.7}$$

$$2H^+ + 2OH^- \leftrightarrow 2H_2O \tag{3.8}$$

The hydroxyl ions necessary for the dissolution of silicon must be produced on the silicon surfaces. It has been proposed that water dissociates, forming OH^- ions. In comparison to KOH solutions, hydrogen evolution in TMAH solutions is negligible on $\langle 100 \rangle$ etching fronts with low or no hillock formation.

Table 3.3
Selectivity (silicon $\langle 100 \rangle$/dielectric) in 4 wt% TMAH Solution
and Silicon-Doped Solution (13.5 g silicon/l) at 80°C

Material	Selectivity in Pure TMAH–H₂O Solution	Selectivity in Si-Doped TMAH–H₂O Solution
Thermal oxide	5.3×10^3	34.7×10^3
Low temperature (CVD) oxide	1.3×10^3	4.2×10^3
PECVD–SiO$_x$:H,N	1.4×10^3	4.3×10^3
LPCVD–Si$_3$N$_4$	24.4×10^3	49.3×10^3
PECVD–SiN$_x$:H	9.2×10^3	18.5×10^3

Note: The silicon etch rate was determined to be 39 μm/hr.

Tetramethylammonium hydroxide [55] can be tailored to etch at rates similar to KOH at 95°C at 20 wt% (Figure 3.12). Etch rates for dielectrics and silicon in 20 wt% TMAH at 95°C are shown in Table 3.4. Based upon the values in Table 3.4, the calculated thicknesses required of masking materials to etch through a 20 mil wafer are shown in Table 3.5. TMAH is very attractive as an etchant, because

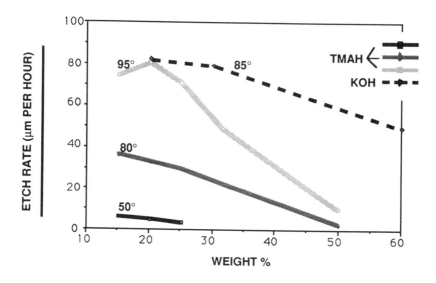

Figure 3.12 Silicon etch rate in TMAH and KOH at various compositions and temperatures.

Table 3.4
Etch Rates of Dielectrics and Silicon
in 20% TMAH at 95°C

	Rate
Dielectric:	
Thermal oxide	124 Å/hr
High temperature oxide	281 Å/hr
Low temperature oxide (360°C)	228 Å/hr
Low temperature oxide (425°C)	190 Å/hr
Thermal nitride	17 Å/hr
Plasma nitride	179 Å/hr
Silicon:	
Polysilicon	77 μm/hr
Silicon $\langle 100 \rangle$	64 μm/hr

Table 3.5
Masking Dielectric Thickness to Etch
a 20 mil Silicon Wafer

Masking Material	Å
Thermal oxide	921
High temperature oxide	2,086
Low temperature oxide (360°C)	1,693
Low temperature oxide (425°)	1,411
Thermal nitride	126
Plasma nitride	1,329

a single, relatively thin film can be used as a mask. Also, TMAH more residue free than other etchants because it contains no metal ions (like the potassium present in KOH).

3.2.3.4 Cesium Hydroxide (CsOH)

This base was selected as an anisotropic silicon etch candidate by Clark, Lund, and Edell [50] at MIT while searching for better selectivity between tantalum and silicon. Weaker bases such as sodium hydroxide and ammonium hydroxide were less selective than stronger bases; therefore, cesium hydroxide was selected for evaluation because of its higher base strength. However, care must be taken when using cesium hydroxide because it is very hygroscopic. Solutions prepared from the solid base must be evaluated to determine the true base content. For example, in Clark et al.'s work [50], one lot was 80.4 wt% CsOH and another was 78.8 wt% CsOH.

It has also been observed [56] that SiO_2 etches more slowly in CsOH than any other group 1 hydroxide. For example, CsOH etches silicon dioxide 5500 times more slowly than it does $\langle 110 \rangle$ silicon. This is an improvement over typical ethylenediamine and pyrocatechol, water-based etchants that attack silicon dioxide about 500 times more slowly than $\langle 110 \rangle$ silicon and KOH, which attacks silicon dioxide only about 200 times more slowly than it does $\langle 110 \rangle$ silicon. This data suggest that a mask of 1000 Å of silicon dioxide could be used to etch 550 μm of $\langle 110 \rangle$ silicon.

Clark et al. [50] used silicon test structures for etch rate determinations. Each wafer was given a 30-s dip in 10:1 H_2O:HF solution just prior to etching to remove any native oxide on the silicon. The presence of a thin oxide layer would have resulted in erroneous silicon etch rate determinations.

The etch rates for $\langle 110 \rangle$ and $\langle 111 \rangle$ oriented silicon, as well as those for silicon dioxide, were studied from 10 wt% to 76 wt% CsOH over the temperature range 25°C to 90°C. The maximum etch rate at 60°C was observed between 50 wt% and 60 wt% CsOH. The maximum selectivity at 60°C for $\langle 110 \rangle$ etch rate versus either

$\langle 111 \rangle$ silicon or silicon dioxide also occurred in this concentration range. Selectivity for $\langle 110 \rangle/\langle 111 \rangle$ silicon etch was 200/1. Selectivity for $\langle 110 \rangle$ Si/SiO$_2$ at 50°C was approximately 8000/1. This latter comparison contrasts to that using 60 wt% KOH of 200/1 and using an ethylenediamine–pyrocatechol–H$_2$O (EDP) etch of 600/1. Thus, the masking ability of oxide to anisotropically etch selected patterns in $\langle 110 \rangle$ silicon is about 40 times better than when using KOH and about 13 times better than when using EDP.

Etched surface smoothness was determined to be a function of the cesium hydroxide concentration. With lower base concentrations, the peak-to-peak surface roughness was more than 10 μm. Peak-to-peak roughness values of less than one micron were seen at the higher base concentrations. The observation of smoother etched surfaces with increasing etchant concentration had been reported earlier for the anisotropic etchants hydrazine [57] and KOH/H$_2$O [58].

The activation energy for the process was determined to be 1.00 eV, which is in good agreement with the value of 0.95 eV reported by Hooley [56]. Hooley proposed the following sequence of reaction steps for silicon dioxide etching:

$$SiO_2(s) + 2H_2O(l) \rightarrow SiO_2 \cdot 2H_2O(s) \tag{3.9}$$

$$SiO_2 \cdot 2H_2O(s) + 2OH^- \rightarrow [Si(OH)_6]^{-2} \tag{3.10}$$

Clark et al. observed that the rate of SiO$_2$ dissolution as a function of CsOH concentration passed through a maximum.

Cesium hydroxide is an attractive alternative to many anisotropic etchants because of its relatively nontoxic nature, especially when compared to hydrazine and ethylenediamine. However, it is a relatively expensive chemical in relationship to potassium hydroxide.

3.2.3.5 General Anisotropic Etching Mechanism

An elegant discussion of the mechanism of anisotropic etching of silicon has been provided by Seidel and his coworkers [54,59]. In the first report, experimental data on the anisotropy, selectivity, and voltage dependence of anisotropic silicon etchants are given. The general unifying electrochemical model, (3.11)–(3.18), to describe the reaction mechanism follows. The main features of all alkaline anisotropic etchants for silicon have also been outlined [59].

$$\begin{array}{c} Si \\ \diagdown \\ Si: + 2OH^- \rightarrow \\ \diagup \\ Si \end{array} \quad \begin{array}{c} Si \quad OH \\ \diagdown \diagup \\ Si \\ \diagup \diagdown \\ Si \quad OH \end{array} + 2e^-_{cond} \tag{3.11}$$

$$\begin{matrix} \text{Si} & \text{OH} \\ & \diagdown \ \diagup \\ & \text{Si} \\ & \diagup \ \diagdown \\ \text{Si} & \text{OH} \end{matrix} \rightarrow \begin{matrix} \text{Si} \\ \diagdown \\ \\ \diagup \\ \text{Si} \end{matrix} \left[\begin{matrix} \text{OH} \\ \diagup \\ \text{Si} \\ \diagdown \\ \text{OH} \end{matrix} \right]^{++} + 2e^-_{\text{cond}} \tag{3.12}$$

$$\begin{matrix} \text{Si} \\ \diagdown \\ \\ \diagup \\ \text{Si} \end{matrix} \left[\begin{matrix} \text{OH} \\ \diagup \\ \text{Si} \\ \diagdown \\ \text{OH} \end{matrix} \right]^{++} + 2\text{OH}^- \rightarrow \text{Si(OH)}_4 + \text{Si}_{\text{Solid}} \tag{3.13}$$

$$\text{Si(OH)}_4 \xrightarrow{\text{pH}>12} \text{SiO}_2(\text{OH})_2^{--} + 2\text{H}^+ \tag{3.14}$$

$$2\text{H}^+ + 2\text{OH}^- \rightarrow 2\text{H}_2\text{O} \tag{3.15}$$

$$4\text{H}_2\text{O} + 4e^- \rightarrow 4\text{H}_2\text{O}^- \tag{3.16}$$

$$4\text{H}_2\text{O}^- \rightarrow 4(\text{OH})^- + 4\text{H} \rightarrow 4(\text{OH})^- + 2\text{H}_2 \tag{3.17}$$

$$\text{Si} + 2\text{OH}^- + 2\text{H}_2\text{O} \rightarrow \text{SiO}_2(\text{OH})_2^{--} + 2\text{H}_2 \tag{3.18}$$

Two hydroxide ions can bind with two dangling bonds of a silicon atom on a $\langle 100 \rangle$ surface, injecting two electrons into the conduction band, (3.11). Next the Si–Si backbones of the Si(OH)_2 neighboring lattice atoms have to be broken to obtain a soluble silicon complex, which is positively charged, (3.12). The silicon–hydroxide complex reacts further with two more hydroxide ions to give orthosilicic acid, (3.13). The Si(OH)_4 can leave the solid surface by diffusion, but in the bulk electrolyte, it is unstable due to the high pH value of the solution. In this environment, the complex in (3.14) can form. The excess electrons in the conduction band can be transferred to water molecules producing hydroxide ions and hydrogen. The overall reaction is represented in (3.18).

When using ethylenediamine–pyrocatechol-based solution, it is thought that the primary role of the pyrocatechol is to convert Si(OH)_4 into a more complex anion, increasing the solubility of the etch product:

$$\text{Si(OH)}_4 + 2\text{OH}^- + 3\text{C}_6\text{H}_4(\text{OH})_2 \rightarrow \text{Si}(\text{C}_6\text{H}_4\text{O}_2)_3^{--} + 6\text{H}_2\text{O} \tag{3.19}$$

3.2.4 Etch-Stop Mechanisms

Given the nature of bulk micromachining technology that relies on anisotropic etching, it is quite clear that etching time should play an important role. If that role were the dominant factor then the control of the etching process would be tedious. Fortunately, other factors important in anisotropic etching can be used effectively to control the etching process. One may take advantage of the fact that etch rates are also a function of the silicon dopant concentration. This, as well as electrochemically assisted anisotropic etching, permits one to control the thickness of diaphragms and determine when to stop in the etching process. Dopant concentrations and electrochemical-assisted anisotropic etching are discussed next.

3.2.4.1 High Boron Concentration

The etch rate of doped silicon has been found to depend on boron concentration [60–65]. Figure 3.13 shows the change in etch rate of silicon as a function of incorporated

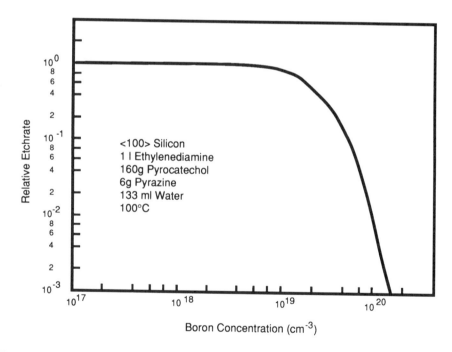

Figure 3.13 Silicon etch rate as a function of incorporated boron concentration.

boron [24, 63]. It can be seen that up to a "critical" boron concentration of approximately 2.5×10^{19} cm^{-3} the etch is independent of boron content. Beyond the "critical" concentration there is a steady drop of the etch rate and a final reduction in the etch rate by three orders of magnitude. This drop is inversely proportional to the fourth power of boron concentration [63]. The explanation of this behavior involves phenomena related to the degeneration of silicon [66].

The reduction of the etch rate in heavily boron-doped silicon can be used very effectively to control etching as an etch-stop mechanism [3,7,9,17,18]. Figure 3.14 shows an example of a silicon wafer doped with boron, which creates the p^+ region (concentration higher than 10^{20} cm^{-3}). This p^+ region is used effectively in anisotropic etching as an etch stop mechanism. When the etching front (proceeding from the back of the wafer) hits the p^+ region then, for all practical purposes, the etch stops because the etch rate of the p^+ is 1,000 times lower than the etch rate of lightly doped silicon. Thus, the p^+ region effectively determines the thickness of the diaphragm.

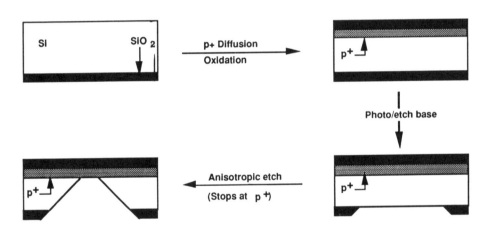

Figure 3.14 The p^+ region used as an etch-stop in anisotropic etching. The thickness of the diaphragm is determined by the thickness of the p^+ region.

3.2.4.2 Electrochemically Controlled Anisotropic Etching

Electrochemical etching is an alternative technique for controlling the shape of silicon microstructures. The coupling of anisotropic etching with electrochemical etch stops has received a great deal of attention particularly with the dawn of micromachining and its application to various sensor devices. A number of papers address this subject [25,67–70]. One of the earliest papers is a brief communication by Wagener that discusses the use of hot alkaline (anisotropic) etches where unwanted silicon is removed chemically and the regions to be kept are passivated electrochemically [67]. Both *n*- and *p*-type layers can be passivated. Silicon orientations other than ⟨111⟩ are needed for processing because ⟨111⟩-oriented silicon etches very slowly in these etches. The etchants themselves are characterized by a relatively sharp active-passive transition voltage between etch and passivation and by a large ratio of silicon etch rates between the active and passive states. Ratios (active/passive) of 200:1 are readily obtained. The ratio of active etch rate to passive etch rate is very important because the thickness uncertainty is determined in large part by this ratio.

Uhlir [71] described electrolytic shaping of germanium and silicon. The wafer being etched was made the anode of the system and current flowed from the semiconductor to the electrolyte. The etchant was a 10 wt% KOH solution, and the rate was considerably enhanced by electrochemically controlled etching (ECE).

A process for electrochemically thinning n/n^+ epitaxial silicon was also described by R.L. Meek [72]. His paper describes a circuit to maintain constant cell potential. This appears to be a good technique for the fabrication of other thin films for experimental purposes. It was also suggested for use in thinning simple diode arrays and manufacturing dielectrically isolated ICs.

Tabata et al. [48] reported that TMAH was effective as an etch stop in electrochemical etching by applying a potential that would passivate the *n*-layer of the epilayer, only the *p*-type substrate was etched.

Jackson, Tischler, and Wise [68] studied the electrochemical etch stop technique for the formation of silicon microstructures, using ethylenediamine-based etchants. Dissolution rates for the passivated areas were only approximately 5 Å/min, which was 3000 times less than for unpassivated areas. Applications for this process can be found in the areas of pressure sensors, accelerometers, and other devices where a thin silicon area must be formed. Figure 3.15 illustrates how a thin silicon diaphragm that might be used as a piezoelectric pressure sensor could be made.

Silicon cantilever beams made by electrochemically controlled etching were reported by Sarro and van Herwaarden [69]. After processing, the wafer with the

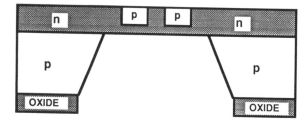

Figure 3.15 Anisotropically etched thin silicon diaphragm. The thickness of the diaphragm is determined by the thickness of the *n*-region.

n-type epilayer containing devices, was mounted face down on a stainless steel holder with black wax. Anodic contact was made to the n-type epilayer, the side to be preserved, with a spring loaded contact. A platinum foil served as the cathode for the process. A schematic of the experimental apparatus is shown (Figure 3.16). Once the wafer is immersed into the solution, etching begins and continues until the n-type epilayer is reached. If there is no applied potential on the n-type back of the wafer, etching continues into the n-region. With an applied potential of 0.8 to 1.0 V, etching stops at the beginning of the reverse biased n-layer. Hot (85°C) KOH was used for the anisotropic etch, and cantilever beams 10 μm thick were prepared by this technique. At this temperature, silicon etched at 1.4 μm per minute. Silicon nitride, 750 Å, was shown to be an excellent masking material for this process.

Silicon pressure sensors with carefully controlled diaphragm thicknesses were studied by Hirata, Suwazono, and Tanigawa [70]. Using a 50–50 hydrazine–water solution at 90°C, p-type silicon was etched with an anodic 5 V bias on the n-epi-material. In this configuration, the authors constructed piezoelectric silicon diaphragms with thickness control of 20 ± 2 μm. They consider the etch stop mechanism to be anodic oxide growth on the n-type epilayer, terminating etching when the p-type material is etched away. Figure 3.17 shows the silicon wafer structure used in this work. Figure 3.18 shows measured anodic current during diaphragm formation. Initially a current flow of 1 mA was observed with p-type material etching. Gas bubbling accompanied the silicon etch. After a 110-min etch, the current increased to 11 mA rather sharply. Then anode current decreased to less than 1 mA and bubbling ceased. The final thickness was essentially determined at this point. After an additional 2-hr etch time, the thickness change was less than 1 ± 1 μm. The etch-stop

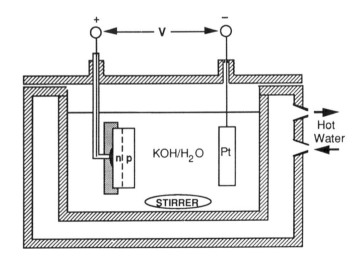

Figure 3.16 Electrochemical etching experimental setup.

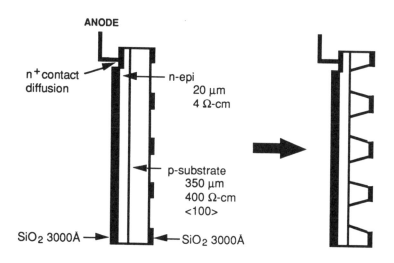

Figure 3.17 Etch-stop mechanism used to form diaphragms with controlled thickness: (a) before etch; (b) after etch.

mechanism is considered to be the following. As the *p-n* junction between the *p*-type substrate and the *n*-type epitaxial layer is reverse biased, only a small leakage current flows from the anode to the *p*-substrate. In this situation, the *p*-type substrate etches normally in the hydrazine–water solution. When the *p*-type substrate in the developing diaphragm region is anisotropically etched away, large currents begin to flow from the *n*-type layer directly into the solution because there is

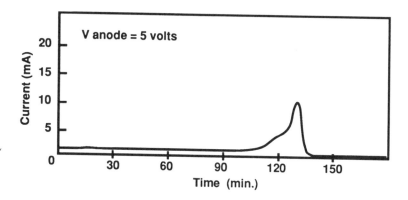

Figure 3.18 Measured anode current during diaphragm formation. A current recorder was set between the anode electrode and the silicon wafer.

now no reverse-biased *p-n* junction barrier. The anode current from the constant voltage source next causes a thin oxide film to grow on the exposed *n*-type silicon in the diaphragm area. This is a good protective coating from the hydrazine–water etch, and the silicon etching stops, leaving only the thin, *n*-type diaphragm for the pressure sensor. Note that the *p-n* junction had to be reverse biased in this work to etch off the *p*-type substrate. Otherwise the *p*-type silicon would have been oxidized and protected from etching. The silicon etch rate in this setup at 2 V was 3.2 μm/ min; while in the zero bias condition, the etch rate was only 1.7 μm per minute, thus indicating that the bias voltage does enhance the etch rate.

High precision thickness control of silicon membranes was also studied by Kloeck et al. [73], using several electrode arrangements. Figure 3.19(a) shows a typical three-electrode etch-stop system. The *p-n* junction is reverse biased, and the *p*-type silicon remains at essentially the open circuit potential, thus allowing it to be etched. Once the *p*-type silicon is completely removed and the diode is destroyed, the *n*-type silicon will be exposed directly to the etch solution. The increase in current density causes passivation and the cessation of the etching process. With only two electrodes, the etch solution potential is ill-defined with respect to the epilayer potential. Addition of a reference electrode (RE) allows precise control of the potential between the solution and the working electrode (WE). This can be accomplished with the aid of a potentiostat by adjusting the current flow through the counterelectrode (CE). In

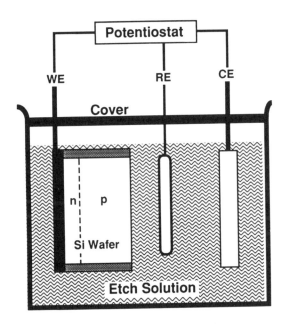

Figure 3.19(a) Electrochemical etching: three-electrode etch setup.

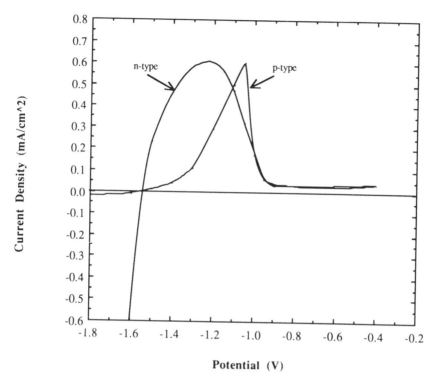

Figure 3.19(b) Electrochemical etching: I/V characteristics of ⟨100⟩ silicon in 40% aqueous KOH solution at 60°C.

this way, the *n*-type epilayer can be maintained at a well-defined passivation potential. Kloeck et al. used KOH as an anisotropic silicon etch with electrochemical assist as an etch stop to carefully control silicon membrane thickness on epiwafers for piezoresistive pressure sensors. The resultant current-voltage characteristics for *p*- and *n*-type ⟨100⟩ silicon in 40% aqueous KOH solution at 60°C, using a system similar to that depicted in Figure 3.19(a), are presented in Figure 3.19(b). As can be seen, the peak current for *n*-type and *p*-type silicon occurs at different potentials.

McNeil et al. [74] obtained similar results in KOH and CsOH solutions. They also determined that both *p*- and *n*-type silicon passivation peaks shift cathodically with increasing etchant concentration at a given temperature for both KOH and CsOH. However, the *n*-type silicon shifted at a faster rate than the *p*-type. Moreover, at a fixed etchant concentration, a decrease in temperature resulted in a cathodic shift for the *n*-type and an anodic shift for the *p*-type silicon.

Finally, it should be noted that the ECE process may have an advantage over using heavily doped boron (5×10^{19} cm^{-3}) as a method of controlled etch stop.

Heavily doped silicon is typically highly strained due to the high boron concentrations required to achieve an effective stop.

3.3 CMOS TECHNOLOGY AND BULK MICROMACHINING

In previous sections the basic concept of bulk micromachining was outlined. In this section, we will discuss processing issues relevant to integrated sensor systems, which means that the discussion has to include both micromachining and standard IC processing. We will start the discussion with a brief description of the basic aspects of CMOS processing and then look into compatibility of CMOS processing and bulk micromachining.

3.3.1 CMOS Processing

Today, CMOS processing can meet the requirements of both digital and analog applications [75]. It is optimized to accomplish superb performance of integrated circuits based both on n-channel and p-channel devices. Since its appearance in the early 1960s it has been a tradition to use an elemental inverter cell when describing the specifics of CMOS processing. We will use the same approach here to describe briefly the twin-tub process [76]. A typical inverter is shown in Figure 3.20. It is fabricated using $\langle 100 \rangle$ p^+-substrate silicon wafers. The p-tub and n-tub regions are formed by ion implantations and drive-in diffusions into a Si–epitaxial layer. Then,

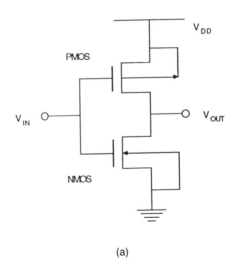

(a)

Figure 3.20 CMOS inverter: (a) equivalent circuit; (b) cross section, where 1 is thermal oxide, 2 is cvd-oxide, 3 is cvd-oxide, 4 is polysilicon, and 5 is metal.

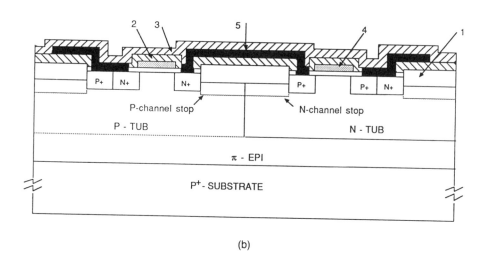

(b)

Figure 3.20 continued.

through the subsequent steps of field oxidation, gate oxidation, polysilicon deposition, p^+- and n^+-ion implantations and diffusions, metallization and passivation, both n-channel and p-channel devices are fabricated. A cross section of the final structure is shown in Figure 3.20(b). It should be noted that many variations of the CMOS process exist and the one described here serves just as an illustration.

3.3.2 Micromechanical Structures in CMOS Processing

Can the standard process described previously meet micromachining requirements? The topic has been already discussed extensively in [36], and the answer is yes. The key element for use of CMOS process is the use of ⟨100⟩ orientation. Everything else is a matter of design. We emphasize again that the particular twin-tub process briefly described in the previous paragraphs has no bearing on the conclusion as long as ⟨100⟩ orientation is used. Many other variations of CMOS and BiCMOS process have equal potential for fabrication of micromechanical structures. The benefit of using IC process is that it enables merging of IC circuits and micromechanical structures on the same chip. The integration of these technologies can be divided in two distinctive steps: IC processing itself followed by micromachining as a postprocessing step. For example, the integrated pressure sensor requires fabrication of the circuit (first step) and micromachining from the backside of the wafer to create the Si diaphragm (postprocessing as a second step). By the same approach, silicon cantilevers and other micromechanical structures could be made. It is important to mention that the selectivity of the etch that is used for anisotropic etching plays a critical role in this procedure.

Merger of CMOS process and bulk micromachining also adds new possibilities for microfabrication of micromechanical structures. In addition to silicon it is possible to utilize silicon dioxide, polysilicon, metal, as well as other thin films used in IC processing for micromechanical structures. The selectivity of anisotropic etches with respect to silicon dioxide (very low etch rate) can be used very successfully in this matter. For example, CVD oxide is typically deposited on IC devices for passivation. This passivating layer can serve as a mask for anisotropic etching. At the same time, polysilicon and metal can be used as active parts of the micromechanical devices. Field oxide and nitride can be used as a basic holder of mechanical devices. There are many possible combinations of the films for fabrication of multilayer dynamic micromechanical structures [77]. The devices fabricated using this approach can measure gas flow [78] or be used as actuators [79]. Here, we will describe just a few basic structures as an illustration of the versatility of this approach.

3.3.2.1 Bridges and Suspended Mass

The simple method for designing of a microbridge is presented in Figure 3.21. The two triangular shaped areas (on either side of the diagonally shaped field oxide–polysilicon–metal–CVD oxide layers) represent silicon exposed to anisotropic etch. The edges of the rectangles are aligned in $\langle 110 \rangle$ directions that will define the cavity. As can be seen, the edges of the diagonal are misaligned with respect to the $\langle 110 \rangle$ directions. This will cause undercutting during the etching, which will create a microbridge on the top of the truncated pyramidal pit (cavity).

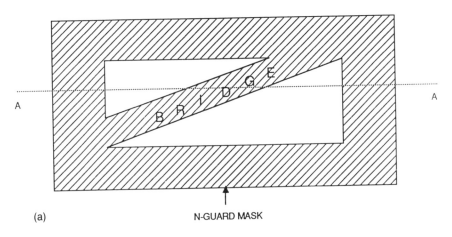

(a)   N-GUARD MASK

Figure 3.21 Design of a microbridge: (a) top view; (b) cross section; (c) SEM micrograph of a microbridge, where 1 is thermal oxide, 2 is cvd-oxide, 3 is cvd-oxide, 4 is polysilicon, and 5 is metal.

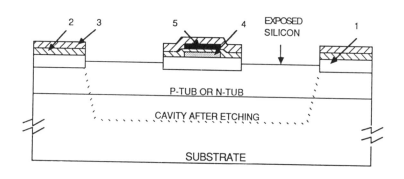

(b)

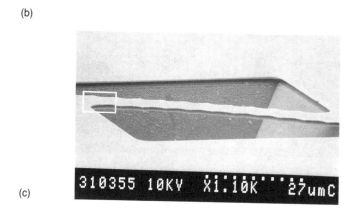

(c)

Figure 3.21 continued.

Following this method bridges have been designed and manufactured using standard twin-tub CMOS process [36,80]. The postprocessing step of anisotropic etching is done by EDP and KOH, creating a cavity beneath the suspended bridge as shown in Figure 3.21(b). Using the same approach, many other complex structures including a suspended mass can be fabricated. An example of the suspended mass is shown in Figure 3.22. The structure was also fabricated in the twin-tub CMOS process using field oxide, polysilicon, and CVD oxide. The arms that keep the mass suspended are spiral like. Micromachining is again done as a postprocessing step.

3.3.2.2 Cantilevers

The method for designing a cantilever beam is based on the property of anisotropic etch to undercut convex corners. All edges of the mask shown in Figure 3.23 are

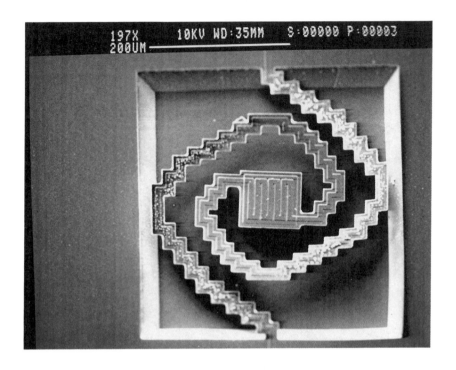

Figure 3.22 SEM micrograph of a suspended mass structure.

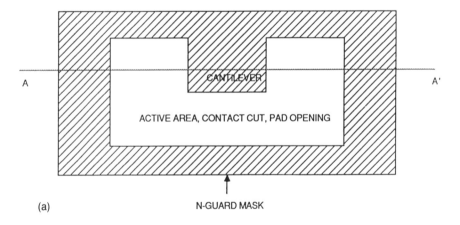

(a)

N-GUARD MASK

Figure 3.23 Design of a cantilever beam: (a) top view, (b) cross section, and (c) SEM micrograph of a cantilever array, where 1 is thermal oxide, 2 is cvd-oxide, 3 is cvd-oxide, 4 is polysilicon, and 5 is metal.

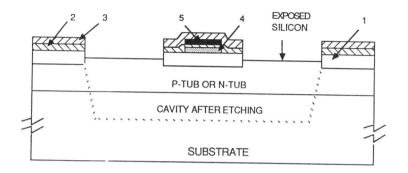

(b)

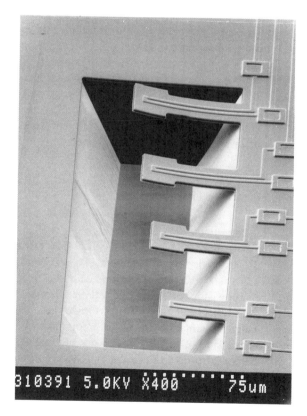

(c)

Figure 3.23 continued.

aligned with the $\langle 110 \rangle$ directions, which define a truncated pyramidal pit. When silicon is exposed to anisotropic etch, undercutting will start from both convex corners and will stop when the $\langle 111 \rangle$ surface is reached. This will leave a suspended cantilever over the cavity.

The combination of layers for an array of cantilevers shown in Figure 3.23(b) included field oxide, polysilicon, and CVD oxide, where polysilicon is basically sealed between the two oxides. The cantilevers are also fabricated using the twin-tub CMOS process and micromachined as a postprocessing step. An early version of this approach dealing only with oxides was published in 1988 [81].

3.4 APPLICATIONS OF BULK MICROMACHINING

There are many useful practical examples of bulk micromachining applications for the manufacture of sensors and actuators. Therefore, choosing only a few representative devices for illustration is not an easy decision. In this section we have selected three examples, which, in our opinion, are distinctive enough and at the same time demonstrate the capability of bulk micromachining technology described in previous sections. The first example is a bulk micromachined accelerometer which utilizes piezoresistive sensing. The second and third examples are representative of what may be expected in the future in the form of integrated systems.

3.4.1 Bulk Micromachined Accelerometer

A bulk micromachined accelerometer is shown in Figure 3.24 [82]. The entire structure consists of a sensing device and two caps. The sensing device has a 3.6 mm square silicon proof mass suspended by four tethers. The accelerometer has overall dimensions of $7.7 \times 7.2 \times 1.2$ mm and is designed for the range of acceleration from 1–500 g. It uses piezoresistors placed in the tethers (flexures) to sense flexure strain induced by deflection of the silicon proof mass under acceleration. The piezoresistors are produced using ion implantation. The proof mass is created by anisotropic etching from the back side of the wafer and dry etching from the front side of the wafer. The dry etching also determines the final tether dimensions.

The axis of sensitivity is in the direction perpendicular to the chip surface. Because the center of the proof mass is not on the same plane as the flexures, acceleration in any direction other than the desired sensing direction tends to induce torsion about a perpendicular axis. To compensate for cross-axis sensitivity, two resistors are placed on each flexure (one at the mass end and one at the frame end). They are all connected in a Wheatstone bridge arrangement with two sensing resistors

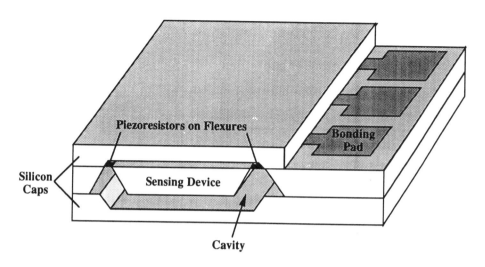

Figure 3.24 Bulk micromachined accelerometer.

in each leg of the bridge. Using this configuration, off-axis sensitivity can be reduced to less than 1% of full-scale response.

As already mentioned, the sensing device is capped from both sides. The shallow cap cavities are etched at the wafer level with a depth control of ± 1 μm. The top and bottom caps are bonded to the central wafer for several reasons: The caps provide an isolated space for the proof mass to move, they protect the sensing device during sawing, and they act as overtravel stops for large acceleration inputs. At the same time, the cavity created by the caps determines the damping characteristics of the device. That is, the cavity restricts the movement of air around the mass, which increases the damping and reduces oscillations. A range of damping levels can be obtained by varying internal cavity size, which could be accomplished by varying the depth of the cavity. Figure 3.25 shows the response as a function of frequency using the three dominant modes of oscillation of the proof mass: vibration in the sensing direction and torsion about the two axes perpendicular to the sensing direction. The cases for critically damped devices (curve for optimum depth), overdamped devices (+2 and +5 μm curves), and underdamped devices (-2 and -5 μm curves) are shown (± 2 and ± 5 μm are referred with respect to the depth of the cavity for critical damping). As can be seen, a flat frequency response can be achieved by appropriate damping.

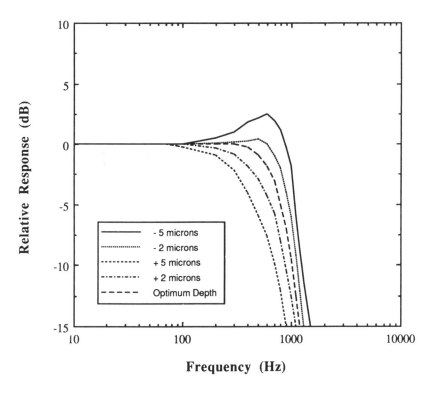

Figure 3.25 Damping as a function of frequency for various cavity sizes. Curves for cavities smaller than the cavity with optimum depth are marked with a minus sign; for the cavities larger than the cavity with optimum depth, the curves are designated with a plus sign.

3.4.2 Thermal Infrared Sensor

The second example in this section deals with an infrared sensing array based on silicon thermopiles [83]. Figure 3.26 shows a schematic diagram of an infrared sensing array. The array consists of a group of 10-μm thick, 3-mm long, and 440-μm wide cantilever beams surrounded by a supporting rim that acts as a heat sink. The basic structure is fabricated using a combination of standard IC technology and bulk micromachining processes. A 10-μm thick n-type epilayer was used to define the thickness of the cantilevers. A subsequent p-type diffusion was used to create the thermopile strips in the cantilever area. The thermopiles consist of a p–Si/Al combination. A backside silicon nitride mask was used for the anisotropic etch. Elec-

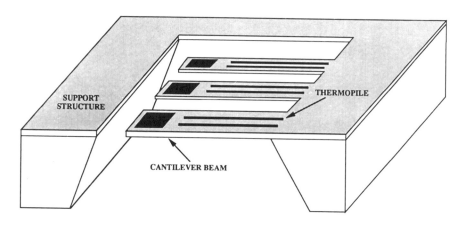

Figure 3.26 Infrared sensing array; a 10-μm epilayer determines the thickness of the cantilevers.

trochemical etching was employed to create the cavity from the back side of the wafer with the *n*-type epilayer used as an etch stop. Silicon plasma etching from the front side was used to define the cantilevers. Finally, the beams were coated with absorbent material to produce the structure shown in Figure 3.26. Each cantilever is half covered with the absorbent material and the other half hosts the *p*–Si/Al thermopile.

The principle of operation of the thermopiles is based on the Seebeck effect. The doping of *p*-type strip was chosen to be approximately 35 Ω/square in order to get a high Seebeck coefficient (0.7 mV/K). When exposed to radiation, absorbed energy creates a temperature gradient in the cantilever that can be detected by the thermopile array as the heat is distributed via conduction along the length of the beam. The output voltage of one of the thermopiles is shown as a function of black-body temperature in Figure 3.27. The device is characterized in vacuum over a temperature of 300K to 400K using a blackbody source at a distance of 5 cm from the sensor. Good agreement with Stefan's law was found. The output voltage rises approximately with the fourth power of the blackbody temperature. Values of 4 μV output voltage per 1° temperature difference between the source and the thermopiles were measured.

This device is a good example of the possibilities for future integrated systems. Because the bulk micromachining is done as a postprocessing step following the standard IC processing, it is obvious that full integration of the sensing array and signal processing has become a reality. Similar devices sitting on a silicon diaphragm instead of cantilevers were designed and fabricated in the early 1980s [84].

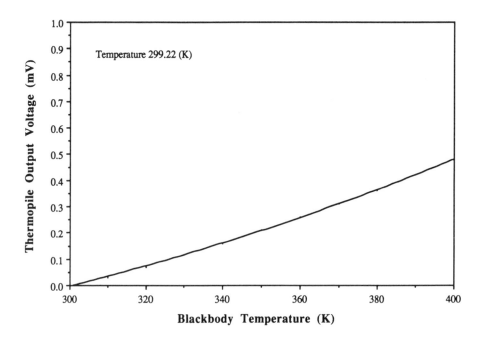

Figure 3.27 Thermopile output as a function of blackbody temperature.

3.4.3 Inductors in CMOS Technology

The next example illustrates many aspects of future integrated systems made possible by the merger of CMOS technology and bulk micromachining. CMOS technology is used to fabricate a monolithic IC amplifier, and bulk micromachining techniques are employed to produce inductors with self-resonance at 3 GHz. The net result is a fully integrated RF amplifier on a single chip [85].

The growing interest in monolithic RF amplifiers stems from the current need to develop miniature components for wireless communication in the 1 GHz band. The most challenging aspect of this endeavor involves fabrication of monolithic inductors, because RF amplifiers use a tuned load to obtain a desired gain and to filter band signals and noise. Previous attempts to fabricate spiral inductors on silicon substrates have failed primarily for two reasons: High parasitic capacitance to the substrate precludes use of inductors at high frequency, and the series spreading resistance in the substrate severely limits the quality factor (Q) of the inductors. These problems can be overcome if the substrate area under the inductor acts as an insulator. This could be done if the silicon beneath the inductor were etched away using bulk micromachining techniques [4].

Figure 3.28 shows the inductor designed as an aluminum spiral with 20 turns. A 100 nH inductor was obtained using the second-layer aluminum, 4 μm wide, with a separation between the metal lines of 4 μm. First-layer aluminum is used to bring the contact outside from the inner end of the inductor. The inductors together with an integrated circuit were fabricated in a standard 2 μm n-well CMOS process. A bulk micromachining step is performed afterward, using the approach described in

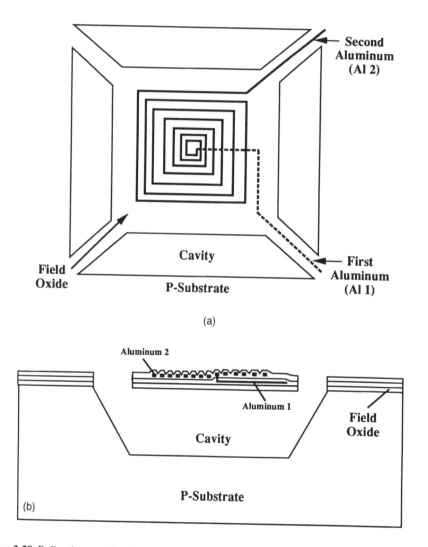

Figure 3.28 Bulk micromachined inductor: (a) top view; (b) cross section.

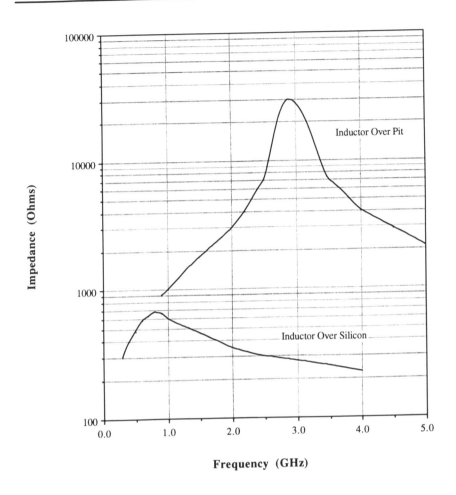

Figure 3.29 Inductor impedance as a function of frequency.

Section 3.3.2, leaving the inductors suspended over the pit, Figure 3.28(b). By elim-inating the substrate beneath the inductors the self-resonance was increased from 800 MHz to 3 GHz, Figure 3.29. As a consequence, a fully integrated RF amplifier was realized operating from a 3V power supply, at 770 MHz with a peak gain of 14 dB, and 7 mW power dissipation, Figure 3.30. In contrast, the performance of the RF amplifier with the inductors sitting on the silicon was much poorer, 400 kHz with a peak gain of 5 dB. This achievement is indeed an excellent example of what the future holds for integrated systems with the merger of micromachining and IC processing.

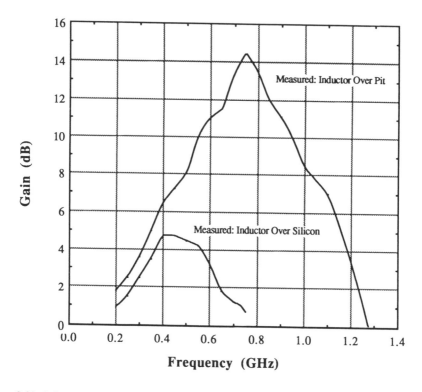

Figure 3.30 Gain of RF amplifier with the inductor loads as a function of frequency.

3.5 SUMMARY

The emphasis of the chapter was on basic elements of bulk micromachining technology. An effort has been made to describe key features. These included silicon as a mechanical material, wet anisotropic etching and its dependence on crystallographic orientation, patterning, and selectivity. Also, the most important anisotropic etchants were described, as well as a mechanism of anisotropic etching. The discussion has included etch-stop mechanisms, which are important for reproducible control of the dimensions of micromechanical structures. Further, a special section was devoted to the compatibility of CMOS technology and bulk micromachining. Finally, several examples of practical applications of bulk micromachining were included, showing that the merging of the IC and micromechanical worlds can lead to the on-chip integrated system.

REFERENCES

[1] E. Bassous, "Nozzles Formed in Mono-crystalline Silicon," U.S. Patent 3,921,916, 1975.

[2] E. Bassous, L. Kuhn, A. Reisman, and H. Taub, "Inkjet Nozzle," U.S. Patent 4,007,464, 1987.

[3] E. Bassous, H. Taub, and L. Kuhn, "Inkjet Printing Nozzle Arrays Etched in Silicon," *Appl. Phys. Lett.*, Vol. 31, 1975, p. 135.

[4] C. Hu and S. Kim, "Thin Film Dye Laser with Etched Cavity," *Appl. Phys. Lett.*, Vol. 29, 1976, p. 582.

[5] S. C. Terry, "A Gas Chromatography System Fabricated on Silicon Wafer Using Integrated Circuit Technology," Ph.D. dissertation, 1975.

[6] S. C. Terry, J. H. Jerman, and J. B. Angel, "A Gas Chromatograph Air Analyzer Fabricated on a Silicon Wafer," *IEEE Trans. Electron. Devices*, Vol. ED-26, 1979, p. 1880.

[7] E. Bassous, "Fabrication of Novel Three-Dimensional Microstructures by the Anisotropic Etching of (100) and (110) Silicon," *IEEE Trans. Electron. Devices*, Vol. ED-25, 1978, p. 1178.

[8] O. N. Tufte, P. W. Chapman, and K. Long, "Silicon Diffused Element Piezoresistive Diaphragms," *J. Appl. Phys.*, Vol. 33, 1962, p. 3322.

[9] H. Guckel, S. Larsen, M. G. Lagally, B. Moore, J. B. Miller, and J. D. Wiley, "Electromechanical Devices Utilizing Thin Si Diaphragms," *Appl. Phys. Lett.*, Vol. 31, 1977, p. 618.

[10] L. K. Clark and K. D. Wise, "Pressure Sensitivity in Anisotropically Etched Thin Diaphragm Pressure Sensors," *IEEE Trans. Electron. Devices*, Vol. ED-26, 1979, p. 1887.

[11] E. Obermeier, "Polysilicon Layers Lead to a New Generation of Pressure Sensors," *Tech. Digest*, Transducers 1985, Int. Conf. on Solid State Sensors and Actuators, Philadelphia, 1985, p. 430.

[12] H. I. Chau and K. D. Wise, "Scaling Limits in Batch Fabricated Silicon Pressure Sensors," *Tech. Digest*, Transducers 1985, Int. Conf. on Solid State Sensors and Actuators, Philadelphia, 1985, p. 174.

[13] R. T. Howe and R. S. Muller, "Polycrystalline Silicon Micromechanical Beams," *J. Electrochem. Soc.*, Vol. 130, 1983, p. 1420.

[14] R. T. Howe, "Integrated Silicon Electrochemical Vapor Sensor," Ph.D. thesis, University of California, Berkeley, 1981.

[15] Y. C. Tai, R. S. Muller, and R. T. Howe, "Polysilicon Bridges for Anemometer Applications," *Tech. Digest*, Transducers 1985, Int. Conf. on Solid State Sensors and Actuators, Philadelphia, 1985, p. 354.

[16] R. T. Howe, "Resonant Microsensors," *Tech. Digest*, Transducers 1987, Fourth Int. Conf. on Solid State Sensors and Actuators, Tokyo, 1987, p. 843.

[17] K. E. Petersen, "Micromechanical Light Modulator Array Fabricated on Silicon," *Appl. Phys. Lett.*, Vol. 31, 1977, p. 521.

[18] K. E. Petersen, "Dynamic Micromechanics on Silicon: Techniques and Devices," *IEEE Trans. Electron. Devices*, Vol. ED-25, 1978, p. 1241.

[19] H. C. Nathanson and R. A. Wickstrom, "A Resonant Gate Silicon Surface Transistor with High Q Bandpass Properties," *Appl. Phys. Lett.*, Vol. 7, 1965, p. 84.

[20] H. C. Nathanson, W. E. Newell, R. A. Wickstrom, and J. R. Davis, Jr., "The Resonant Gate Transistor," *IEEE Trans. Electron. Devices*, Vol. ED-14, 1967, p. 117.

[21] P. Chen, R. S. Muller, T. Shiosaki, and R. M. White, "Silicon Cantilever Beam Accelerometer Utilizing a PI-FET Capacitive Transducer," *IEEE Trans. Electron. Devices*, Vol. ED-26, 1979, p. 1857.

[22] R. D. Jolly and R. S. Muller, "Miniature Cantilever Beams Fabricated by Anisotropic Etching of Silicon," *J. Electrochem. Soc.*, Vol. 127, 1980, p. 2750.

[23] P. M. Zavracky and R. H. Morrison, "Electrically Actuated Micromechanical Switches with Hysteresis," *Tech. Digest*, IEEE Solid State Conf., Hilton Head Island, SC, 1984, p. 50.

[24] L. Csepregi, "Micromechanics: A Silicon Microfabrication Technology," *Microelectronic Eng.*, Vol. 3, 1985, p. 221.

[25] K. E. Petersen, "Silicon as a Mechanical Material," *Proc. IEEE*, Vol. 70, 1982, p. 420.

[26] *Tech. Digest*, 1984 IEEE Solid State Sensor Conference, Hilton Head Island, SC, 1984.

[27] L. S. Fan, Y. C. Tai, and R. S. Muller, "*Pin* Joints, Gears, Springs, Cranks, and Other Novel Micromechanical Structures," *Tech. Digest*, Transducers 1987, Fourth Int. Conf. on Solid State Sensors and Actuators, Tokyo, 1987, p. 849.

[28] K. J. Gabriel, W. S. N. Trimmer, and M. Mehregany, "Microgears and Turbines Etched From Silicon," *Tech. Digest*, Transducers 1987, Fourth Int. Conf. on Solid State Sensors and Actuators, Tokyo, 1987, p. 853.

[29] W. S. N. Trimmer, K. J. Gabriel, and R. Mahadevan, "Silicon Electrostatic Motors," *Tech. Digest*, Transducers 1987, Fourth Int. Conf. on Solid State Sensors and Actuators, Tokyo, 1987, p. 857.

[30] M. Mehregany, K. J. Gabriel, and W. S. N. Trimmer, "Integrated Fabrication of Polysilicon Mechanisms," *IEEE Trans. Electron. Devices*, Vol. ED-35, 1988, p. 719.

[31] L. S. Fan, Y. C. Tai, and R. S. Muller, "Integrated Movable Micromechanical Structures for Sensors and Actuators," *IEEE Trans. Electron. Devices*, Vol. ED-35, 1988, p. 719.

[32] T. A. Lober and R. T. Howe, "Surface-Micromachining Process for Electrostatic Microactuator Fabrication," *Tech. Digest*, IEEE Solid State Sensor and Actuator Workshop, Hilton Head Island, SC, 1988, p. 59.

[33] J. H. Lang and S. F. Bart, "Toward the Design of Successful Electric Micromotors," *Tech. Digest*, IEEE Solid State Sensor and Actuator Workshop, Hilton Head Island, SC, 1988, p. 127.

[34] L. S. Fan, Y. C. Tai, and R. S. Muller, "IC-Processed Electrostatic Micromotors," *Tech. Digest*, IEDM, San Francisco, 1988, p. 666.

[35] C. S. Smith, "Piezoresistance Effect in Germanium and Silicon," *Phys. Rev.*, Vol. 94, 1954, p. 42.

[36] Lj. Ristic, "CMOS Technology: A Base for Micromachining," *Microelectronics J.*, Vol. 20, 1989, p. 153.

[37] J. M. Crishal and A. L. Harrington, "A Selective Etch for Elemental Silicon," *Electrochemical Society Extended Abstracts*, Vol. 109, Abst # 89, Spring Meeting, Los Angeles, 1962, p. 71C.

[38] R. M. Finne and D. L. Klein, "A Water Soluble Amine Complexing Agent System for Etching Silicon," *J.E.C.S.*, Vol. 114, 1967, p. 965.

[39] D. B. Lee, "Anisotropic Etching of Silicon," *J. Appl. Phys.*, Vol. 40, 1969, p. 4569.

[40] H. Muraoka and T. Y. Sumitomo, "Controlled Preferential Etching Technology," in *Semiconductor Silicon*, ed. H.R. Huff and R.R. Burgess, The Electrochemical Society, 1973, p. 327.

[41] K. D. Bean, "Anisotropic Etching of Silicon," *IEEE Trans. Electron. Devices*, Vol. ED-25, 1978, p. 1185.

[42] M. Declercq, L. Gerzberg, and J. Meinol, "Optimization of the Hydrazine–Water Solution for Anisotropic Etching Silicon in Integrated Circuit Technology," *J. Electrochem. Soc.*, Vol. 122, 1975, p. 545.

[43] X. P. Wu and W. H. Ko, "A Study on Compensating Corner Undercutting in Anisotropic Etching of ⟨100⟩ Silicon," *Tech. Digest*, Transducers 1987, Fourth Int. Conf. on Solid State Sensors and Actuators, Tokyo, 1987, p. 126.

[44] A. Reisman, M. Berkenbilt, S. A. Chan, F. B. Kaufman, and D. C. Green, "The Controlled Etching of Silicon in Catalyzed Ethylenediamine–Pyrocatechol–Water Solutions," *J. Electrochem. Soc.*, Vol. 126, 1979, p. 1406.

[45] M. P. Wu, Q. H. Wu, and W. H. Ko, "A Study on Deep Etching of Silicon Using Ethylenediamine–Pyrocatechol–Water," *Sensors and Actuators*, Vol. 9, 1986, p. 333.

[46] A. I. Stoller, "The Etching of Deep, Vertical-Walled Patterns of Silicon," *RCA Rev.*, Vol. 31, 1970, p. 271.

[47] D. L. Kendall, "On Etching Very Narrow Grooves in Silicon," *Appl. Phys. Lett.*, Vol. 26, 1975, p. 195.

[48] O. Tabata, R. Asahi, H. Funabashi, and S. Sugiyama, *Digest of Technical Papers, Transducer '91*, IEEE Int. Conf. on Solid-State Sensor and Actuator, 1991, p. 811.

[49] U. Schnakenberg, W. Benecke, and P. Lange, *Digest of Technical Papers, Transducer '91*, IEEE Int. Conf. on Solid-State Sensor and Actuator, 1991, p. 815.

[50] L. D. Clark, Jr., J. L. Lund, and D. J. Edell, "Cesium Hydroxide (CsOH): A Useful Etchant for Micromachining Silicon," *Tech. Digest*, IEEE Solid State Sensor and Actuator Workshop, Hilton Head Island, SC, 1988, p. 5.

[51] M. M. Abu-Zeid, "Corner Undercutting in Anisotropically Etched Isolation Contours," *J. Electrochem. Soc.*, Vol. 131, 1984, p. 2138.

[52] J. B. Price, "Anisotropic Etching of Silicon with KOH-H_2O-Isopropyl Alcohol," in *Semiconductor Silicon*, ed. H. R. Huff and R. R. Burgess, Princeton, NJ, Electrochemical Society Proceedings, 1973, p. 339.

[53] T. D. Mok and C. A. T. Salama, "A V-Groove Schottky-Barrier FET for UHF Applications," *IEEE Trans on Elec. Dev.*, Vol. ED-25, No. 10, 1978, p. 1235.

[54] H. Seidel, "The Mechanism of Anisotropic, Electrochemical Silicon Etching in Alkaline Solution," IEEE Solid-State Sensor and Actuator Workshop, Hilton Head Island, SC, 1990, p. 86.

[55] H. G. Hughes and P. C. Lue, Motorola, unpublished data.

[56] J. G. Hooley, "The Kinetics of the Reaction of Silica with Group I Hydroxide," *Canadian J. of Chem.*, Vol. 39, 1961, p. 1221.

[57] D. B. Lee, "Anisotropic Etching of Silicon," *J. Applied Physics*, Vol. 40, No. 11, 1969, p. 4569.

[58] D. L. Kendall and G. R. deGuel, "Orientation of the Third Kind: The Coming of Age of ⟨110⟩ Silicon," in *Micromachining and Micropackaging of Transducers*, ed. C. D. Fung, P. W. Chung, W. H. Ko, and D. G. Fleming, 1985, p. 107.

[59] H. Seidel, L. Csepregi, A. Heuberger, and H. Baumgartel, "Anisotropic Etching of Crystalline Silicon in Alkaline Solutions—I. Orientation Dependence and Behavior of Passivation Layers," *J. Electrochem. Soc.*, Vol. 137, 1990, p. 3612.

[60] J. C. Greenwood, "Ethylenediamine–Catechol–Water Mixture Shows Preferential Etching of *p-n* Junction," *J. Electrochem. Soc.*, Vol. 116, 1969, p. 1325.

[61] A. Bohg, "Ethylenediamine–Catechol–Water Mixture Shows Etching Anomaly in Boron-Doped Silicon," *J. Electrochem. Soc.*, Vol. 118, 1971, p. 401.

[62] J. W. Faust, Jr., E. D. Palik, H. F. Gray, and R. F. Greene, "Study of the Etch-Stop Mechanism in Silicon," *J. Electrochem. Soc.*, Vol. 129, 1982, p. 2051.

[63] N. F. Raley, Y. Sugiyama, and T. Van Duzer, "⟨100⟩ Silicon Etch-Rate Dependence on Boron Concentration in Ethylenediamine–Pyrocatechol–Water Solutions," *J. Electrochem. Soc.*, Vol. 131, 1984, p. 161.

[64] H. Seidel and L. Csepregi, "Etch-Stop Mechanism of Highly Boron-Doped Silicon Layers in Alkaline Solutions," *Electrochem. Soc. Ext. Abs.*, 1985, p. 839.

[65] E. D. Palik, V. M. Bermudez, and O. J. Glembocki, "Ellipsometric Study of the Etch-Stop Mechanism in Heavily Doped Silicon," *J. Electrochem. Soc.*, Vol. 132, 1985, p. 135.

[66] H. Seidel, L. Csepregi, A. Heuberger, and, H. Baumgartel, "Anisotropic Etching of Crystalline Silicon in Alkaline Solution-Influence of Dopants," *J. Electrochem. Soc.*, Vol. 137, 1990, p. 3626.

[67] H. A. Waggener, "Electrochemically Controlled Thinning of Silicon," *Bell Syst. Tech. J.*, Vol. 50, 1970, p. 473.

[68] T. N. Jackson, M. A. Tischler, and K. D. Wise, "An Electrochemical P-N Junction Etch-Stop for the Formation of Silicon Microstructures," *IEEE Electron. Dev. Lett.*, Vol. EDL-2, 1981, p. 44.

[69] P. M. Sarro and A. W. van Herwaarden, "Silicon Cantilever Beams Fabricated by Electrochemically Controlled Etching for Sensor Applications," *J. Electrochem. Soc.*, Vol. 133, 1986, p. 1724.

[70] M. Hirata, S. Suwazono, and H. Tanigawa, "Diaphragm Thickness Control in Silicon Pressure Sensors Using an Anodic Oxidation Etch-Stop," *J. Electrochem. Soc.*, Vol. 134, 1987, p. 2037.

[71] A. Uhlir, Jr., "Electrolytic Shaping of Germanium and Silicon," *Bell Syst. Tech. J.*, Vol. 36, 1956, p. 333.

[72] R. L. Meek, "Electrochemically Thinned N/N+ Epitaxial Silicon-Method and Applications," *J. Electrochem. Soc.*, Vol. 118, 1971, p. 1240.

[73] B. Kloeck, S. D. Collins, N. F. Rooh, and R. L. Smith, "Study of Electrochemical Etch-Stop for High-Precision Thickness Control of Silicon Membranes," *IEEE Trans. on Elec. Dev.*, Vol. ED-36, No. 4, 1989, p. 663.

[74] V. M. McNeil, S. S. Wang, K. Y. Ng, and M. A. Schmidt, "An Investigation of the Electrochemical Etching of (100) Silicon in CsOH and KOH," *Tech. Digest*, IEEE Solid State Sensor and Actuator Workshop, Hilton Head Island, SC, 1990, p. 92.

[75] J. Y. Chen, "CMOS—The Emerging VLSI Technology," *IEEE Circuits and Devices Magazine*, Vol. 2, 1986, p. 16.

[76] J. Agraz-Guerena, R. A. Ashton, W. J. Bertram, R. C. Melin, R. C. Sun, and J. T. Clements, "Twin-Tub CMOSIII-a Third Generation CMOS Technology," *IEDM Tech. Digest*, 1984, p. 63.

[77] Lj. Ristic, A. C. Dhaded, R. Chau, and W. Allegretto, "Edges and Corners of Multilayer Dynamic Microstructures," *Sensors and Actuators*, Vol. A21–A23, 1990, p. 1042.

[78] M. Parameswaran, A. M. Robinson, Lj. Ristic, K. Chau, and W. Allegretto, "A CMOS Thermally Isolated Gas Flow Sensor," *Sensors and Materials*, Vol. 2, 1990, p. 17.

[79] M. Parameswaran, Lj. Ristic, A. M. Robinson, K. Chau, and W. Allegretto, "Electrothermal Microactuators in Standard CMOS Process," *Sensors and Materials*, Vol. 2, 1990, p. 197.

[80] M. Parameswaran, Lj. Ristic, A. C. Dhaded, and H. P. Baltes, "Fabrication of Microbridges in Standard Complementary Metal Oxide Semiconductor Technology," *Can. J. Phys.*, Vol. 67, 1989, p. 184.

[81] M. Parameswaran, Lj. Ristic, A. C. Dhaded, and H. P. Baltes, "Sandwiched Oxide Cantilever Beams in Standard CMOS Technology," *Proc. Canadian Conf. on Electrical and Comp. Eng.*, Vancouver, 1988, p. 781.

[82] S. Terry, "A Miniature Silicon Accelerometer with Built-in Damping," *Tech. Digest*, IEEE Solid. St. Sensor and Actuator Workshop, Hilton Head Island, SC, 1988, p. 114.

[83] P. M. Sarro, H. Yashiro, A. W. van Herwaarden, and S. Middelhoek, "An Infrared Sensing Array Based on Integrated Silicon Thermopiles," *Transducers '87*, Tokyo, 1987, p. 227.

[84] G. R. Lahiji and K. D. Wise, "A Batch-Fabricated Silicon Thermopile Infrared Detector." *IEEE Trans. Electron Dev.*, Vol. ED-29, 1982, p. 14.

[85] J. Y. C. Chang, A. A. Abidi, and M. Gaitan, "Large Suspended Inductors on Silicon and Their Use in a 2 μm CMOS RF Amplifier," *IEEE Electron. Dev. Lett.*, Vol. EDL-14, 1993, p. 246.

Chapter 4
Surface Micromachining Technology

Lj. Ristic, F.A. Shemansky, M.L. Kniffin, and H. Hughes

Motorola

4.1 INTRODUCTION

It has been pointed out in Chapter 3 that micromachining technology can be divided into two distinct categories: bulk micromachining and surface micromachining. The important aspects of bulk micromachining were covered in Chapter 3, and this chapter will cover basics of surface micromachining technology. The advantages of surface micromachining technology result from the ability to produce smaller sensors as well as from the fact that this approach shares many common features with current IC technology. It should be pointed out that progress in this field in the last couple of years has been unprecedented, and a single chapter is not enough to cover all the results in this rapidly expanding area. With this in mind, we will concentrate on surface micromachining dealing with polysilicon as a core material. The study of polysilicon for surface micromachining applications represents a primary effort in the sensor arena, with polysilicon processing technology verging on manufacturability. This chapter covers the key aspects of surface micromachining with polysilicon, including deposition, doping, annealing, sacrificial layers, and sacrificial etching, as well as characterization of polysilicon films. Some examples of practical applications are also included.

In addition to polysilicon technology the chapter contains a section on the LIGA process because the process shares two features with surface micromachining: reliance on orientation-independent etching and use of sacrificial layers for making three-dimensional structures. Another reason for the inclusion of LIGA process is our belief that this process, and particularly its evolution, will be one of the technologies of the future; as such it deserves our attention and a place in the book.

4.2 BASIC CONCEPT OF SURFACE MICROMACHINING TECHNOLOGY

Surface micromachining is a technique for fabricating three-dimensional microme-chanical structures from multilayer stacked and patterned thin films. The basic con-cept of surface micromachining was originally demonstrated with metal films by Nathanson et al. in the late 1960s [1]. The renewed interest in this technique grew after the presentation of two papers on polysilicon machined structures in 1987 at Transducers '87 in Tokyo [2,3]. Since that time, the rush has been on to develop this technology further.

4.2.1 Layer Stacking

The basic concept of surface micromachining is depicted in Figure 4.1. First, an isolation layer is deposited on a silicon substrate for electrical isolation or as a sub-strate protective layer. Then, a spacer layer is deposited on top of the isolation layer and patterned (this layer is also referred to as a *sacrificial layer*). The next step is the deposition of the structural layer (body or core layer) over the sacrificial layer followed by its patterning. Finally, the sacrificial layer is etched, using layer-selective etching. As a result the free-standing micromechanical structure is produced.

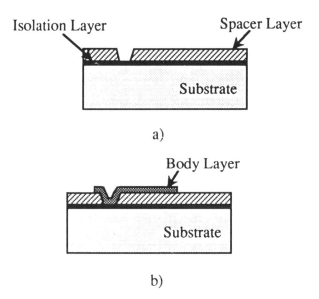

Figure 4.1 Surface micromachining: (a) patterned sacrificial layer; (b) patterned structural layer; (c) suspended beam after sacrificial etching.

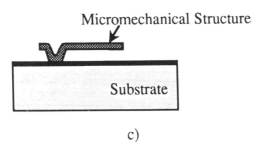

Micromechanical Structure

Substrate

c)

Figure 4.1 continued.

The surface micromachining approach is attractive because smaller structures can be produced with far better dimensional control compared to bulk micromachining. Other differences between bulk and surface micromachining are outlined in Table 4.1.

Surface micromachining can be used with a variety of combinations of thin films. The structural layer may be polysilicon [2,3], silicon nitride [4], silicon dioxide [5], polyimide [6], tungsten [7], molybdenum [8], amorphous silicon carbide [9], TiNi alloy [10], nickel–iron permalloy [11], or composite films such as polysilicon–ZnO [12], or polysilicon–silicon nitride–polysilicon [13]. The sacrificial layer composition depends on the choice of structural layer. Table 4.2 provides a summary of the various combinations of sacrificial and structural layers described in the literature. This is certainly an illustration of the potential of surface micromachining technology. There is no doubt that future research in this field will bring new materials and many new possibilities.

Table 4.1
Bulk vs. Surface Micromachining

Features	Bulk Micromachining	Surface Micromachining
Core Material	Silicon	Polysilicon
Sacrificial layer	————	PSG, (SiO$_2$)
Size	Large (typical cavity dimensions are several hundred μm)	Small (high precision controlled by thickness of the film; typical dimensions are several μm)
Processing factors	Single or double side processing (front and back side)	Single side processing (front side)
	Selectivity in material etching	Selectivity in material etching
	Etching: anisotropic (depends on crystal orientation)	Etching: isotropic
	Etch stops	Residual stress in films (depends on deposition, doping, annealing)
	Patterning	

Table 4.2
Combination of Structural and Sacrificial Layers in Surface Micromachining

	Structural Layer		Sacrificial Layer	
Material	Typical Thickness (μm)	Material		Typical Thickness (μm)
Polysilicon	1–4	PSG, SiO$_2$		1–7
Si$_2$N$_4$	0.2–2	PSG, SiO$_2$		2
SiO$_2$	1–3	Polysilicon		1–3
Polyimide	10	Al		1.5–3
W	2.5–4	SiO$_2$		8
Mo	0.5	Al		0.7
SiC	1.5	SiO$_2$		1.5
TiNi	8	Polyimide or		3
		Au		2
NiFe	2.5	Al or		7
		Cu		7
PolySi–ZnO	2–0.95	PSG		0.6
PolySi–Si$_3$N$_4$–PolySi	1–0.2–1	PSG		2

4.2.2 Sealing

In addition to free-standing micromechanical devices, surface micromachining technology offers the unique capability of forming hermetically sealed cavities. The cavities themselves are formed by the sacrificial etching, and sealing can be done in two ways: by a reactive sealing [14] and by a deposition process using polysilicon [14] or silicon nitride [4]. In reactive sealing, openings are sealed via thermal oxidation, which follows sacrificial etching. The internal cavity pressure may be varied by adjusting the partial pressure of trapped oxygen inside the cavity. Reactive sealing is shown schematically in Figure 4.2(b). Sealing by deposition is accomplished by

a)

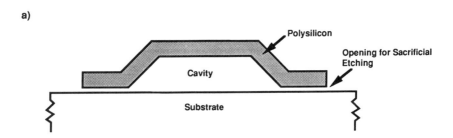

Figure 4.2 Sealing: (a) cavity after sacrificial etching; (b) reactive sealing; (c) sealing by deposition.

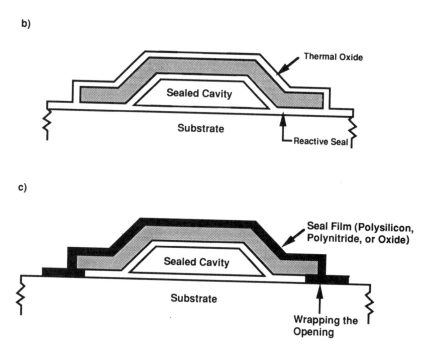

Figure 4.2 continued.

wrapping the openings with a deposited film, Figure 4.2(c). A good example of the application of sealing by deposition can be found in the work of the Berkeley group on incandescent microlamps [15].

4.3 POLYSILICON FOR SURFACE MICROMACHINING

Polysilicon has rapidly become the material of choice for a wide variety of surface micromachined devices. Polysilicon's popularity in this area is a direct result of its mechanical properties and its relatively well-developed deposition and processing technologies. These characteristics along with the capability of utilizing established integrated circuit (IC) processing techniques make it a natural selection. However, the use of polysilicon for micromechanical structures is not without its inherent difficulties. Producing structures that meet the required electrical and mechanical specifications of a design can be a very challenging task. The art of film engineering through judicious selection of deposition parameters and subsequent thermal processing is still maturing. The following sections focus on critical processing issues, such as deposition techniques, doping methods, annealing, and the effects of these processes on film microstructure and electrical characteristics. The discussion also includes sacrificial layers and sacrificial etching.

4.3.1 Deposition of Polysilicon

Polycrystalline silicon can be deposited by many different methods using a variety of source gases including silane, disilane, and the chlorosilanes (mono-, di-, tri-, and tetra-). Films with various properties have been generated by low-pressure chemical vapor deposition (LPCVD) [16–37], atmospheric pressure (APCVD) [37–39], plasma-enhanced chemical vapor deposition (PECVD), and molecular-beam deposition. However, the most predominant method for surface micromachining applications has been LPCVD in hot-wall batch reactors using silane as the silicon source gas.

The LPCVD silane pyrolysis process is moderately activated with an apparent activation energy $E_a = 32$–50 kcal/mol [16,19,20,22,40]. Deposition conditions range from 530–700°C with total pressures from 10^{-3}–10 torr. Through the variation of process parameters, temperature and silane partial pressure, amorphous, pseudo-amorphous (partially crystalline), or polycrystalline silicon can be deposited. Transition temperatures between amorphous and crystalline material ranging from 560–600°C have been reported [17,19,21–23,26,29] for LPCVD films.

Kinsbron, Sternheim, and Knoell [29] have shown that the incubation time for nucleation and the crystallization rate are most likely responsible for this transition-regime temperature spread. If the total time required to deposit the film is shorter than the incubation period, then a completely amorphous film will result. On the other hand, if deposition proceeds beyond the nucleation time, crystallization will commence and a fully polycrystalline film will eventually result, provided the crystal front moves more rapidly than the film deposition rate. Implicit in Kinsbron's formulation is the effect of reactant species partial pressure on the observed transition range, because incubation time for nucleation will be a function of this parameter. The phase transition generally begins heterogeneously at the substrate-film interface then proceeds along a uniform growth front to the surface of the film, as shown schematically in Figure 4.3(a) [29,41]. Kinsbron developed a simple expression to model the in-situ crystallization of LPCVD silicon films [29].

A plot of the crystalline volume fraction (f_x) of as-deposited film as a function of film thickness for the growth conditions modeled in Kinsbron's work is shown in Figure 4.3(b). Increasing the film thickness or decreasing the deposition rate leads to an increase in the deposition time, resulting in an increased extent of crystallization in the film. A similar effect can be achieved by increasing the substrate temperature.

Joubert et al. [23] tracked polysilicon-preferred orientation as a function of deposition temperature and silane partial pressure. Their experiments, and a com-

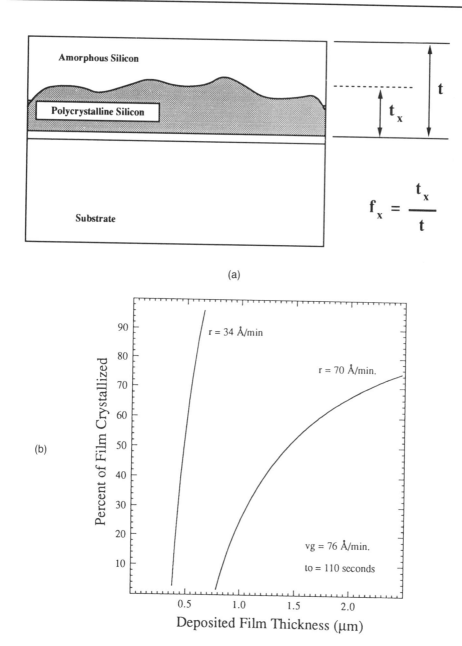

(a)

(b)

Figure 4.3 Microstructure of LPCVD silicon films deposited at temperatures near the amorphous-crystalline transition temperature: (a) schematic representation of the as-deposited polysilicon microstructure; (b) extent of crystallization in an as-deposited film as a function of both the film thickness and the deposition rate (r).

pilation of data from previous investigators, indicates at least a qualitative relationship between process parameters and film microstructure. The transition temperature from amorphous to crystalline material generally increases with increasing silane partial pressure. These observations are also consistent with those of Kamins and Cass [37], who showed a transition temperature of 680°C for APCVD silicon.

The transition regime is an important consideration in the fabrication of micromachined structures because the ultimate average residual stress and stress gradient through the thickness of the film will be strongly influenced by the initial film morphology. For example, if a constrained structure such as a diaphragm or a bridge is required, then a tensile stress film would be desirable to assure that the structure is free standing and not distorted. Postdeposition thermal processing of an as-deposited amorphous film could produce the desired characteristics. On the other hand, the appropriate stress condition may be difficult to achieve with an as-deposited polycrystalline film.

An attempt to quantify the relationship between deposition parameters and microstructure was presented by Voutsas and Hatalis [42]. Polycrystalline silicon films were deposited at temperatures as low as 530°C with pressures ranging from 2×10^{-3}–4×10^{-3} torr. This work showed that polysilicon microstructure can be altered by varying the deposition rate while maintaining the same deposition temperature. Guidelines were also set forth for predicting grain size of as-deposited polysilicon as a function of deposition parameters.

Hatalis and Greve conducted a detailed investigation of the kinetics of crystallization and grain growth in amorphous LPCVD silicon films [43,44], and some of their results are summarized in Figure 4.4. They found that the average grain size in the crystallized film was a function of both the deposition conditions and the temperature of the postdeposition anneal.

Both low deposition temperatures and high deposition rates were determined to lead to an increase in the average grain size of the annealed polysilicon film. A similar trend has been reported by French and Evans [45]. It is argued that this occurs because low substrate temperatures and fast growth rates suppress the formation of stable subcritical nuclei, thus decreasing the concentration of available nucleation sites at the initial stages of crystallization. Hatalis and Greves also found that reducing the temperature at which crystallization takes place results in a significant increase in the average polysilicon grain size.

LPCVD silicon films deposited at substrate temperatures above about 620°C are invariably polycrystalline [17,22,27,45,46]. A typical cross-sectional TEM micrograph of an undoped LPCVD polysilicon film deposited at 630°C (290 mtorr) is shown in Figure 4.5. The film consists of large columnar grains that grow out of a fine-grained transition region at the polysilicon-substrate interface. The columnar structure develops as the grains that have their fastest growing plane oriented parallel to the substrate dominate the growth of the film.

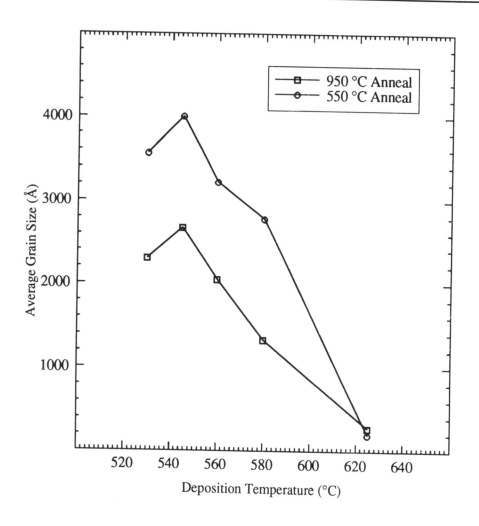

Figure 4.4 The effect of deposition temperature and postdeposition anneal on the average grain size of LPCVD polysilicon thin films.

Similar microstructures are observed for LPCVD polysilicon films that have been deposited at substrate temperatures up to 700°C. However, the dominant growth direction changes from the $\langle 110 \rangle$ direction to the $\langle 100 \rangle$ direction in the vicinity of 650°C [17,22,27,46]. The average grain size also tends to increase as the deposition temperature is raised.

Figure 4.5 Cross-sectional TEM dark-field micrograph of a 2-μm thick LPCVD polysilicon film deposited at 630°C. The film has a columnar structure and exhibits a preferred ⟨110⟩ texture.

4.3.2 As-Deposited Film Stress

One of the foremost considerations in producing micromechanical structures from polysilicon has to be the film stress. Both the average residual stress and the stress gradient are dependent on deposition conditions. The initial microstructure will contribute significantly to ultimate film properties regardless of subsequent processing steps and therefore must be considered carefully.

The average film stress can be determined using Stoney's equation, which relates stress to wafer curvature. In (4.1), E_s is the substrate elastic Young's modulus and ν_s is Poisson's ratio:

$$\sigma_f = E_s \frac{t_s^2}{6(1 - \nu_s)t_f R_c} \tag{4.1}$$

The radius of curvature is represented by R_c and t_s and t_f are substrate and film thicknesses, respectively. By measuring the wafer radius of curvature before and after deposition (or before and after film removal), an estimate of average film stress can be obtained.

Typically both fully amorphous and fully polycrystalline as-deposited LPCVD silicon films are in a state of compression [24,33,47] with stress magnitudes ranging up to *ca.* 700 MPa. Residual compressive stress values have also been reported for amorphous PECVD and APCVD silicon films ranging from about -300 to -500 MPa [38,41,48].

For LPCVD silicon films deposited under conditions that fall within the amorphous-crystalline transition region, the magnitude and sign of the as-deposited residual stress is extremely sensitive to the deposition conditions. The as-deposited stresses for these films typically range from $+500$ MPa tension to -500 MPa compression [24,32,38,41,42,46,49]. The magnitude of the residual stress can probably be correlated with the extent to which the amorphous-crystalline phase transition has gone to completion during the deposition process.

Work on average residual polysilicon stress as a function of deposition temperature at atmospheric pressure was presented by Adamczewska and Budzynski [38]. They showed that as-deposited films exhibit a peak in stress at around 630°C and rapidly drop off to a low stress condition at higher deposition temperatures. Similar results have been shown for LPCVD films [50]. For example, an order of magnitude reduction in compressive film stress (196 MPa to 20 MPa) has been observed by increasing the deposition temperature from 630°C (290 mtorr) to 700°C (290 mtorr). In these experiments, the silane flow rate was adjusted to maintain a constant reactant conversion level between runs in order to avoid obscuring results with a shift in deposition mechanism. Qualitatively similar results were obtained by Krulevitch et al. [26,41,51]; however, a direct comparison would not be appropriate, due to variations in reactor conditions, deposition parameters, and film thicknesses.

Extraction of absolute reactant conversion levels from the data presented by Kru-levitch is not possible, but calculated relative conversions (assuming identical deposition area for all runs) indicate as much as a fourfold variation between runs.

A stress gradient over the film thickness results in a bending moment manifested in deflection of a cantilever beam–type structure. Figure 4.6 is an SEM photo of a series of 2-μm thick polysilicon cantilever beams deposited at 630°C and 290 mtorr. The film stress gradient has resulted in an upward deflection. Figure 4.7 shows the resultant deflection as a function of beam length for both the 630°C film and polysilicon deposited at 700°C (290 mtorr). Also included in this figure are results from depositions performed at 630°C and at a slightly reduced total pressure of 180 mtorr. As previously indicated, a considerable reduction in average stress can be obtained by increasing the deposition temperature to 700°C, and Figure 4.7 demonstrates that the stress gradient can also be reduced. Similar trends are observed when reducing the deposition pressure at a constant temperature.

Both the average stress and stress gradient appear to be a function of the dominant orientation in fully crystalline films, with the $\langle 110 \rangle$ orientation producing the

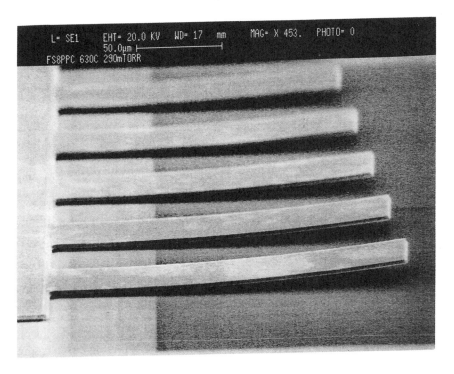

Figure 4.6 SEM micrograph of 2.0-μm thick polysilicon film deposited at 630°C and 290 mtorr total pressure.

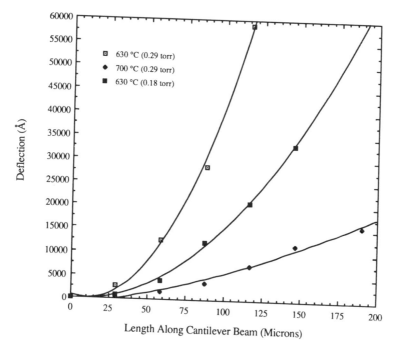

Figure 4.7 Polysilicon cantilever beam deflection as a function of processing conditions.

highest magnitude and randomly oriented films, the lowest [27,38,41]. The $\langle 100 \rangle$ fiber texture is expected in polysilicon films deposited at 700°C between 100 mtorr and 1 torr [23], which results in considerable stress reduction. Randomly oriented films have been demonstrated at reduced pressures (10^{-2}–10^{-3} torr) between 600°C and 700°C [23] and also at atmospheric pressure and temperatures greater than 750°C [38].

4.3.3 Annealing of Undoped Films

Numerous authors have examined the effects of furnace annealing on recrystalliza-tion and grain growth in undoped LPCVD polysilicon films deposited at substrate temperatures around 630°C [17,27,46]. The grain structure of these films is relatively stable with little change in microstructure expected during subsequent processing, even upon annealing in an N_2 furnace ambient at temperatures as high as 1050°C. However, after annealing at temperatures in excess of 1100°C significant recrystal-lization and grain growth is observed [17,27].

Guckel and coworkers [24,32,47,49] have used micromechanical test structures to characterize the effects of annealing on the properties of LPCVD silicon films deposited at 580°C. They have found that annealing leads to significant changes in the sign and magnitude of the average residual strain field in these films. Annealing the films at temperatures close to their deposition temperature leads to a gradual reduction in the magnitude of the compressive strain present in the as-deposited film, with the polysilicon approaching a strain free state for long annealing times.

Chang et al. have observed similar behavior with PECVD silicon films [48]. Postdeposition anneals at temperatures between 650 and 950°C cause the initially compressive amorphous or pseudo-amorphous films to become tensile. This phenomenon has been attributed to the volume contraction associated with the crystallization of the amorphous silicon layer located at the surface of the as-deposited film [41,46,49]. The magnitude of the average residual tensile strain decreases slightly with increasing anneal temperature in this range. After annealing at temperatures in excess of 1000°C, complete relaxation of the internal strain field is observed. Krulevtich [41,46] noted a similar stress relaxation in polysilicon thin films deposited at 605°C. TEM micrographs of the annealed films indicate that strain relaxation is not associated with recrystallization as has been previously argued.

As previously indicated, CVD films deposited at temperature-pressure combinations that lie above the amorphous-crystalline transition region tend to be compressive with the magnitude of the internal strain decreasing significantly with increasing deposition temperature. A plot of the average residual stress versus annealing temperature for films deposited at 630°C and 290 mtorr is shown in Figure 4.8. Significant reductions in the residual strain field can be achieved by moderate increases in anneal temperature. TEM micrographs have shown the relaxation in residual stress is not associated with significant changes in the microstructure of the film.

4.3.4 In-Situ Doping

In-situ doping refers to the procedure of impurity incorporation during the deposition step. In-situ doping, as opposed to an ex-situ doping process, offers a reduction in the total number of required processing steps, provides a flat concentration profile over the thickness of the film, and allows precise dopant concentration control.

Arsine (AsH_3), phosphine (PH_3), and diborane (B_2H_6) are often used as source gases for in-situ doping of polycrystalline silicon. The addition of diborane to silane tends to accelerate deposition and may reduce the apparent activation energy slightly ($E_a = 20$ kcal/mol) in the temperature range 620–900°C. Conversely, phosphine and arsine suppress the deposition rate significantly, and the addition of phosphine appears to slightly increase the apparent activation energy ($E_a = 44$ kcal/mol) [52,53].

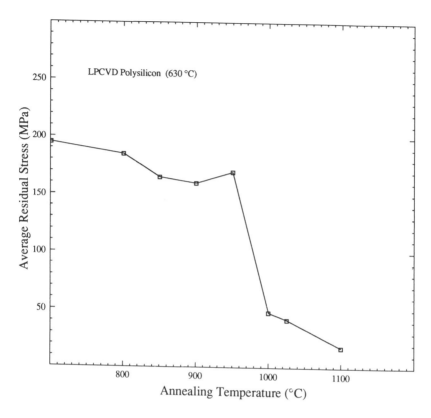

Figure 4.8 Average compressive stress in an undoped LPCVD polysilicon film deposited at 630°C, as a function of postdeposition annealing temperature.

Phosphine is probably the most commonly used source gas for doping during deposition. Unfortunately, films used for surface micromachining are often considerably thicker than those used in conventional integrated circuit applications, and the depressed deposition rate results in a throughput reduction. Introduction of phosphine also facilitates a degradation in film thickness uniformity, resulting in the need for specialized equipment. Gas injectors may be used to distribute the flow evenly throughout the boat of a batch system, and recently single wafer reactors have been employed to improve film thickness uniformity.

Deposition rate and uniformity reductions are generally attributed to preferential adsorption of phosphine on the substrate surface [54,55]. Using a Langmuir model, Meyerson and Yu [55] calculated sticking coefficients for phosphine and for silane of $S_{PH_3} = 1.0$ and $S_{SiH_4} \leq 0.025$, respectively. The preferential phosphine adsorption is thought to suppress a heterogeneous reaction mechanism, with residual polysilicon growth resulting from a gas phase intermediate formed by the homogeneous decomposition of silane. The highly reactive intermediate, formed in low quantities, would be rapidly depleted resulting in large concentration gradients and nonuniform deposition. A reactor scale model developed by Yeckel, Middleman, and Hochberg [56,57] also indicates that the homogeneous decomposition pathway is necessary to explain the deposition nonuniformity observed in this system.

Other effects of the addition of phosphine to the reaction system include a promotion of crystallization resulting in a lowering of the transition temperature between amorphous and crystalline material [52]. Harbeke et al. [52] also showed that the texture of as-deposited in-situ doped polysilicon films differs markedly from undoped films. At a deposition temperature of 580°C and 500 mtorr total pressure, a preferred orientation of $\langle 311 \rangle$ is indicated, switching to $\langle 111 \rangle$ at 600°C, a combination of $\langle 111 \rangle$ and $\langle 311 \rangle$ at 620°C, and strongly $\langle 311 \rangle$ at 640°C [52]. Effects of film texture on mechanical properties were not discussed. In fact, a dearth of published information exists concerning the mechanical properties of in-situ doped polysilicon films.

Average residual stress values for 2.0-μm thick in-situ phosphorus-doped polysilicon films have been shown to be a function of the input phosphine-to-silane flow ratio. Films deposited in an amorphous state at 575°C and 600°C, at atmospheric pressure, and annealed at 700°C for 1 hr, with an input phosphine-to-silane flow ratio ranging from 0.02 to 0.08 show a considerable reduction in tensile stress magnitude as PH_3/SiH_4 ratio increases [50]. Stress magnitudes ranged from *ca.* 500 MPa ($PH_3/SiH_4 = 0.02$) to *ca.* 100 MPa ($PH_3/SiH_4 = 0.08$). Corresponding resistivities for these films decrease monotonically with increasing phosphine-to-silane ratio from 100 Ω/square to less than 10 Ω/square. Similar resistivity trends were observed by Kurokawa [58] over the same phosphine–silane range. Kurokawa's polysilicon films were deposited at either 630°C or 650°C (1 torr total pressure) and annealed at 1000°C for 30 min. A resistivity saturation at a PH_3/SiH_4 ratio of 0.04 was also noted.

As previously discussed, introduction of phosphine into a reaction system utilizing silane as the silicon source results in a suppression in polysilicon growth rate. In order to circumvent this problem, disilane (Si_2H_6) has been employed by several investigators [59,60]. Nakayama, Yonezawa, and Murota [59] showed no reduction in silicon deposition rate in the Si_2H_6–PH_3–He system at 520–665°C and 3 torr total pressure. In fact, at high phosphine concentrations, deposition rates were enhanced several orders of magnitude relative to the silane-based in-situ phosphorus doping process. Whidden, Lindlow, and Stiles [60] confirmed these results while depositing amorphous silicon at temperatures as low as 480°C.

Figure 4.9 shows cross-sectional TEM micrographs of a 1.5-μm silicon film deposited from disilane and phosphine at 510°C and 800 mtorr total pressure. The as-deposited film is amorphous with little change in microstructure following anneals at 700°C, 800°C, and 900°C. The film annealed at 900°C had an average residual tensile stress of *ca.* 100 MPa and a resistivity of 10 Ω/square.

1.0 μm

(a)

Figure 4.9 Cross-section TEM micrographs of silicon films deposited at 510°C and 800 mtorr total pressure from disilane and phosphine: (a) as-deposited, (b) following 1 hr 700°C annealing, (c) following 1 hr 800°C annealing, and (d) following 1 hr 900°C annealing.

0.5 μm

(b)

Figure 4.9 continued.

0.5 µm

(c)

Figure 4.9 continued.

0.5 μm

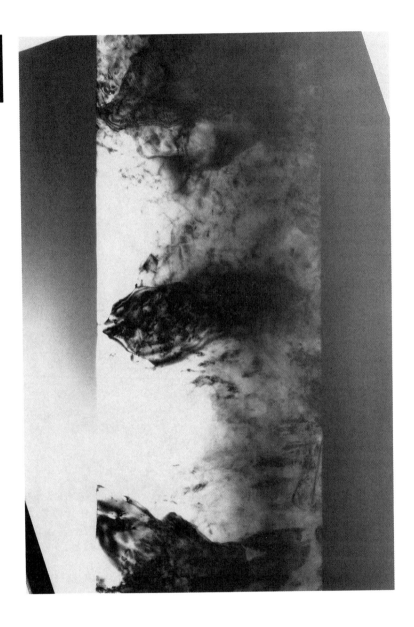

(d)

Figure 4.9 continued.

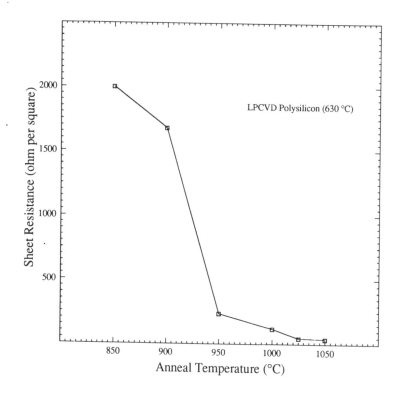

4.3.5 Ex-Situ Doping

Polysilicon films for surface-micromachining applications are typically deposited onto a sacrificial oxide layer. This oxide is usually doped with phosphorus to increase its etch rate in hydrofluoric acid solutions. During postdeposition thermal processing the doped glass can serve as a diffusion source, resulting in impurity incorporation into adjacent polysilicon films.

The sheet resistance of a 2-μm thick polysilicon film deposited at 630°C (290 mtorr) is plotted in Figure 4.10 as a function of annealing temperature. Considerable doping of the polysilicon film from the sacrificial PSG layer (8 wt% phosphorus) occurs at temperatures in excess of 950°C.

The processes used to dope polysilicon films can have a significant impact on electrical, chemical, and mechanical properties. Consequently there is considerable interest in understanding these processes and the resultant changes in the properties

Figure 4.10 Variation in the sheet resistance of an LPCVD silicon film ($T_{dep} = 630$°C) as a function of postdeposition annealing temperature. Doping of the film is achieved via phosphorous diffusion from the PSG spacer layer.

of the film. However, to date, little information has been published relating ex-situ doping of polysilicon films to the physical properties of the material.

For ex-situ doping, via predeposition and drive-in diffusion, phosphorous-containing compounds such as phosphorous oxychloride ($POCl_3$) or phosphine (PH_3) are reacted with oxygen to form phosphorous pentoxide (P_2O_5). The phosphorous containing oxide then serves as a source for the diffusion of phosphorous into the polysilicon film. The concentration of phosphorous at the polysilicon surface is a function of both the gas mixture and the furnace temperature. The phosphorous content of the film used as a diffusion source in a standard predeposition cycle is typically higher than the phosphorous content of a sacrificial PSG spacer layer. Consequently significantly higher doping levels can be achieved at only moderate processing temperatures at the expense of an additional processing step.

The moderate doping cycles commonly used in surface-micromachining applications tend to have no effect on the microstructure of polysilicon films deposited at temperatures near 630°C. Comparative analysis of a 630°C deposited LPCVD polysilicon film before and after doping for 30 min at 875°C in a PH_3/O_2 atmosphere shows no evidence of recrystallization or grain growth. Heavier doping cycles (higher temperatures or longer times) can lead to significant recrystallization and grain growth [61]. This behavior has been attributed to an enhancement of the silicon vacancy concentration in the presence of a large concentration of n-type dopants.

For LPCVD polysilicon films deposited at temperatures below about 630°C, one would expect crystallization of any amorphous material present in the film to occur during a doping oxide deposition process, and this has been observed experimentally.

For 2-μm thick LPCVD films deposited at temperatures above 630°C, a moderate phosphorous predeposition tends to lead to a slight increase in the average residual compressive stress [27,61]. Heavier doping cycles (higher temperatures or longer times) can lead to a significant reduction in the average residual stress [61]. The reduction in film stress at higher doping temperatures may be associated with the recrystallization and grain growth typically observed upon annealing heavily doped n-type polysilicon films.

The residual stress gradient tends to be significantly more sensitive to phosphorous predeposition than the average residual film stress [63,64]. The effect of a moderate phosphorous predeposition and drive-in on the residual stress gradient in a 630°C LPCVD polysilicon film is illustrated in Figure 4.11, where variations in the curvature of doped and undoped polysilicon cantilever beams are plotted as a function of the anneal temperature. Similar results have been observed for films deposited at other temperatures [65]. A moderate phosphorous predeposition cycle causes the released film to bend toward the substrate, which represents a reversal in the sign of the residual stress gradient in the undoped film. This indicates that the phosphorous predeposition process contributes to a significant increase in the magnitude of the compressive stress at the polysilicon surface. The mechanism for this

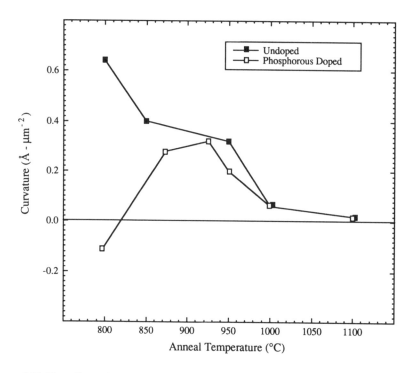

Figure 4.11 The effect of a phosphorous predeposition and drive-in on the curvature of surface-micromachined polysilicon (T_{dep} = 630°C) cantilever beams. Data on cantilevers fabricated from polysilicon films that did not receive a phosphorous predeposition is plotted for comparison.

is not entirely clear, although it has been suggested that the precipitation of a silicon-phosphide phase at the grain boundaries is responsible [17,27,66]. The properties of the doped film gradually approach those of the undoped film as the annealing times and temperatures are increased. This phenomenon is probably due to the diffusion of phosphorous away from the top surface of the polysilicon film into the bulk.

The sheet resistance of a diffusion-doped polysilicon film is a complex function of polysilicon microstructure, film thickness, composition of the sacrificial oxide layers in contact with the film, predeposition conditions, and time and temperature of the annealing. Some of the key points in determining the sheet resistance follow.

Increasing both the time and temperature of the predeposition process leads to a decrease in the sheet resistance of the film. This is consistent with the trends observed in single crystal silicon, although the sheet resistances of the polysilicon tend to be substantially higher. Subsequent annealings lead to a gradual reduction in sheet resistance, with the temperature of the annealing having a much greater impact than the anneal time.

The physical properties of the polysilicon film also play an important role in determining the resistivity of the film [67,68]. The effect of substrate deposition temperature on the sheet resistance of phosphorous diffusion-doped polysilicon films is illustrated in Figure 4.12. Here the sheet resistances of 2-μm thick polysilicon films are shown. These films were doped during a 30 min PH_3/O_2 predeposition at 875°C, and their sheet resistances are plotted as a function of the polysilicon deposition temperature. The sheet resistance goes through a maximum for films deposited at 630°C. Kamins and coworkers have observed similar results [67,68], with this behavior being attributed to variations in the average grain size, finer grained materials yielding higher sheet resistances.

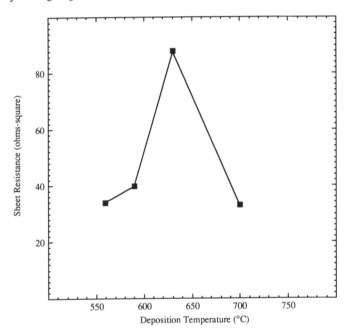

Figure 4.12 Polysilicon sheet resistance following a PH_3/O_2 predeposition as a function of the polysilicon deposition temperature.

4.3.6 Sacrificial Layers and Sacrificial Etching

Sacrificial film layers can be used as a scaffolding material in the construction of three-dimensional structures. These layers may also serve as a dopant source. The use of phosphorus-doped oxide as the sacrificial layer is most common. Etch rate and reflow properties can be engineered by controlling dopant incorporation. Moreover, the etch selectivity of the oxide in dilute hydrofluoric acid (HF) solution over other materials such as polysilicon and silicon nitride is generally good.

Phosphosilicate glass (PSG) deposition for surface micromachining applications is typically accomplished by thermal LPCVD using SiH_4, O_2, and PH_3. PECVD using SiH_4, N_2O, and PH_3 or even tetraethylorthosilicate (TEOS) and trimethylphosphite (TMP) are other possibilities. Regardless of the deposition technique, blanket etch rates in HF solution exhibit a monotonic increase with increasing phosphorus content. PECVD PSG etch rates are consistently higher than LPCVD PSG over a range (0–14 wt%) of phosphorus concentrations [69,70]. For example, Shioya and Meada [69] show that at 6 wt% phosphorus LPCVD PSG etches at 5,500 Å/min in 1.25% HF (20°C) solution whereas PECVD PSG etches at 10,000 Å/min. The plasma deposition conditions were not specified; however, an increase in etch rate is typically observed with a decrease in deposition temperature [70].

Although these blanket etch rates are important, the PSG sacrificial etching process, in many surface micromachining applications involving complex structures, quickly becomes diffusion limited as opposed to reaction rate limited. Figure 4.13 shows a schematic representation of the sacrificial etching sequence with seven distinctive steps identified. Monk, Soane, and Howe [71] state that external mass transfer limitations have been proven insignificant, and they model the etch process as a steady-state lumped-parameter diffusion-reaction system. The diffusive flux ($\mathcal{I}_{diff}$), or delivery rate of reactant to the fluid-solid interface, is set equal to the reactive flux ($\mathcal{I}_{rxn}$) and the resulting equation is solved numerically,

$$\mathcal{I}_{diff} = \mathcal{D}_{HF} \frac{C^b - C^s}{\delta} \tag{4.2a}$$

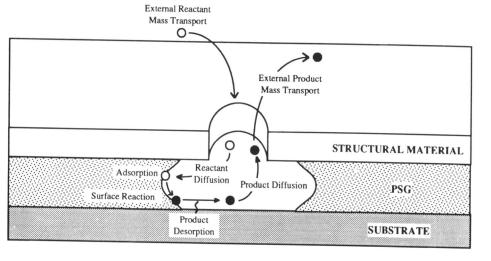

Figure 4.13 Schematic representation of PSG sacrificial etching mechanism.

$$\mathcal{P}_{\text{rxn}} = k_{\text{eff}}(C^s)^n \tag{4.2b}$$

In this analysis, adsorption, surface reaction, and desorption rate constants are combined and represented by k_{eff} and n is the reaction order. C^b and C^s are bulk and surface concentrations of reactant, respectively. $\mathcal{D}_{\text{HF}}$ is the diffusivity of HF in water and δ is the diffusion length. Experimental results for concentrated HF solutions (49%) in conjunction with model predictions indicate that, after only 10 min in the etch solution, reactant diffusive flux becomes rate determining. The experimental results of Monk et al. also indicate that the etch-channel geometry affects the etch time with the distance etched into a rectangular channel increasing with increasing channel width. Since liquid-phase reactant diffusivities are directly proportional to temperature [72], PSG etch time may be reduced by increasing the etch solution temperature. However, with etching reaction rates having an exponential temperature dependence, selectivities of other materials in solution (for example, silicon nitride, metals) may be adversely affected. Thus, sacrificial etch conditions for a particular application will vary and should be optimized for each situation.

Increasing the film phosphorus content to increase blanket PSG etch rates may be an overriding consideration, particularly with short diffusion lengths and when etch selectivities are a major concern. Moreover, extended exposure of polysilicon to HF solution may degrade mechanical properties, thus providing more incentive to reduce sacrificial etch time. Higher PSG phosphorus content can however result in PSG reflow [73] or dewetting manifested in structural layer deformation during high temperature stress relaxation cycles.

This problem may be circumvented by employing a composite sacrificial film [34]. The composite depicted in Figure 4.14 was used for experiments conducted on

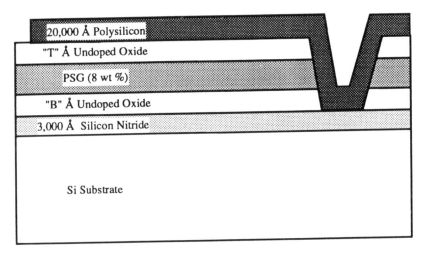

Figure 4.14 Composite PSG structure.

2-μm thick (total for both doped and undoped films) sacrificial oxide films. RTA conditions ranging 1,050–1,100°C and 60–300s were employed, which was sufficient to produce flat cantilever beam structures. The doped portion of the composite oxide contained 8 wt% phosphorus and the top and bottom undoped layers ranged in thickness from 0–5,000Å. With no underlying undoped layer ($B = 0$Å), none of the films survived the anneal cycle required to alleviate the stress in the polysilicon structural material. An underlying undoped layer ($B = 1,000$–5,000Å; $T = 0$Å) produced structures that were borderline in terms of structural integrity while all PSG sandwich films produced structures with no distortion. Thus, even if the annealing temperature exceeds the reflow temperature for the PSG [73], undistorted structures can be produced at the expense of only a small compromise in etch time. Similar results were obtained by Putty et al. using a composite 16,000Å underlying silicon dioxide layer covered by a 4,000Å 6% PSG film that was annealed for 3 min at 1150°C [34].

Finally, step coverage of the sacrificial films is becoming an increasingly important issue as microstructure complexity increases, critical dimensions decrease, and multiple structural layers become commonplace. Typically structural films used in micromachining applications are several micrometers thick, with spacing between adjacent structures on the same order. This can result in difficult to fill areas and unwanted surface topographies that are propagated through subsequently deposited layers.

Figure 4.15 compares SEM micrographs of PSG (8 wt%) deposited in rectangular channels etched into single crystal silicon with a nominal aspect ratio of 2.0.

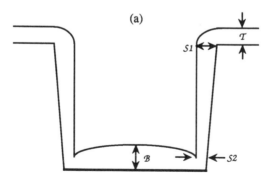

Figure 4.15 (a) SEM micrographs comparing PSG step coverage in rectangular trenches from (b) LPCVD silane, oxygen, and phosphine with (c) PECVD tetraethoxysilane, trimethylphosphite, and oxygen.

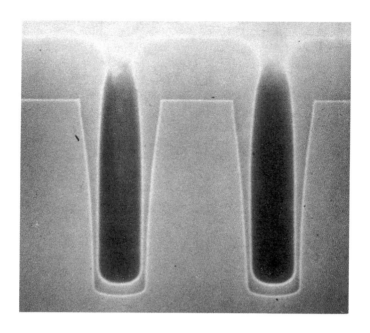

LPCVD Process

39
10

(b)   25

Figure 4.15 continued.

The blanket film thickness in each case is approximately 2.0 μm. Figure 4.15(b) shows the results from a standardly employed thermal LPCVD process using SiH_4, O_2, and PH_3 at 425°C and 280 mtorr total pressure. The PSG in Figure 4.15(c) was deposited from TEOS, TMP, and O_2 in a PECVD system. The step coverage values tabulated under each micrograph show that the PSG film-thickness uniformity is greatly enhanced with the PECVD process. Potential difficulties associated with the

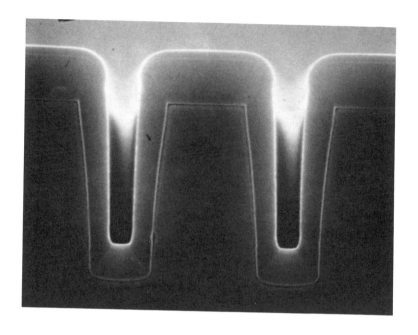

PECVD Process

57
41
(c)   70

Figure 4.15 continued.

poor surface coverage of the LPCVD PSG are many and range from mechanical obstruction to electrical shorting. Figure 4.16 shows the step coverage (bottom sidewall/top) as a function of aspect ratio (initial feature height/initial feature width) for both processes. In each case, the conformality decreases with increasing aspect ratio. The LPCVD process degrades more rapidly than the PECVD process, dropping to nearly 0 at aspect ratios above approximately 3.0.

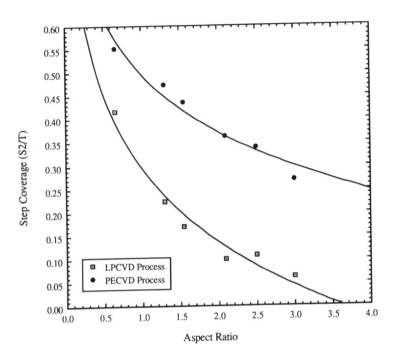

Figure 4.16 Step coverage of PSG as a function of rectangular trench aspect ratio for PECVD and LPCVD processes.

4.3.7 Stiction

During the rinse and dry cycle that follows the sacrificial etch process, the evaporating rinse solution creates a large capillary force that tends to draw the micromechanical device structure toward the underlying substrate [74,75]. Should the device structure come in contact with the substrate as a result of this capillary action, substantial attractive forces can develop between the two surfaces. For large-area or low-spring-constant devices this attractive force can easily exceed the restoring force of structure. As a consequence, the micromechanical device will remain adhered to the wafer surface after the rinse and dry cycle is completed. Once stuck together, the two surfaces are difficult to separate, and the device is rendered useless. Obviously this can lead to significant reductions in device yield, hence considerable effort has been devoted to understanding and solving the problem of stiction. In this section we will briefly review some of the techniques developed to solve this problem.

The simplest way to circumvent the stiction problem is to prevent the micromechanical device from coming in contact with the underlying substrate at any point during the rinse and dry cycle. One means of attaining this goal is to minimize the forces acting on the device by the withdrawing liquid. This can be achieved by utilizing low surface tension solutions such as methanol or *n*-hexane, rather than water, for the final rinse bath [74]. This approach has met with some success. However, it is not always completely effective when applied to large-area or low-spring-constant devices.

The surface tension forces can be completely eliminated by supplanting the rinse bath with a solid phase that can then be removed through a vapor phase process such as sublimation or reactive ion etching. For example, Orpana and Korhonen have demonstrated that solutions of acetone and photoresist can be used to form a solid spacer layer, which is then removed using an oxygen plasma to leave a freestanding structure [76]. Others have shown that micromachined devices can be immersed in a fluid, such as a methanol:water mixture or cyclohexane, which is then frozen to form a solid spacer layer that is subsequently sublimed in vacuum to release the device [77]. Techniques of this nature show promise but need additional work to be made compatible with a low-cost, high-volume manufacturing environment.

Several authors have designed temporary structural modifications that can be used to stiffen the micromechanical device during the sacrificial etch process, thus preventing contact with the substrate during drying and subsequent handling. The additional structural support is then removed prior to assembly or testing, in order to allow the device to operate freely. Examples of this approach are fusible polysilicon links [78] and plasma-etchable parylene support columns. In general, techniques of this type are quite effective methods of reducing yield loss due to stiction. However, they can require the integration of a number of additional steps into the device process flow.

An alternative approach to solving the stiction problem involves minimizing the magnitude of the attractive forces that develop between the substrate and the movable mechanical parts should they come in contact with one another. By reducing the magnitude of these interfacial forces, one increases the probability that the device's mechanical restoring forces will be sufficient to overcome the forces causing it to adhere to the underlying substrate. Several methods of modifying the silicon surface termination to achieve this goal have been examined. The simplest processes involve treating the device in an oxidizing bath such as H_2O_2 or subjecting it to an RCA cleaning prior to the final rinse [75]. Others involve terminating the silicon surface with hydrophobic functional groups using hexamethyldisilazane [74] or octadecyltrichlorosilane [75]. Both techniques have been shown to reduce yield loss due to stiction, though the former is most easily integrated into a conventional IC manufacturing process.

Structural modifications to the micromechanical device have also been employed to reduce the attractive forces between the device and substrate. Standoff bumps have been designed to reduce the area of contact between the device and the underlying substrate, thus minimizing the net force acting on the device [79]. This approach can be particularly effective in combination with some of the aforementioned techniques for reducing stiction; however, it does require the introduction of several additional fabrication steps into the process flow.

4.4 MECHANICAL CHARACTERIZATION OF POLYSILICON

With the advent of surface micromachining technology, numerous nontraditional methods for the characterization of thin-film mechanical properties on the micrometer and submicrometer scales have been developed. These techniques involve the fabrication of micromechanical test structures that exhibit a controlled response to an internally or externally applied load that can be analytically related to various mechanical properties of the thin film. The micromechanical test structures can be placed at different locations across the wafer surface, thus providing an accurate picture of the on-wafer variation in these properties. Many of these devices can also be fabricated in parallel with actual device structures to serve as in-situ process control monitors.

Thin films used in the fabrication of surface-micromachined devices tend to have large residual stress fields that are extremely sensitive to both the deposition conditions of the film and its postdeposition process history. These residual stresses can affect the load-response behavior of the device, the frequency dependence of its output, and many other critical operating parameters; therefore they must be fully understood and carefully controlled if one is to develop a manufacturable surface-micromachining process. In the following section we will review some of the micromechanical test structures and discuss their application to the mechanical characterization of thin films.

4.4.1 Test Structures for In-Situ Characterization

One class of test structures used for evaluation of internal strain fields is based on the measurement of dimensional changes that occur upon the release of a stressed film from the underlying substrate. After release from the substrate, a stressed film is free to relax to its unstressed state, and the resultant contraction or elongation of the film is directly proportional to the magnitude of the internal strain field. These test structures have been devised to provide both direct and indirect measurements of the stress relaxation [80–83].

4.4.1.1 Cantilevers

Direct measurement of the strain relaxation can be achieved by measuring the contraction or elongation of a thin cantilever beam after release from the substrate [81].

The residual strain in the film is related to the dimensional change in the beam by the following expression:

$$\varepsilon = \frac{L_0 - L}{L_0} \tag{4.3}$$

Where ε is the residual strain, L_0 is the initial length of the cantilever beam, and L is the length of the beam after release from the substrate. The difference between L_0 and L is typically determined by comparing the end of the cantilever beam to fiducial marks patterned on the wafer surface [81].

This measurement technique can prove problematic when applied to films with relatively high elastic moduli, as is often the case in surface-micromachined devices. That is, the dimensional changes in the film tend to be quite small and difficult to measure, even when large aspect ratio structures are utilized. Senturia and coworkers have developed [80,82,83] several structures for the characterization of tensile films that effectively amplify the stress relaxation. However, for practical device dimensions the amplication factors are on the order of one-and-a-half to two; therefore devices of this type are still best suited to the measurement of films that exhibit large internal stress, such as polyimide or other polymeric materials.

4.4.1.2 Bridges

Guckel et al. [24,25,47,84] have developed structures for the measurement of localized strain fields based on the tendency of slender members to buckle when a compressive load acting on its ends exceeds a critical value ε_{CR}. The Euler buckling strain for a thin beam under an axial load is given by [84–86]:

$$\varepsilon_{CR} = \frac{\pi^2 t^2}{KL^2} \tag{4.4}$$

where K is a constant related to the geometrical boundary conditions, t is the thickness of the thin film, and L is the length of the beam. For surface-micromachined bridges, the values for K lie between those of a simply supported beam ($K = 12$) and those of a doubly clamped beam ($K = 3$).

The upper and lower bounds for the residual strain in a compressively stressed thin film can be obtained by fabricating an array of bridge structures of varying lengths and observing the point at which the critical buckling strain is exceeded. An example of this technique is given in Figure 4.17, where an optical interference micrograph of a polysilicon bridge array is shown. Interference fringes on the unbuckled beams are not straight while those on the buckled beams are straight. In this array all bridges that exceed 150 μm in length have bowed, whereas all those that

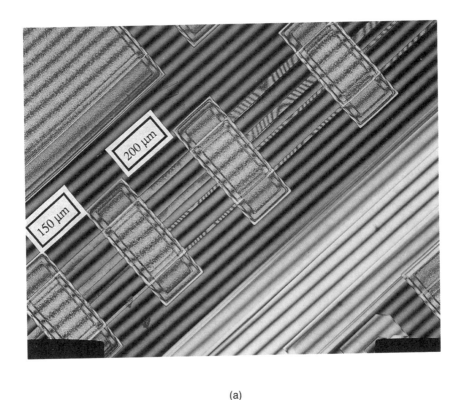

(a)

Figure 4.17 Surface-micromachined bridge array for the evaluation of residual compressive strains in thin films: (a) optical interference micrograph of an array of polysilicon bridges; (b) theoretical buckling strains for the bridges in this array.

are shorter than 150 μm remain straight. One can see from Figure 4.17(b) that this indicates that the magnitude of the internal strain lies between -3.2×10^{-4} and -5.6×10^{-4}. This is consistent with substrate curvature measurements taken on similarly processed films.

4.4.1.3 Ring and Beam Structures

A similar technique can be applied to the characterization of tensile films if one utilizes the ring and beam structure [24,25,47] shown schematically in Figure 4.18. When this structure is released from the substrate, the relaxation of the internal stress in the film will result in the application of a tensile load to the ring at the two opposing anchor points. As a result of this tensile load, the ring will want to contract

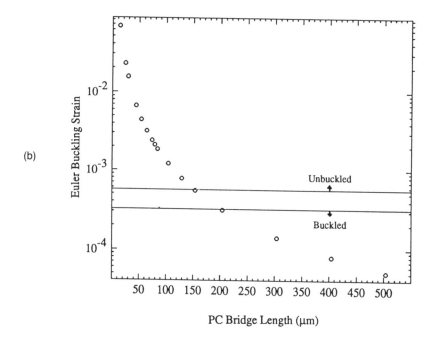

(b)

Figure 4.17 continued.

radially at the point where the ring joins the internal crossbar [87]. This radial contraction will introduce compressive stress in the crossbar that is related to the average residual strain in the film by the following expression:

$$\varepsilon_{crossbar} = G\varepsilon_{film} \tag{4.5}$$

where G is determined by the specific geometry of the ring and beam structure [24,25,47]. Combining (4.4) and (4.5) one obtains the following expression for the critical value of the residual strain required to buckle the crossbar:

$$\varepsilon_{film}^{CR} = \left(\frac{\pi^2 t^2}{3(2R_{CR})^2}\right)\left(\frac{1}{G}\right) \tag{4.6}$$

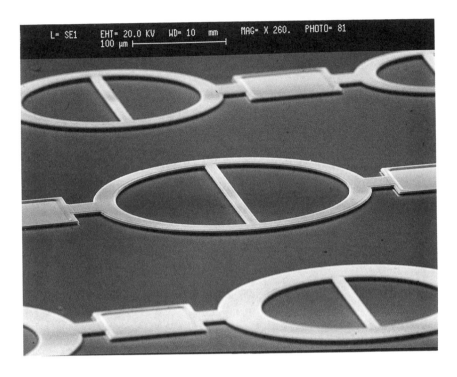

Figure 4.18 Schematic drawing of a ring-and-beam micromechanical test structure for the measurement of residual tensile stress in thin films (after [47]).

As in the bridge structures, upper and lower bounds for the residual strain in the film can be obtained by fabricating and analyzing an appropriate array of these test structures.

4.4.2 Gradient of Residual Stress

It should be noted that the expressions for the critical buckling stresses are truly valid only for films in which the residual stresses do not vary through the thickness of the film, as a residual strain gradient introduces an eccentricity to the axial load that will tend to alter the elastic stability of the beam [86]. In-plane stresses typically vary through the thickness of a thin film, particularly if the film has been subjected to substantial postdeposition processing [88]. This variation in the film stress results in an internal bending moment M whose magnitude is given by [88]:

$$M = \int_{-t/2}^{t/2} \sigma_x(y)y\,dy \qquad (4.7)$$

Where t is the thickness of the thin film and $\sigma_x(y)$ is the magnitude of the in-plane stress at a distance y from the center of the film. This effective bending moment will cause the film to curl when it is released from the substrate, resulting in a nonplanar device structure. Several test structures have been devised to evaluate the magnitude of this internal stress gradient [88–90], the simplest being the surface-micromachined cantilever beam.

Relaxation of the internal stress gradients in the film will cause a surface-micromachined cantilever to bend out of the plane of the substrate upon release from the sacrificial oxide layer, as illustrated in Figure 4.6. The magnitude of this deflection can be measured using either a SEM or a phase-contrast microscope [64]. Such data can then be used to estimate the magnitude of the internal stress field.

For a residual stress gradient that remains constant along the length of the beam, the vertical deflection $\delta(x)$ at a point x can be described by [85]:

$$\frac{\delta(x)}{x} = K + \frac{(1 - v^2)}{2EI} Mx \qquad (4.8)$$

Where x is the position along the length of the cantilever, K is a constant that depends upon the boundary conditions at the support point, $E/(1 - v^2)$ is the biaxial modulus of the film, and I is the moment of inertia of the beam about the z-axis. The first term in (4.8) is related to the geometrical boundary conditions at the constraint end; the second term describes the warping of the cantilever beam due to the residual strain gradient in the film. Figure 4.19 shows a plot of $\delta(x)/x$ versus x whose slope is proportional to $M(1 - v^2)/E$, which is process sensitive.

Fan et al. [90] have developed a spiral microstructure for the characterization of residual strain gradients in thin films. When fabricated from a thin film that exhibits a residual strain gradient, spiral structures tend to contract or expand upon release from the substrate. The resultant deformation provides three different response variables, which can be directly related to the residual strain gradient: endpoint rotation, endpoint height, and lateral contraction. The advantage of the Archimedean spiral structure is that the strain gradient can be estimated from a single data point, and two of the response variables (end rotation and lateral contraction) do not require the use of a SEM or phase contrast microscope to obtain the measurement. This substantially reduces the time required to characterize the thin film. However, the relatively large area required to fabricate Archimedean spirals with the requisite sensitivity is somewhat prohibitive. As an illustration, a SEM micrograph of an Archimedean spiral structure is shown in Figure 4.20. This particular structure is fabricated from a polysilicon film with a small residual stress gradient.

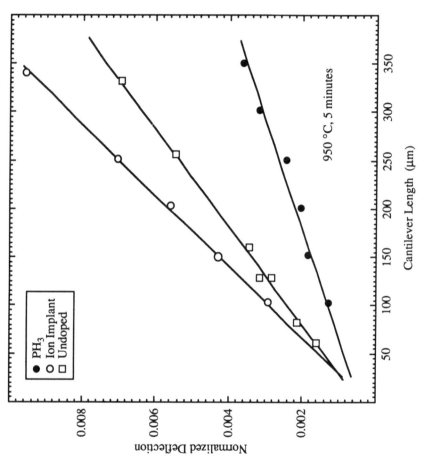

Figure 4.19 Plot of normalized deflection versus cantilever length for polysilicon cantilever beams that have been subjected to different process conditions.

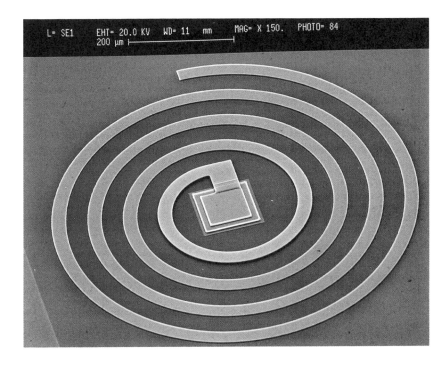

Figure 4.20 SEM micrograph of a polysilicon Archemedian spiral structure used for the characterization of internal stress gradients.

4.4.3 Load-Response Characterization

The preceding sections have discussed methods of measuring the magnitude and homogeneity of residual stresses in thin films. Numerous other thin film mechanical properties, such as Young's modulus, Poisson's ratio, and the yield strength of the material, play an equal if not greater role in determining the characteristics of a micromechanical device. These properties can also be quite sensitive to the nature of the film and therefore must be carefully monitored. Measurements designed to evaluate these properties typically involve characterizing the response of a micromechanical structure to an externally applied load. The variation in the response of the structure with the magnitude of the applied load can then be directly related to the mechanical properties of the materials from which it is fabricated. The remainder of this section will review some of these methods and their application to the evaluation of thin film mechanical properties.

134

Several authors have developed characterization techniques based on the load-response behavior of cantilever beams. The mechanics of a thin cantilever beam are well defined and closed-form solutions to the load-response behavior of this structure are available for a variety of different loading configurations [85], therefore the measured response of the cantilever beam to an applied load can easily be related to the mechanical properties of the material from which the cantilever is fabricated.

Weihs et al. [91] have developed a technique for characterizing the mechanical response of thin film cantilever beams using a load controlled submicrometer indentation instrument called a *Nanoindenter*. The Nanoindenter has a diamond tipped stylus, which is used to mechanically deflect the beam while continuously monitoring the applied load and the deflection of the cantilever beam. This is shown schematically in Figure 4.21(a). The load-response characteristics of the cantilever beam can be extracted from these data and used to obtain information about the elastic and plastic properties of the cantilever material.

Brittle materials such as polysilicon will exhibit linear-elastic behavior up to their breaking points. A typical load-deflection curve for a polysilicon cantilever beam is shown in Figure 4.21(b). In the linear-elastic regime the deflection of the beam $\delta(x)$ is related to the applied load P by:

$$\delta(x) = \frac{4Px^3(1 - \nu^2)}{Ewh^3} \tag{4.9}$$

(a)

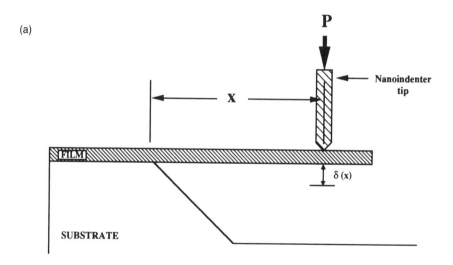

Figure 4.21 Mechanical characterization of thin films using the Nanoindenter technique: (a) schematic drawing of the measurement technique (after [92]); (b) load-deflection data for a surface-micromachined polysilicon cantilever beam.

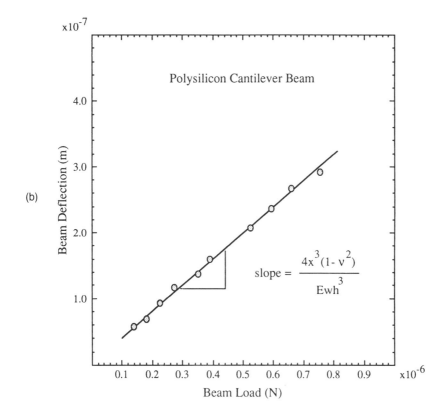

Figure 4.21 continued.

where x is the point at which the load is being applied and w is the width of the cantilever beam. $E/(1 - \nu^2)$ is the biaxial modulus of the material, which can be calculated from the slope of this line to the accuracy that the dimensions of the cantilever beam can be determined.

Under small loads, beams fabricated from metals or other nonbrittle materials will also exhibit linear-elastic behavior. However, the onset of plastic flow will cause the shape of the load-response curve to become nonlinear at higher loads. The load at which the response of the cantilever deviates from linear elastic behavior is defined as the yield load P_y. The yield strength σ_y of the material can be estimated from P_y by calculating the maximum stress in the beam under these conditions [91].

Dynamic properties of cantilever beams have also been used to evaluate the elastic properties of thin films [93–95], although this technique cannot be extended to the characterization of plastic deformation. The beam is electrostatically vibrated by the application of a sinusoidally varying voltage between the cantilever beam and the substrate while the amplitude of oscillation of the cantilever beam is measured optically using the reflected laser light, as illustrated schematically in Figure 4.22(a). Variations in the vibrational amplitude of the cantilever beam are then monitored as a function of the frequency of applied voltage.

As shown in Figure 4.22(b), the vibrational amplitude of an underdamped cantilever beam goes through a maxima at some critical frequency f_{cr} [93]. This frequency is related to the mechanical resonance frequency of the cantilever beam by:

$$f_{cr} = f_r \left(1 - \frac{1}{4Q^2} \right)^{1/2} \qquad (4.10)$$

where Q reflects the energy loss due to damping of the mechanical vibrations of the cantilever beam. Q can be extracted from the width of the resonance frequency or from the ratio of peak vibrational amplitude to the amplitude of vibration at very low frequencies. The resonant frequency of a cantilever beam is a function of both the geometry of the device and the properties of the material or materials from which it is fabricated. It is related to these parameters by:

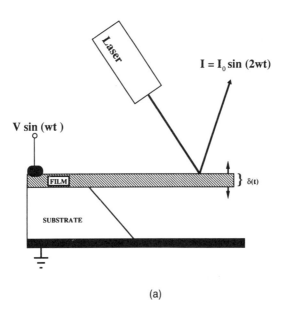

(a)

Figure 4.22 Mechanical characterization of thin films using resonant frequency measurements: (a) schematic drawing of the experimental setup; (b) typical frequency response of an SiO₂ cantilever beam; (c) resonant frequency versus beam length for an array of SiO₂ beams.

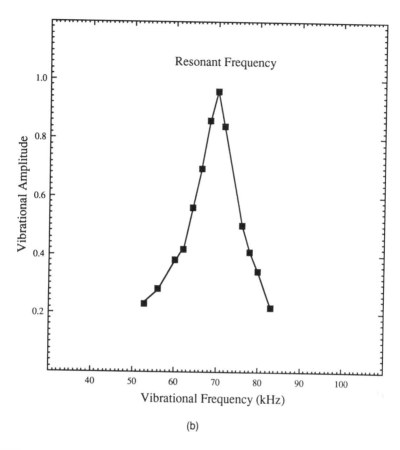

(b)

Figure 4.22 continued.

$$f_r^2 = \frac{3.52}{2\pi} \left(\frac{I}{AL^4} \frac{E}{\rho} \right) \tag{4.11}$$

where L is the length of the cantilever beam, A is its cross-sectional area, and I is its moment of inertia about the z-axis. E is the elastic modulus of the material from which the beam is fabricated, and ρ is the material density. A plot of $1/L^2$ versus f_r for an array of cantilever beams will yield a straight line whose inverse slope is proportional to E/ρ [93]. E/ρ can be calculated from the slope of this line using (4.11), and E can then be estimated from this value by using published values of the material density. Modulus values obtained in this manner are typically in reasonable agreement with bulk values of the elastic modulus.

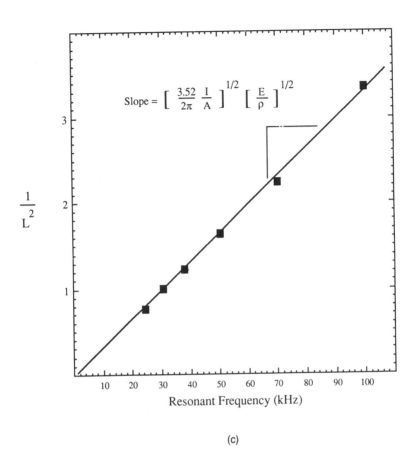

(c)

Figure 4.22 continued.

Guckel and coworkers have used a similar technique to evaluate both the average residual stress field and the elastic properties of polysilicon thin films [24]. In this case, the resonating structure was a micromachined beam that was clamped at both ends rather than a cantilever beam. Constraining the beam at both ends causes the beam to bear an axial load proportional to the magnitude of internal stress field in the film. The resonant frequency of the structure varies with the magnitude of an applied load much like the pitch of a guitar string changes as the tension on the string is varied [24,96].

The response of micromachined diaphragms to an applied pressure can also be used to evaluate the mechanical properties of thin films [80,82,83,97,98]. To obtain information on the mechanical properties of the material, a pressure differential is created across the diaphragm, causing the diaphragm to bow outward. The vertical displacement at the center of the diaphragm is then measured using an optical interferometer or other optical measurement technique. The center displacement, d, can be related to the applied pressure differential $(P_1 - P_0)$ by the following expression:

$$P_1 - P_0 = A \frac{t}{a^2} \sigma_0 d + Bf(\nu)\left(\frac{t}{a^4}\right)\left(\frac{Ed^3}{1 - \nu}\right) \tag{4.12}$$

where $2a$ is the width of the diaphragm and t is the thickness of the thin film [97]. The term σ_0 is the average residual stress in the film, and $E/(1 - \nu)$ is the biaxial modulus. A and B are constants that depend upon the geometry of the device, and $f(\nu)$ is a function of Poisson's ratio for the diaphragm material [80,82,83,97,98]. With the appropriate geometrical constants, the load-deflection data can be fit to yield values for the average residual film stress as well as the biaxial modulus of the thin film, as illustrated in Figure 4.23, [80].

As micromachining technology matures, the level of understanding of the mechanical behavior of thin-film materials that is required increases tremendously. To address these issues, micromechanical versions of a wide range of bulk mechanical characterization techniques have been developed. A survey of the recent literature reveals discussions of devices for the characterization of friction and wear, fracture strain and fracture toughness [79,99,100], and many other thin-film mechanical properties. Because of space limitations we refer the reader to the available literature on these topics.

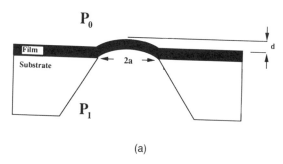

(a)

Figure 4.23 Mechanical characterization of thin films with the membrane-deflection technique: (a) schematic drawing of the test structure; (b) pressure-deflection data for polyimide membranes.

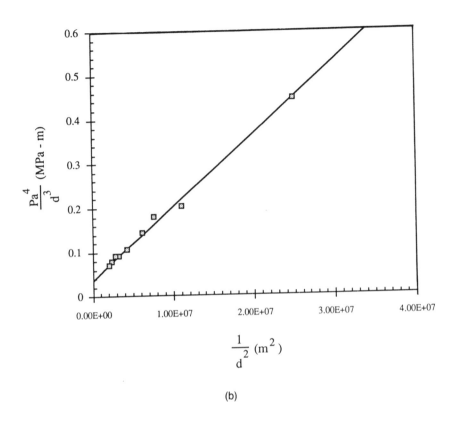

(b)

Figure 4.23 continued.

4.5 APPLICATION OF SURFACE MICROMACHINING

A wide variety of different sensors have already been produced using surface mi-cromachining technology. A complete description would require a chapter in itself. Therefore, we have limited our choice of examples to two specific applications: the accelerometer and the electrostatically driven actuator. We feel that these examples illustrate the basic technology concepts quite adequately. In addition, this section will include discussion concerning integrated surface micromachined systems. This issue is becoming increasingly important with the current trend toward integration of sensor components with signal processing circuits.

4.5.1 Surface Micromachined Accelerometer

The development of a reliable accelerometer for automotive applications currently represents one of the most challenging tasks in the field of sensor technology. To meet this challenge the designers must weigh the pros and cons of each different approach, such as piezoresistive, capacitive, or piezoelectric. Each approach has its inherent limitations.

The technology chosen for description here is a capacitive accelerometer manufactured by surface micromachining [101,102]. The basic structure is a double-pinned differential capacitor utilizing three polysilicon layers. The generic term *double-pinned* refers to a structure that is anchored to the substrate at locations on opposite sides of the suspended mass. Similar devices can be constructed using a cantilever-type design [101]. There are potentially significant differences in processing between the cantilever-type and the double-pinned structures. For example, for double-pinned designs it is desirable that the resultant residual stress in the polysilicon film be tensile to produce a flat undistorted structure. However, with the cantilever design the stress gradient through the thickness of the polysilicon is the critical constraint that must be minimized to avoid generation of a bending moment. An example of a double-pinned device is shown in Figure 4.24. The movable part (polysilicon layer 2) is composed of a suspended mass and four tethers that are anchored to the substrate. The differential capacitor in Figure 4.24 is made of three polysilicon layers:

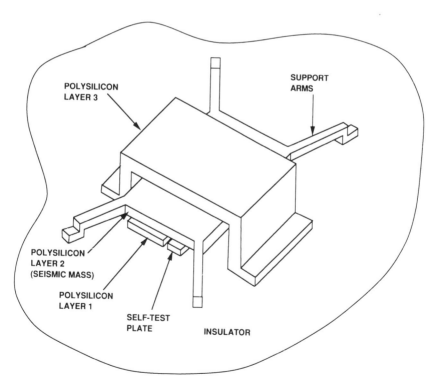

Figure 4.24 Polysilicon surface micromachined accelerometer.

the bottom capacitor is formed between polysilicon layer 2 and polysilicon layer 1, whereas the top capacitor is formed between the polysilicon layer 3 and polysilicon layer 2. The poly-1 and poly-3 layers are fixed. The basic principle of operation involves movement of the polysilicon-2 layer (suspended mass) under acceleration. This movement occurs in a direction perpendicular to the surface of the chip, and it is translated into a change of capacitance.

An additional feature of this device is a built-in self-test capability. This is achieved relatively easily by partitioning the poly-1 layer. That is, in addition to the poly-1 plate that makes up the bottom capacitance there is a self-test plate beneath poly-2. The self-test plate is electrically isolated from the bottom plate of the sense capacitor. When a voltage is applied to the self-test plate with respect to poly-2, electrostatic force will act on poly-2, causing a deflection that results in a change of capacitance, which in turn causes a change in the output voltage. The applied self-test voltage can be adjusted to correspond to a desired level of acceleration.

The differential capacitors of the micromachined structure can be coupled to either open- or closed-loop detector circuitry. An example of an open-loop config-uration is shown in Figure 4.25. C_t and C_b are the top and bottom sensing capaci-tances of the micromechanical structure. The oscillator and reference blocks determine the internal clock frequency and reference voltages, respectively. The control logic contains digital circuitry that provides the pulses for proper operation of the switched-

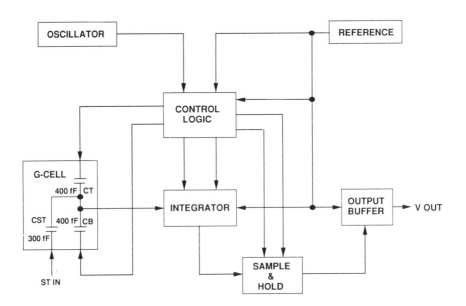

Figure 4.25 Block diagram of an open-loop accelerometer. C_T and C_B are sensing capacitors, and C_{ST} is a self-test capacitor.

capacitor networks. These networks include an integrator and a sample-and-hold block. The switched-capacitor networks provide the output signal that is converted into analog form in the output buffer.

The response of a typical open loop system is shown in Figure 4.26. As can be seen, the output voltage is linear with a sensitivity of about 100 mV/g.

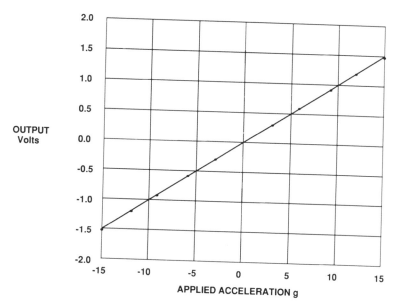

Figure 4.26 Output voltage as a function of acceleration for the open-loop system. The supply voltage is 5V. The sensitivity is 100 mV/g.

4.5.2 Electrostatically Driven Resonators

A particularly interesting group of surface micromachined devices is electrostatically driven actuators. They can be realized as micromotors [103,104] or comb-drive actuators [105]. For this class of devices the driving force is the electrostatic force created by a high electric field developed between the electrodes. The spacing between the electrodes is on the order of a few micrometers and the voltage between the electrodes is small, on the order of 10V. The spacing between the electrodes can also be adjusted into the submicrometer region [106].

In this section we focus on the electrostatic-comb-drive actuators that can be excited to work as resonators. The importance of resonators comes from their potential use for measurement of such quantities as pressure, acceleration, or vapor concentration. In all of these applications physical or chemical parameters alter the potential or kinetic energy of the resonator, thus shifting the resonance frequency.

An example of a polysilicon resonator is shown in Figure 4.27. The resonator is made of three electrodes: a resonant plate (B-electrode), and two electrostatic comb drivers (A and C electrodes). The resonant structure is suspended by folded beams attached to the anchors. Fingers of the B-electrode are placed between the fingers of comb drivers A and C. The A and C electrodes are stationary and used for driving or sensing. To excite the resonant structure both a dc and an ac drive voltage are needed. Usually one stationary electrode is used for driving (A, for example) and the other is used for sensing (C-electrode). When the B-electrode is excited, it moves in a lateral direction, causing the change in comb capacitance. The change of capacitance is a function of finger dimensions as well as of the comb gap (space between the fingers). The current at the sense comb caused by the motion of the B-electrode is proportional to the change of capacitance and thus to the velocity of the B-electrode. This current can be measured by an electromechanical amplitude modulation scheme [105]. The resonance frequency of the structures is found to be in the range between 20 and 75 KHz, depending on structure dimensions. The quality factor depends on damping and is in the range between 20 and 100. Typical dc driving voltage is about 50V, and the amplitude of the ac driving voltage around 10V. It should be pointed out that resonators are also suitable for characterization of thin films. For example, for structures like the one shown in Figure 4.27, Young's modules of polysilicon can be estimated [107]. Further, the resonant structures can be designed with straight beams instead of folded beams [107], in which case the intrinsic stress in the film remains in the structure causing axial forces. The existence of axial forces is the source of nonlinearity in the deflection motion of the structure and can be used to estimate the intrinsic stress.

4.5.3 Integration of Surface Micromachined Structures—Processing Issues

Integration of sensors and actuators with signal processing circuitry on the same chip offers numerous advantages, including higher sensitivity and accuracy, compensation of parasitic effects, faster data acquisition, improved reliability, and lower cost However, integration is not an easy task, particularly when surface micromachined devices and the control circuitry have conflicting processing requirements. Therefore, in this section, we will discuss specific processing requirements of the surface micromachined structures and integrated circuits and outline possible solutions to problems associated with integration.

Analysis of polysilicon surface micromachined structures reveals several potential areas of incompatibility. First, the thickness of the polysilicon film used in surface micromachining is almost an order of magnitude greater than those used for the gate in IC technology. The thicker polysilicon is required to achieve structurall

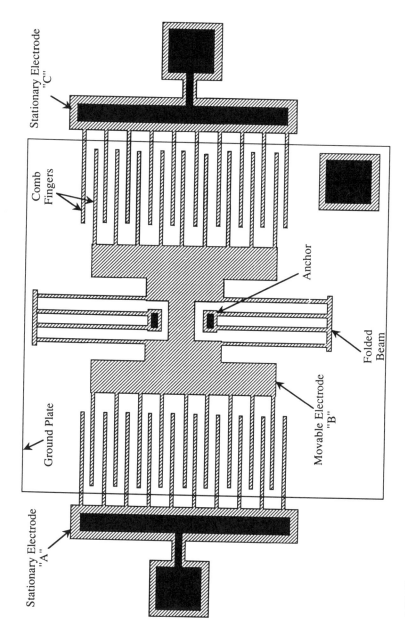

Figure 4.27 Electrostatic comb-drive actuator.

sound designs [35]. To distinguish this structural polysilicon from the gate polysilicon we will call it *thick polysilicon* in further discussion in this section. Second, the mechanical properties of polysilicon are extremely dependent on the deposition and annealing processing scheme. Finally, the thick polysilicon films have to be conductive because they are used simultaneously as mechanical and electrical elements.

The logical question, when it comes to integration, is how polysilicon surface micromachined structures should be integrated together with the circuitry using a full IC process, and at the same time, how can the specific process requirements of both technologies be met? Should one complete the structural layer processing followed by IC processing or vice versa?

Actually separating structural layer and IC processing completely is a difficult proposition. Full IC processing typically includes several high-temperature steps, such as initial oxidation, *p*-well drive-in (diffusion), *n*-well drive-in, field oxidation, gate oxidation, $n+$ source-drain drive-in, and $p+$ source-drain drive-in, or corresponding predeposition steps. All of these processes require temperatures in excess of 900°C. The mechanical properties of polysilicon are extremely sensitive to post-deposition thermal processing and required design specifications may not be achievable if the structural layers are deposited prior to high-temperature IC steps. If the thick polysilicon is deposited after full IC processing, three elements must be considered: (1) influence of polysilicon stress relaxation annealing on IC devices, (2) protection of IC devices during sacrificial etching, and (3) effects of doping intended for polysilicon on IC devices.

Annealing temperatures, if above 900°C for extended time periods, can modify circuit parameters resulting in catastrophic failure. One possible method to circumvent this problem is to achieve a high temperature quickly and maintain this temperature for only a very short time, using rapid thermal annealing (RTA) procedures [34,104]. The IC designs can also be more readily compensated to tolerate short (on the order of seconds) RTA cycles. Protection of the IC devices during sacrificial etching may be accomplished with a passivating layer of silicon nitride [34]. Finally, appropriate dopant levels in the thick polysilicon may be achieved either by ion implantation [34] or in-situ doping [104]. Both are relatively low temperature process steps and should not change parameters of already fabricated IC devices. In-situ doping offers advantages over ion implantation for certain surface micromachining applications because ion implantation can induce structural alterations in the polysilicon films.

4.6 LIGA PROCESS

The LIGA process is a technique for fabrication of three-dimensional micromechanical structures with high aspect ratios (height/width) having heights of several hundred micrometers. LIGA, an acronym for, lithography, electroforming, and micromolding (in German, lithographe, galvanoformung, abformung) was first demonstrated in Germany [109]. The process combines X-ray lithography with thick resist layers and electroplated metal layers to form three-dimensional structures. It also can be combined with sacrificial layers [110,111], in which case it is called *sacrificial LIGA* (SLIGA) [111]. The basic steps in SLIGA are presented in Figure 4.28.

Processing starts with the deposition and patterning of the sacrificial layer. Requirements for the sacrificial layer include good adhesion to the substrate, good coverage of the sacrificial layer by a plating base, resistance to X-ray damage, and ease of removal during etching. Because the LIGA process involves temperatures no greater than 200°C, polyimide films meet all of the sacrificial layer requirements. Following deposition, the polyimide is patterned and then covered by a plating base that will be used latter as a seed for electroplating of core material. The plating base (seed) for the example shown in Figure 4.28 consists of 150Å of titanium (adhesive metal) and 150Å of nickel. Both of these films are sputtered. The next step involves application of a thick photoresist layer (the photoresist can be as thick as several hundred micrometers, depending on the type of structure desired). The photoresist should have high selectivity between the exposed and unexposed areas, which, in turn, should produce vertical walls. This requirement can be satisfied with poly-methylmethacrylate (PMMA) combined with aqueous developing system. After the application of PMMA, it is exposed to X-ray photons from a synchrotron. The synchrotron generates high-energy X-ray collimated photons needed to achieve complete exposure of the thick photoresist. The X-ray mask is also specially designed and constructed of a basic material (membrane) that is X-ray transparent. Silicon nitride or tensile polysilicon films can be used for this purpose. In addition, a patterned layer of gold on the membrane serves as an absorber. The combination of the membrane and absorber allows for locally exposed patterns that produce vertical photoresist walls after the development of PMMA.

Development of PMMA is followed by electroplating of a core material (Ni in this case) and subsequent removal of PMMA and the plating base in selective areas. The final step in SLIGA is etching of the sacrificial layer thus producing a suspended structure (bridge in Figure 4.28).

Note that further evolution of the LIGA process has led to polyimide-based processing [11], which does not require use of the synchrotron. This simplification of the technology will play an important role in future applications.

It should be emphasized that the LIGA process greatly expands micromachining capabilities, making possible vertical cantilevers, coils, microoptical devices, microconnectors, actuators, and so forth. One possible application of the SLIGA process is for constructing actuators. Guckle et al. provide an example of such an application [112]. Magnetic or thermal actuation mechanisms are accomplished by passing a current through the actuator. The actuator geometry determines the magnitude of sensitivity to one mechanism or the other, and the geometries can be tailored to emphasize one effect and suppress the other. The example in Figure 4.29

148

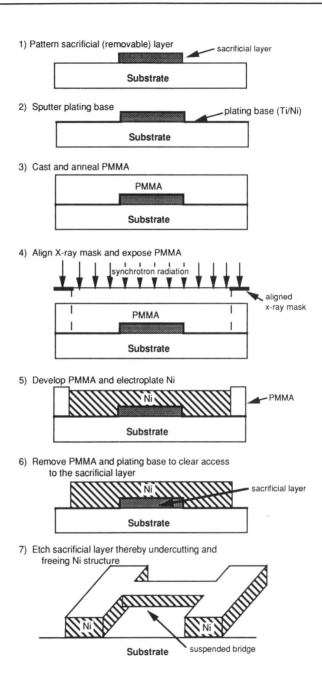

Figure 4.28 The SLIGA process.

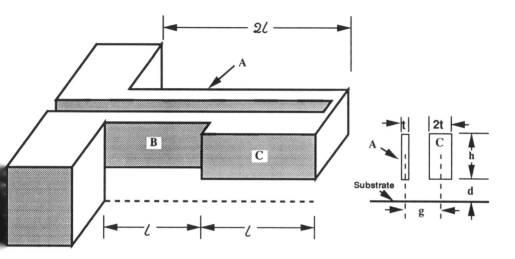

Figure 4.29 Thermally driven actuator: $l = 1000\ \mu m$, $g = 10\ \mu m$, $t = 4\ \mu m$, $r = 50\ \mu m$.

shows a cantilever designed to explore the thermal actuation mechanism. Sections A and B are identical in thickness but differ in length. Section C is thicker than A and B, resulting in a lower resistance. The length of A is equal to the length of B and C together. By passing a current through the device, dissipated power will raise its temperature. The lower resistance of section C will cause lower power dissipation in that area, which in turn will cause a smaller temperature increase in C relative to sections A and B. As a consequence, section A will expand more than sections B and C combined. The final result of the thermal actuation mechanism is a deflection of the beam. This is a good example of how geometric differences (asymmetry) can be used as a driver for the thermal actuation mechanism. It is interesting to note that the measured deflection of the tip of the beam, made of nickel with overall length of 1 mm and height of 50 μm, exceeds 100 μm. This type of actuator can be used either as a microswitch or for position sensing.

4.7 SUMMARY

The focus of this chapter was on polysilicon surface micromachining technology. The basic concept of surface micromachining was presented as were the key processing issues pertinent to polysilicon. The discussion included deposition conditions, doping, annealing, and stress of the films. Furthermore, attention was paid to sacrificial layers and sacrificial etching including stiction. Also, a section was devoted to test structures for mechanical characterization of polysilicon. Finally, several examples of practical applications of surface micromachining were cited. In addition, this chapter contains a brief section describing the LIGA process.

REFERENCES

[1] H. C. Nathanson, W. E. Newell, R. A. Wickstrom, and J. R. Davis, Jr., "The Resonant Gate Transistor," *IEEE Trans. Electron Dev.*, Vol. ED-14, 1967, p. 117.

[2] L. S. Fan, Y. C. Tai, and R. S. Muller, "Pin Joints, Gears, Springs, Cranks, and Other Novel Micromechanical Structures," *Tech. Digest*, Transducers '87, 1987, p. 849.

[3] K. J. Gabriel, W. S. N. Trimmer, and M. Mehregany, "Microgears and Turbines Etched From Silicon," *Tech. Digest*, Transducers '87, 1987, p. 853.

[4] S. Sugiyama, T. Suzuki, K. Kawahata, K. Shimaoka, M. Takigawa, and I. Igarashi, "Micro-Diaphragm Pressure Sensor," *Tech. Digest*, IEDM, Los Angeles, 1986, p. 184.

[5] R. Gutteridge, J. Schmiesing, and Lj. Ristic, "Characterization of Poly–SiO_2: Application for Dynamic Micromechanical Structures," unpublished results.

[6] G. Blackburn and J. Janata, "The Suspended Mass Ion Selective Field Effect Transistor," *Electrochem. Soc. Spring Mtg*, Vol. 82-1, Montreal, 1982, p. 196.

[7] W. Yun, "CMOS Metallization For Integration With Micromachining Processes," M.S. thesis, University of California, Berkeley, 1989.

[8] R. B. Brown, M. L. Ger, and T. Nguyen, "Characterization of Molybdenum Thin Films for Micromechanical Structures," *Proc. IEEE MEMS '90*, Napa Valley, CA, 1990, p. 77.

[9] L. Tong and M. Mehregany, "Silicon Carbide as a New Micromechanics Material," *Tech. Digest*, Solid-St. Sensor and Actuator Workshop, Hilton Head Island, SC, 1992, p. 198.

[10] J. A. Walker, K. J. Gabriel, and M. Mehergany, "Thin-Film Processing of TiNi Shape Memory Alloy," *Sensors and Actuators*, Vols. A21–23, 1990, p. 243.

[11] C. H. Ahn and M. G. Allen, "A Fully Integrated Micromagnetic Actuator With a Multilevel Meander Magnetic Core," *Tech. Digest*, IEEE Solid-St. Sensor and Actuator Workshop, Hilton Head Island, SC, 1992, p. 14.

[12] T. Tawagawa, P. Schiller, H. Yoon, and D. L. Polla, "Micromachined Zinc Oxide Thin Film Sensors," Abs. no. 762, ECS Fall Mtg, Seattle, 1990, p. 1084.

[13] F. Shemansky, Lj. Ristic, D. Koury, and H. Hughes, "Mechanical Properties of Polysilicon Based Composite Films," unpublished results.

[14] H. Guckel and D. W. Burns, "Planar Processed Polysilicon Sealed Cavities for Pressure Transducer Arrays," *Tech. Digest*, IEDM, 1984, p. 223.

[15] C. H. Mastrangelo and R. S. Muller, "Vacuum-Sealed Silicon Micromachined Incandescent Light Source," *Tech. Digest*, IEDM, 1989, p. 503.

[16] R. S. Rosler, "Low Pressure CVD Production Processes for Poly, Nitride, and Oxide," *Solid State Technology*, 1977, p. 63.

[17] T. I. Kamins, M. M. Mandurah, and K. C. Saraswat, "Structure and Stability of Low Pressure Chemically Vapor Deposited Silicon Films," *J. Electrochem. Soc.*, Vol. 125, No. 6, 1978, p. 927.

[18] C. Linder and N. F. De Rooij, "Investigations on Free Standing Polysilicon Beams in View of Their Application as Transducers," *Sensors and Actuators*, Vol. A21–A23, 1990, p. 1053.

[19] W. A. P. Claassen, J. Bloem, W. G. J. N. Valkenburg, and C. H. J. Van Den Brekel, "The Deposition of Silicon from Silane in a Low Pressure Hot Wall System," *J. Crystal Growth*, Vol. 57, 1982, p. 259.

[20] A. M. Beers and J. Bloem, "Temperature Dependence of the Growth Rate of Silicon Prepared Through Chemical Vapor Deposition from Silane," *Appl. Phys. Lett.*, Vol. 41, No. 2, 1982, p. 153.

[21] G. Harbeke, L. Krausbauer, E. F. Steigmeier, A. E. Widmer, H. F. Kappert, and G. Neugebauer, "High Quality Polysilicon by Amorphous Low Pressure Chemical Vapor Deposition," *Appl. Phys. Lett.*, Vol. 42, No. 3, 1983, p. 249.

[22] G. Harbeke, L. Krausbauer, E. F. Steigmeier, A. E. Widmer, H. F. Kappert, and G. Neugebauer,

"Growth and Physical Properties of LPCVD Polycrystalline Silicon Films," *J. Electrochem. Soc.*, Vol. 131, No. 3, 1984, p. 675.

[23] P. Joubert, B. Loisel, Y. Chouan, and L. Haji, "The Effect of Low Pressure on the Structure of LPCVD Polycrystalline Silicon Films," *J. Electrochem. Soc.*, Vol. 134, No. 10, 1987, p. 2541.

[24] H. Guckel, D. W. Burns, H. A. C. Tilmans, D. W. DeRoo, and C. R. Rutigliano, "Mechanical Properties of Fine Grained Polysilicon the Repeatability Issue," *Tech. Digest*, IEEE Solid State Sensor and Actuator Workshop, Hilton Head Island, SC, 1988, p. 96.

[25] H. Guckel and D. W. Burns, "Polysilicon Thin Film Process," U.S. Patent # 4,897,360, 1990.

[26] P. Krulevitch, R. T. Howe, G. C. Johnson, and J. Huang, "Stress in Undoped LPCVD Polycrystalline Silicon," *Tech. Digest*, Transducers '91, 1991, p. 949.

[27] T. I. Kamins, "Design Properties of Polycrystalline Silicon," *Sensors and Actuators*, Vols. A21–A23, 1990, p. 817.

[28] L. S. Fan and R. S. Muller, "As Deposited Low Strain LPCVD Polysilicon," *Tech. Digest*, IEEE Solid State Sensor and Actuator Workshop, Hilton Head Island, SC, 1988, p. 55.

[29] E. Kinsbron, M. Sternheim, and R. Knoell, "Crystallization of Amorphous Silicon Films During Low Pressure Chemical Vapor Deposition," *Appl. Phys. Lett.*, Vol. 42, No. 9, 1983, p. 835.

[30] H. Guckel, T. Randazszo, and D. W. Burns, "A Simple Technique for the Determination of Mechanical Strain in Thin Films with Applications to Polysilicon," *J. Appl. Phys.*, Vol. 57, No. 5, 1985, p. 1671.

[31] H. Guckel, J. J. Sniegowski, T. R. Christenson, and F. Raissi, "The Application of Fine Grained, Tensile Polysilicon Mechanically Resonant Transducers," *Sensors and Actuators*, Vols. A21–A23, 1990, p. 346.

[32] H. Guckel, D. W. Burns, C. R. Rutigliano, D. K. Showers, and J. Uglow, "Fine Grained Polysilicon and its Application to Planar Pressure Transducers," *Tech. Digest*, Transducers '87, 1987, p. 277.

[33] H. Guckel and D. W. Burns, "Measurement and Control of Mechanical Properties of Thin Films," IV International Workshop on the Physics of Semiconductor Devices, Madras, India, 1987.

[34] M. W. Putty, S. Chang, R. T. Howe, A. L. Robinson, and K. D. Wise, "Process Integration for Active Polysilicon Resonant Microstructures," *Sensors and Actuators*, Vol. 20, 1989, p. 143.

[35] R. T. Howe and R. S. Muller, "Polycrystalline and Amorphous Silicon Micromechanical Beams: Annealing and Mechanical Properties," *Sensors and Actuators*, Vol. 4, 1983, p. 447.

[36] R. T. Howe and R. S. Muller, "Stress in Polycrystalline and Amorphous Silicon Thin Films," *J. Appl. Phys.*, Vol. 54, 1983, p. 4674.

[37] T. I. Kamins and T. R. Cass, "Structure of Chemically Deposited Polycrystalline Silicon Films," *Thin Solid Films*, Vol. 16, 1973, p. 147.

[38] J. Adamczewska and T. Budzynski, "Stress in Chemically Vapor Deposited Silicon Films," *Thin Solid Films*, Vol. 113, 1984, p. 271.

[39] W. A. Bryant, "The Kinetics of the Deposition of Silicon by Silane Pyrolysis at Low Temperatures and Atmospheric Pressure," *Thin Solid Films*, Vol. 60, 1979, p. 19.

[40] F. C. Eversteyn and B. H. Put, "Influence of AsH_3, PH_3, and B_2H_6 on the Growth Rate and Resistivity of Polycrystalline Silicon Films Deposited from a SiH_4-H_2 Mixture," *J. Electrochem. Soc.*, Vol. 120, No. 1, 1973, p. 106.

[41] P. Krulevitch, T. D. Nguyen, G. C. Johnson, R. T. Howe, H. R. Wenk, and R. Gronsky, "LPCVD Polycrystalline Silicon Thin Films: The Evolution of Structure, Texture and Stress," *Mat. Res. Soc. Symp. Proc.*, Vol. 202, 1991, p. 167.

[42] A. T. Voutsas and M. K. Hatalis, "Structure of As-Deposited LPCVD Silicon Films at Low Deposition Temperatures and Pressures," *J. Electrochem. Soc.*, Vol. 139, No. 9, 1992, p. 2659.

[43] M. K. Hatalis and D. W. Greve, "Pressure Chemical Vapor Deposited Amorphous Silicon Films," *J. Appl. Phys.*, Vol. 63, 1988, p. 2260.

[44] M. K. Hatalis and D. W. Greve, "High-Performance Thin-Film Transistors in Low-Temperature

Crystallized LPCVD Amorphous Silicon Films," *IEEE Electron Dev. Lett.*, Vol. EDL-8, 1987, p. 361.

[45] A. G. R. French and P. J. Evans, "Effect of Deposition Temperature on LPCVD Polysilicon," *Electron. Lett.*, Vol. 22, 1986, p. 716.

[46] P. Krulevitch, G. C. Johnson, and R. T. Howe, "Stress and Microstructure in Phosphorus Doped Polycrystalline Silicon," *Mat. Res. Soc. Symp. Proc.*, Vol. 276, 1992.

[47] H. Guckel, D. W. Burns, C. C. G. Visser, H. A. C. Tilmans, and D. Deroo, "Fine Grained Polysilicon Films with Built in Tensile Strain," *IEEE Trans. Electron Dev.*, Vol. ED-35, No. 6, 1988, p. 800.

[48] S. Chang, W. Eaton, J. Fulmer, C. Gonzalez, B. Underwood, J. Wong, and R. L. Smith, "Micromechanical Structures in Amorphous Silicon," *Tech. Digest*, Transducers '91, 1991, p. 751.

[49] H. Guckel, "Surface Micromachined Pressure Transducers," *Sensors and Acuators*, Vol. 28, 1991, p. 133.

[50] F. A. Shemansky, "Mechanical and Electrical Properties of In-situ Phosphorus Doped Polysilicon," unpublished results.

[51] J. Huang, P. Krulevitch, G. C. Johnson, R. T. Howe, and H. R. Wenk, "Investigation of Texture and Stress in Undoped Polysilicon Films," *Mat. Res. Soc. Symp. Proc.*, Vol. 182, 1990, p. 201.

[52] G. Harbeke, L. Krausbauer, E. F. Steigmeier, A. E. Widmer, H. F. Kappert, and G. Neugebauer, "LPCVD Polycrystalline Silicon: Growth and Physical Properties of In-Situ Phosphorus Doped and Undoped Films," *RCA Review*, Vol. 44, 1983, p. 287.

[53] A. J. Learn and D. W. Foster, "Deposition and Electrical Properties of In-Situ Phosphorus Doped Silicon Films Formed by Low Pressure Chemical Vapor Deposition," *J. Appl. Phys.*, Vol. 61, No. 5, 1987, p. 1898.

[54] B. S. Meyerson and W. Olbricht. "Phosphorus Doped Polycrystalline Silicon via LPCVD: I. Process Characterization," *J. Electrochem. Soc.*, Vol. 131, No. 10, 1984, p. 2361.

[55] B. S. Meyerson and M. L. Yu, "Phosphorus Doped Polycrystalline Silicon via LPCVD: II. Surface Interactions of the Silane/Phosphine/Silicon System," *J. Electrochem. Soc.*, Vol. 131, No. 10, 1984, p. 2366.

[56] A. Yeckel and S. Middleman, "A Model of Growth Rate Nonuniformity in the Simultaneous Deposition and Doping of a Polycrystalline Silicon Film by LPCVD," *J. Electrochem. Soc.*, Vol. 134, No. 5, 1987, p. 1275.

[57] A. Yeckel, S. Middleman, and A. K. Hochberg, "The Origin of Nonuniform Growth of LPCVD Films from Silane Gas Mixtures," *J. Electrochem. Soc.*, Vol. 136, No. 7, 1989, p. 2038.

[58] H. Kurokawa, "P-Doped Polysilicon Film Growth Technology," *J. Electrochem. Soc.*, Vol. 129, No. 11, 1982, p. 2620.

[59] S. Nakayama, H. Yonezawa, and J. Murota, "Deposition of Phosphorus Doped Silicon Films by Thermal Decomposition of Disilane," *Jap. J. Appl. Physics*, Vol. 23, No. 7, 1984, p. L493.

[60] T. Whidden, E. Lindow, and S. Stiles, "In-Situ Doped Polysilicon Thin Films Utilizing Disilane in a Vertical Reactor," *Microelectronics Manufacturing Technology*, 1991, p. 35.

[61] S. P. Murarka and T. F. Retajczk, Jr., "Effect of Phosphorus Doping on Stress in Silicon and Polycrystalline Silicon," *J. Appl. Physics*, Vol. 54, No. 4, 1983, p. 2069.

[62] D. Flowers, Lj. Ristic, and H. G. Hughes, "Mechanical and Structural Characterization of In-Situ Doped Plasma Enhanced Alpha Silicon Films," *Tech. Digest*, Transducers '91, San Francisco, 1991, p. 961.

[63] Lj. Ristic, M. L. Kniffin, R. Gutteridge, and H. G. Hughes, "Properties of Polysilicon Films Annealed by RTA Process," *Thin Solid Films*, Vol. 220, 1992, p. 106.

[64] T. Lober, J. Huang, M. Schmidt, and S. Senturia, "Characterization of the Mechanisms Producing Bending Moments in Polysilicon Micro-Mechanical Beams by Interferometric Deflection Measurements," *Tech. Digest*, IEEE Solid St. Sensor and Actuator Workshop, Hilton Head Island, SC, 1988, p. 92.

[65] F. A. Shemansky and M. L. Kniffin, "Mechanical and Electrical Properties of Ex-situ Phosphorus Doped Polysilicon Films," unpublished results.

[66] J. C. Bravman and R. Sinclair, "Transmission Electron Microscopy Studies of the Polycrystalline Silicon–SiO_2 Interface," *Thin Solid Films*, Vol. 104, 1983, p. 153.

[67] T. I. Kamins, "Structure and Properties of LPCVD Silicon Films," *J. Electrochem. Soc.*, Vol. 127, 1980.

[68] T. I. Kamins, J. Manoliu, and R. N. Tucker, "Diffusion of Impurities in Polycrystalline Silicon," *J. Appl. Phys.*, Vol. 43, 1972, p. 83.

[69] Y. Shioya and M. Maeda, "Comparison of Phosphosilicate Glass Films Deposited by Three Different Chemical Vapor Deposition Methods," *J. Electrochem. Soc.*, Vol. 133, No. 9, 1986, p. 1943.

[70] A. Takamatsu, S. Miyoko, H. Sakai, and T. Yoshimi, "Plasma-Activated Deposition and Properties of Phosphosilicate Glass Film," *J. Electrochem. Soc.*, Vol. 131, No. 8, 1984, p. 1865.

[71] D. J. Monk, D. S. Soane, and R. T. Howe, "A Diffusion/Chemical Reaction Model for HF Etching of LPCVD Phosphosilicate Glass Sacrificial Layers," *Tech. Digest*, IEEE, Solid State Sensor and Actuator Workshop, Hilton Head Island, SC, 1992, p. 46.

[72] R. B. Bird, W. E. Stewart, and E. N. Lightfoot, in *Transport Phenomena*, New York, John Wiley & Sons, 1960.

[73] R. A. Levy and K. Nassau, "Viscous Behavior of Phosphosilicate and Borophosphosilicate Glasses in VLSI Processing," *Solid State Technology*, 1986, p. 123.

[74] P. R. Scheeper, J. A. Voorthuyzen, W. Olthius, and P. Bergveld, "Investigation of Attractive Forces Between PECVD Silicon Nitride Microstructures and an Oxidized Silicon Substrate," *Sensors and Actuators*, Vol. A30, 1992, p. 231.

[75] R. L. Alley, G. J. Cuan, R. T. Howe, and K. Komvopoulos, "The Effect of Release-Etch Processing on Surface Microstructure Stiction," *Tech. Digest*, IEEE Solid St. Sensor and Actuator Workshop, Hilton Head Island, SC, 1992, p. 202.

[76] M. Orpana and A. O. Korhonen, "Control of Residual Stress of Polysilicon Thin Films by Heavy Doping in Surface Micromachining," *Tech. Digest*, Transducers '91, 1991, p. 957.

[77] H. Guckel and J. J. Sniewgowski, "Formation of Microstructures with Removal of Liquids by Freezing and Sublimation," U.S. Patent No. 5,013,693.

[78] G. K. Fedder and F. T. How, "Thermal Assembly of Polysilicon Microstructures," Proceedings IEEE Microelectro-Mechanical Systems Workshop, Salt Lake City, 1989.

[79] Y. C. Tai and R. S. Muller, "Fracture Strain of LPCVD Polysilicon," *Tech. Digest*, IEEE Solid-State Sensor and Actuator Workshop, Hilton Head Island, SC, 1988, p. 88.

[80] M. G. Allen, M. Mehregany, R. T. Howe, and S. D. Senturia, "Microfabricated Structures for the *In-Situ* Measurement of Residual Stress, Young's Modulus, and Ultimate Strain of Thin Films," *Appl. Phys. Lett.*, Vol. 51, 1987, p. 241.

[81] L. S. Fan, Y. C. Tai, and R. S. Muller, "Integrated Movable Micromechanical Structures for Sensors and Actuators," *IEEE Trans. Electron Dev.*, Vol. ED-35, 1988, p. 724.

[82] M. Mehregany, R. T. Howe, and S. D. Senturia, "Novel Microstructures for the In-situ Measurement of Mechanical Properties of Thin Films," *J. Appl. Phys.*, Vol. 62, 1987, p. 3579.

[83] Stephen D. Senturia, "Microfabricated Structures for the Measurement of Mechanical Properties and Adhesion of Thin Films," *Tech. Digest*, Transducers '87, Tokyo, 1987, p. 11.

[84] H. Guckel, T. Randazzo, and D. W. Burns, "A Simple Technique for the Determination of Mechanical Strain in Thin Films with Applications to Polysilicon," *J. Appl. Phys.*, Vol. 57, 1985, p. 1671.

[85] R. Roark and W. Young, in *Formulas for Stress and Strain*, New York, McGraw-Hill, 1989.

[86] S. H. Crandall, N. C. Dahl, and T. J. Lardner, in *An Introduction to the Mechanics of Solids*, New York, McGraw-Hill, 1978.

[87] A. Blake, in *Practical Stress Analysis in Engineering Design*, New York, Marcel Dekker, 1990.

[88] D. W. Burns and H. Guckel, "Thin Films for Micromechanical Sensors," *Journal of Vac. Sci. Tech.*, Vol. A8, 1990, p. 3606.

[89] R. Mahadevan, M. Mehregany, and K. J. Gabriel, "Application of Electric Microactuators to Silicon Micromechanics," *Sensors and Actuators*, Vols. A21–A23, 1990, p. 219.

[90] L. S. Fan, R. S. Muller, W. Yun, R. T. Howe, and J. Huang, "Spiral Microstructures for the Measurement of Average Strain Gradients in Thin Films," *Proc. IEEE Conf. on Micro Electro Mechanical Systems*, 1990, p. 177.

[91] T. P. Weihs, S. Hong, J. C. Bravman, and W. D. Nix, "Mechanical Deflection of Cantilever Microbeams: A New Technique for Testing the Mechanical Properties of Thin Films," *J. Mater. Res.*, Vol. 3, 1988, p. 931.

[92] W. D. Nix, "Mechanical Properties of Thin Films," *Metallurgical Trans.*, Vol. 20A, 1989, p. 2217.

[93] K. E. Petersen, "Dynamic Micromechanics on Silicon: Techniques and Devices," *IEEE Trans. Electron. Dev.*, Vol. ED-25, 1978, p. 1241.

[94] K. E. Petersen and C. R. Guarnieri, "Young's Modulus Measurement of Thin Films Using Micromechanics," *J. Appl. Phys.*, Vol. 50, 1979, p. 6761.

[95] L. Kiesewetter, J. M. Zhang, D. Houdeau, and A. Steckenborn, "Determination of Young's Moduli of Micromechanical Thin Films Using the Resonance Method," *Sensors and Actuators*, Vol. A35, 1992, p. 153.

[96] J. D. Zook, D. W. Burns, H. Guckel, J. J. Sniegowski, R. L. Engelstad, and Z. Feng, "Characteristics of Polysilicon Resonant Microbeams," *Sensors and Actuators*, Vol. A35, 1992, p. 51.

[97] J. Y. Pan, P. Lin, F. Maseeh, and S. D. Senturia, "Verification of FEM Analysis of Load-Deflection Methods for Measuring Mechanical Properties of Thin Films," *Tech. Digest*, IEEE Solid State Sensor and Actuator Workshop, Hilton Head Island, SC, 1990, p. 70.

[98] F. Maseeh, M. A. Schmidt, M. G. Allen, and S. D. Senturia, "Calibrated Measurements of Elastic Limit, Modulus and the Residual Stress of Thin Films Using Micromachined Suspended Structures," *Tech. Digest*, IEEE Solid State Sensor and Actuator Workshop, Hilton Head Island, SC, 1988, p. 84.

[99] K. Najafi and K. Suzuki, "Measurement of Fracture Stress, Young's Modulus, and Intrinsic Stress of Heavily Boron-Doped Silicon Microstructures," *Thin Solid Films*, Vol. 181, 1989, p. 251.

[100] L. S. Fan, R. T. Howe, and R. S. Muller, "Fracture Toughness Characterization of Brittle Thin Films," *Sensors and Actuators*, Vol. A21–A23, 1990, p. 872.

[101] Lj. Ristic, R. Gutteridge, B. Dunn, D. Mietus, and P. Bennett, "Surface Micromachined Polysilicon Accelerometer," *Tech. Digest*, IEEE Solid-State Sensor and Actuator Workshop, Hilton Head Island, SC, 1992, p. 118.

[102] Lj. Ristic, R. Gutteridge, J. Kung, D. Koury, B. Dunn, and H. Zunino, "A Capacitive Type Accelerometer with Self-Test Features Based on a Double-Pinned Polysilicon Structure," *Tech. Digest*, Transducers '93, Yokohama, 1993, p. 810.

[103] L. S. Fan, Y. C. Tai, and R. S. Muller, "IC-Processed Electrostatic Micro-Motors," *Tech. Digest*, IEDM, San Francisco, 1988, p. 666.

[104] M. Mehregany, S. F. Bart, L. S. Tavrow, J. H. Lang, S. D. Senturia, and M. F. Schlecht, "A Study of Three Microfabricated Variable-Capacitance Motors," *Sensors and Actuators*, Vols. A21–A23, 1990, p. 173.

[105] W. C. Tang, T.-C. H. Nguyen, and R. T. Howe, "Laterally Driven Polysilicon Resonant Microstructures," *Sensors and Actuators*, Vol. A20, 1989, p. 25.

[106] T. Hirano, T. Furuhata, K. J. Gabriel, and H. Fujita, "Operation of Sub-micron Gap Electrostatic Comb-Drive Actuators," *Tech. Digest*, Transducers '91, San Francisco, 1991, p. 873.

[107] P. I. Pratt, G. C. Johnson, R. T. Howe, and J. C. Chang, "Micromechanical Structures for Thin Film Characterization," *Tech. Digest*, Transducers '91, San Francisco, 1991, p. 205.

[108] Lj. Ristic, R. Gutteridge, and R. Roop, "Integration of Surface Micromachined Dynamic Structures," *Proc. IV International Forum on ASIC and Transducers*, Leuven, 1991, p. 129.

[109] E. W. Becker, W. Ehrfeld, P. Hagmann, A. Maner, and D. Munchmeyer, "Fabrication of Microstructures with High Aspect Ratios and Great Structural Heights by Synchrotron Radiation Lithography, Galvanoforming, and Plastic Molding (LIGA Process)," *Microelectronic Engineering*, Vol. 4, 1986, p. 35.

[110] K. Suzuki, "Single Crystal Silicon Micro-Actuators," *Tech. Digest*, IEDM, 1990, p. 26.

[111] H. Guckel, K. J. Skrobis, T. R. Christensen, J. Klein, S. Han, B. Choi, E. G. Lovel, and T. W. Chapman, "On the Application of Deep X-Ray Lithography with Sacrificial Layers to Sensor and Actuator Construction (The Magnetic Micromotor with Power Take Offs)," *Tech. Digest*, Transducers '91, 1991, p. 393.

[112] H. Guckel, J. Klein, T. Christenson, K. Skrobis, M. Laudon, and E. G. Lovell, "Thermomagnetic Metal Flexure Actuators," *Tech. Digest*, Solid-St. Sensor and Actuator Workshop, Hilton Head Island, SC, 1992, p. 73.

Chapter 5
Silicon Direct Wafer Bonding

F. Secco d'Aragona and Lj. Ristic

5.1 INTRODUCTION

Silicon direct wafer bonding (DWB) technology is a technique of fusing together two Si wafers by heat. Two wafers can also have SiO_2 or some other thin-film layer over the surfaces that are bonded together. DWB has become an attractive process because of its versatile application. For example, silicon-to-silicon bonding can be used as a substitute for thick epitaxial layers for bipolar devices, or two silicon wafers of opposite conductivity can be joined to form a diode with no further processing. Also, two oxidized silicon wafers, or one oxidized wafer and one unoxidized counterpart, can be joined to create a silicon-on-insulator (SOI) structure. Finally, two bonded wafers can be explored for manufacturing different micromechanical structures that find wide use in sensors.

The silicon bonding technique was first mentioned in a patent [1] assigned to NEC, where two *pnp* and *npn* mesa transistors formed a unitary structure. This was achieved by positioning them adjacent to each other and subjecting them to a temperature of 650°C in an atmosphere of oxygen saturated with steam. High-temperature treatment produced the growth of an insulating silicon dioxide layer on the external surfaces of the transistors and also formed the oxide binding layer. The first bonding of entire wafers was reported in 1966 by Cave in a patent assigned to RCA [2] in which a large number of oxidized wafers were stacked and heated at temperatures between 1,200° and 1,250°C. A pressure of 100 to 2,000 psi was also applied to the stack while heating it. Under these conditions of heat and pressure, adjacent oxide layers fused and became bonded in about 3 min. By slicing the stack perpendicular to the surface of the oxide layers, a composite periodic structure consisting of elongated strips of silicon separated by fused silicon dioxide was obtained.

Bonding silicon wafers without an oxide was also suggested by the RCA group [3] in the early 1970s for a fabrication of bipolar field effect transistors. The bonding was done in a vacuum chamber at 1,300°C and under 2,000 psi pressure for about

5 min. In this case a polished polysilicon wafer was bonded to another wafer with an epitaxial layer grown on a heavily doped substrate. The extremely high temperatures and pressures involved in the process made it difficult for manufacturing. It should be mentioned that bonding semiconductor elements without an oxide was also suggested by Dahlberg in 1982 [4,5]. In this work n-doped and p-doped plates or discs were used to form a pn-junctions as a part of an $npnp$ structure. The bonding was attributed to a plastic deformation of the semiconductor lattice.

Silicon-to-silicon (Si//Si) DWB without the intermediate oxide layer and without the application of an external pressure was first reported in 1986 by researchers from Toshiba [6–10]. The authors have fully explored the potential of DWB, applying it to the fabrication of bipolar ICs, by bonding a n^+ doped wafer to a n^- wafer. This was done at temperatures above 200°C. The method is of special interest for power devices where high resistivity epitaxial layers with high thickness are needed. Since the bonding process usually requires a short time at high temperature, during the bonding there is too little time for out diffusion from the heavy doped wafer. Thus the bonding produces sharper $p+/p$ and $n+/n$ transitions than the transitions obtained by growing epitaxial layers of comparable thickness. The same authors also bonded p-type to n-type wafers, fabricating diodes with breakdown voltages higher than 1,600V. They also reported the fabrication of a 1,800V bipolar mode MOSSFETs using DWB technique [7]. Another important application of DWB technique is for manufacturing SOI integrated circuits. Wafer bonding offers reduced defects in the SOI layer. The quality of the layer, its defect density, its final thickness and thickness uniformity—all these are important parameters that determine the characteristics of devices. A comparative analysis of electrical and radiation characteristics of CMOS devices fabricated on 2 μm SOI layers obtained by SIMOX, and Si DWB shows that the wafer-bonding approach provided generally better characteristics because of the better crystal quality of the SOI layer [11].

In addition to all these applications the DWB technology has been playing a key role in the current development of solid state sensors and actuators. Silicon DWB has added a new dimension to the sensor technology, where the combination of micromachining and DWB opens new avenues for sensor devices. Size reduction, superior diaphragm thickness control, dielectrically isolated sensors and circuits, and superb etch stop capabilities are some of the features that we will find in the next generation of sensors, as a direct consequence of the use of DWB.

In the first part of the chapter the mechanism of DWB will be described. The following section discusses the processing considerations, such as flatness, interface integrity, bond strength, shaping of bonded wafers, and micro defects. Finally, the last portion of this chapter will cover application of DWB technology to sensors and actuators. Examples include a threshold pressure switch, a pressure sensor, and a peristaltic membrane pump.

5.2 MECHANISM OF DIRECT WAFER BONDING

For successful bonding of two silicon wafers two basic requirements have to be fulfilled. First, the polished surfaces have to be brought into intimate contact; and second, the Si atoms (or the SiO_4 tetrahedra when SiO_2 is present on the surface of the wafer) have to form a stable bond. The first step is accomplished by joining and pressing together two specularly clean wafer surfaces at room temperature. The second step, the formation of the permanent bond, can be accomplished by subjecting the wafers to heat. Doremus [12] has described the reaction of silica glass with absorbed molecules of water in the following way.

$$H_2O + (\text{-Si-O-Si-}) = 2(\text{-SiOH}) \qquad (5.1)$$

The result of this reaction is a formation of hydroxyl groups that is crucial for bonding. The presence of OH groups on silica surfaces has been confirmed by IR spectroscopy that shows spectral bands related to OH vibrations. Consistent with this interpretation is the observation that the spectral bands related to OH disappear after pumping the silica surfaces at room temperature or heating the wafers to 150°C for a short time. As the temperature is raised above 150°C only the sharp band corresponding to SiOH groups remains. Generally, the following types of hydroxyl groups exists on silica surfaces: isolated SiOH groups, hydrogen bonded SiOH groups, and internal SiOH groups. This is shown schematically in Figure 5.1. The relative amount of these different groups depends on the processing of the silica surface as well as on the temperature and humidity at which it is being observed.

Bonding between two silicon wafers occurs at room temperature, provided the wafer surfaces are mirror polished and free of gross contamination. In that case the interatomic attractive forces act between the opposite wafer surfaces. According to Spierings and Haisma [13], the attractive interfacial forces that play a role in DWB are the van der Waals (dispersion) forces, electrostatic forces, and forces due to chemical interactions. It should be pointed out that the room temperature bonding is a reversible process. At this stage wafers can still be separated. A permanent bond occurs only after a high temperature annealing.

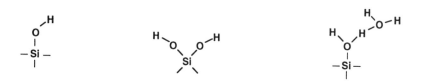

Figure 5.1 Schematic diagrams of OH groups on a silica surface: (a) isolated group; (b) two hydroxyls on one Si atom; (c) hydrogen-bonded groups with an adsorbed H_2O molecule.

5.2.1 SiO₂//SiO₂ Bonding

The original theory for the bonding of two surfaces with thermally grown oxide was developed by Lasky [14,15]. According to this theory the wafer bonding occurs when gaseous oxygen trapped between two wafers is converted to silicon dioxide by the oxidation of silicon. The volume occupied by oxygen decreases by a factor of 4,000 at 1,000°C. The replenishment of gases from the tube is restricted by the small transport through the gap between the wafers. During the processing a partial vacuum is created in the gap between the wafers. Atmospheric pressure, exercised on the outer surfaces of the wafers, pushes the wafers together. Therefore no external applied force is necessary to achieve intimate contact between the wafers.

The bonding mechanism of two oxidized wafers is depicted in Figure 5.2, which shows schematically the sequence of events leading to the formation of the bond. It is well known that the surface of an oxidized silicon wafer terminates at the Si dangling bonds. These bonds react rapidly with water (either atmospheric or during the wafer cleaning) to form SiOH groups. Therefore, the surface of an oxidized wafer normally contains silicon hydroxyl groups (Figure 5.2(a)). When the wafers are pressed together and heated, the two opposite OH groups interact forming H_2O and the Si-O-Si siloxane bond (Figure 5.2(b)). This reaction can be described by the equation

$$\text{Si-OH} + \text{HO-Si} \rightarrow H_2O + \text{Si-O-Si} \tag{5.2}$$

Maszara et al. [16] have analyzed the bond of oxidized wafers over a temperature ranging from 100° to 1,400°C. They suggest the following mechanism. When the wafers are brought together at room temperature a hydrogen bond develops between the opposite wafer surfaces. As the temperature is increased to about 200°C more hydrogen bond bridging develops between the opposite OH groups because of the increased surface mobility of OH groups. As the temperature increase continues the hydrogen bonds are replaced by Si-O-Si bonds and the excess water or hydrogen diffuses away from the bonding interface. This starts to occur at about 300°C. According to Lasky [14] bonding between two thermally grown oxides can be prevented if the wafer surfaces are dried prior to being brought into intimate contact. In one series of experiments wafers were held apart while they were at 1,000°C in an oxygen ambient. After a certain time the separator was withdrawn, allowing the wafers to fall together. Bonding did not occur. Following this finding a similar experiment was designed with a group of wafers oxidized under wet (torch) and dry (O₂) conditions. After the oxidation four groups of wafers were formed. In the first two groups the wafers were joined as soon as they had come out of the furnace. In the second two groups the oxidized wafers were joined only after they were subjected to a treatment designed to produce OH groups on the wafer surface. All wafers were

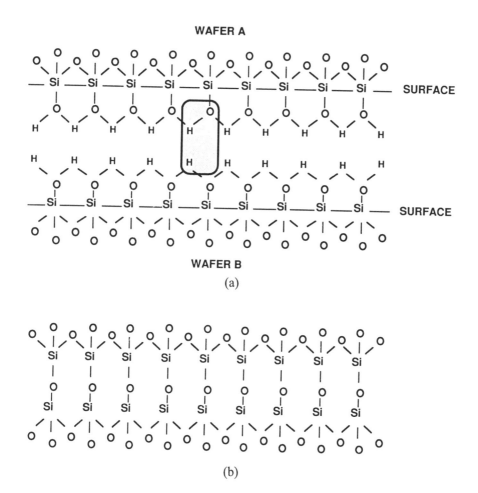

Figure 5.2 (a) Hydroxyl groups on an oxidized silicon wafer. (b) Siloxane bond after bonding.

then annealed at high temperature to achieve permanent bond. Figure 5.3 shows a scanning acoustic tomograph (SAT) image of the bond interface. Both the dry- (*A*) and wet-oxidized (*C*) wafers, joined as dry, show imperfect bonding as evidenced by the dark areas in the pictures. On the other hand, the dry- (*B*) and wet-oxidized (*D*) wafers, joined as wet, show a much improved bond with the exception of scratch- and dirt-induced voids. It must be emphasized that, in practice, it is extremely difficult to remove all traces of H_2O from a dry oxidizing ambient [17]. Consequently, some OH groups will still be present in the dry-oxidized wafers, leading to the partial bond, as shown in Figure 3(a).

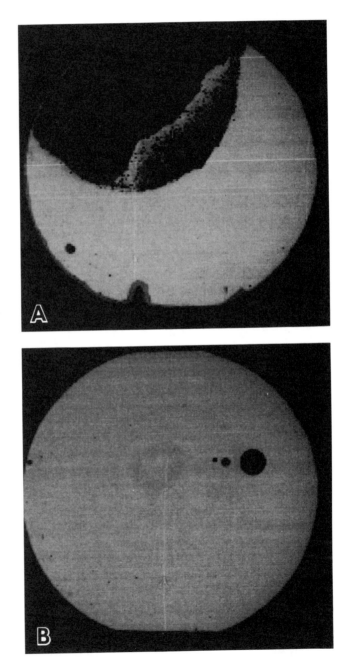

Figure 5.3 SAT images showing the effect of oxide growth conditions and wafer preparation on the quality of the bond: (a) and (c) wafers bonded dry, (b) and (d) wafers bonded wet.

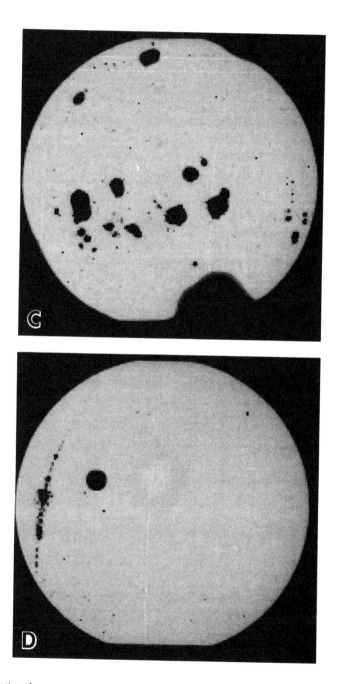

Figure 5.3 continued.

5.2.2 Si//Si Bonding

Silicon-to-silicon bonding depends on the status of the silicon surface preceding the bonding. The surface can be affected by the presence or absence of a native oxide. In the presence of the native oxide the surface is said to be *hydrophilic*, and it represents the stable state of the silicon. The hydrophilic state can be created by using the RCA cleaning solution [18]. In the second case, the absence of the native oxide results in a hydrophobic surface. This is achieved mainly by cleaning the wafers in a HF (hydrofluoric acid) based solution with consecutive water rinsing. The study of hydrophilic and hydrophobic Si surfaces [19] has shown that the hydrophilic state is caused by singular and associated OH groups on the surface, whereas the hydrophobic state is associated mainly with Si-H and $Si-CH_x$ groups, and to a minor extent Si-F groups. A few hydroxyl groups will also exist on the hydrophobic surface due to the water rinse.

The bonding mechanism is directly linked to the state of the surface. For example, the bonding in the presence of the native oxide (the case of hydrophilic surface) is thought to proceed via the hydrogen–siloxane bond in a manner analogous to the one described for oxidized wafers. In the case of hydrophobic surfaces very few OH groups exist, and the same mechanism cannot be invoked to explain the bond. Van der Waals bonding is thought to be the prevalent adhesion mechanism for hydrophobic surfaces. According to Backlund, Ljungberg, and Soderbarg [20], van der Waals forces keep the two wafers together at room temperature. When heating the wafers to about 200°C, the available hydroxyl groups will contribute to the attraction through hydrogen bonds, Figure 5.4(a). At 200°C dehydration of OH groups begins, and the hydrogen bonds are replaced by the Si-O-Si bonds, Figure 5.4(b). Dehydration of Si-H begins at temperatures around 400°C, and very strong covalent Si-Si bonds are formed between hydrophobic surfaces, Figure 5.4(c).

It should be pointed out that the type of bonding, hydrophilic or hydrophobic, determines the strain field in the bonding plane. Examples are shown in Figure 5.5. Transmission electron microscopy analysis of the Si//Si hydrophobic bond shows a strain field, under weak beam diffraction conditions, as a network of dislocations, Figure 5.5(a). This is a cross section of the boundary between two bonded (100)-oriented Si wafers. The dislocations are limited to the boundary plane and do not extend into the silicon lattice on either side of the boundary. To resolve the dislocations, the boundary was tilted about the ⟨110⟩ axis. Consequently, it looks wider than it really is. A view of the boundary "edge on" shows the boundary to be about 100Å wide. The dislocations network is formed because of the rotation of the crystal lattice about an axis normal to the boundary of one wafer with respect to the other wafer. This results in a twist boundary. In contrast, the interface dislocations are absent for the Si//Si hydrophilic bond, as can be seen in Figure 5.5(b). At the same time an interfacial oxide is visible for the case of hydrophilic bond.

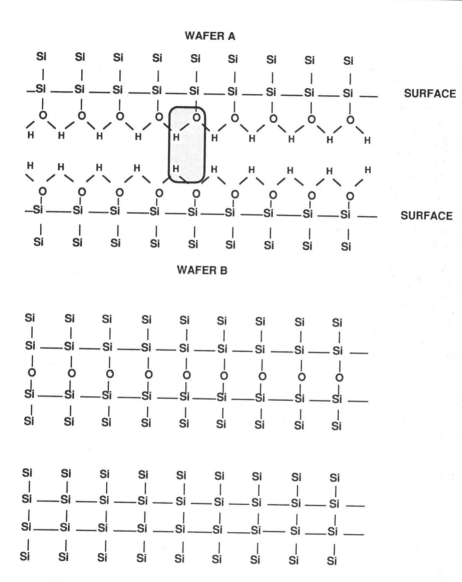

Figure 5.4 Schematic diagram showing the mechanism of Si//Si bonding.

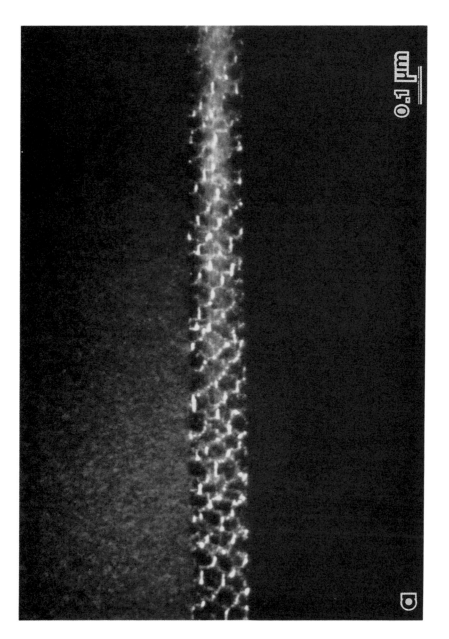

Figure 5.5(a) Transmission electron microscopy of the twist boundary between two hydrophobic silicon wafers.

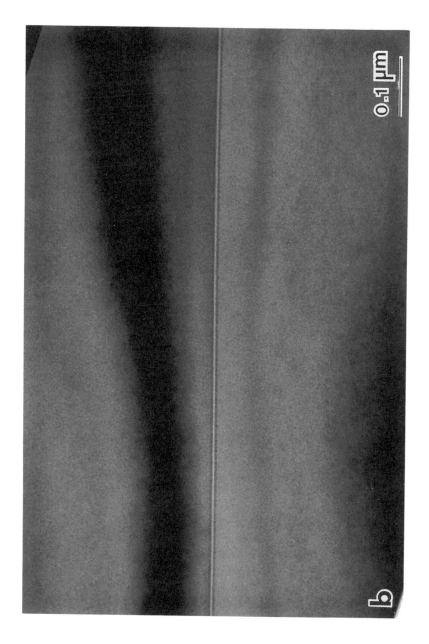

Figure 5.5(b) Boundary between two hydrophilic wafers: interfacial oxide is visible, interface dislocations are absent.

5.3 PROCESSING CONSIDERATIONS

The following section deals with specifics of DWB technology. The material discussed is related to the key processing requirements necessary for successful bonding. The topics include interface integrity and discussions on void formation and bond strength, as well as shaping bonded wafers. The section finishes with a discussion on microdefects in bonded wafers.

5.3.1 Interface Integrity

Interface integrity is one of the key concerns in the bonding process. If the interface integrity is poor that usually means the presence of voids at the interface between the two bonded wafers. Therefore, there is no surprise that every discussion on interface integrity focuses on voids.

According to Mitani, Lehmann, and Gosele [21] there are four reasons for void formation: insufficient wafer flatness, trapped air, particulates, and surface contamination. The same authors also point out that the flatness requirements for the best commercially available wafers are stringent enough to fulfill the first condition. One cause may lead to another. For instance, the presence of particulates will also contribute to pockets of trapped air by preventing the two surfaces from coming in direct contact. Probably the principal cause for large bonding voids, up to several centimeters in diameter, is the presence of particulates and the associated trapped air. The air trapped between the two surfaces causes not only a poor contact at room temperature but also creates a repulsing force between the two bonding wafers at high temperatures. As the wafers heat up, a bond will occur where they touch each other leaving behind a sealed, disclike cavity, filled with air. The pressure of the repulsing force in the cavity is estimated to be 75 psi when the bonding is carried out at 1,100°C. Figure 5.6 shows an example of voids caused by dirt particles on a 100 mm silicon wafer. Figure 5.6(b), shows a particle after removing most of the top wafer by grinding. The voids are several millimeters in diameter, exhibit a circular shape, and almost in all cases a particulate can be recognized at their center. The circular shape is consistent with a minimum energy shape of a gas-filled cavity. Figure 5.7 shows a higher magnification of the particles responsible for the voids. A strain field exists in the vicinity of the particles resulting in slipping of the silicon wafer. The lattice stress around the particle is clearly seen by the bands of slip dislocation aligned in the orthogonal $\langle 110 \rangle$ glide directions. This has been confirmed by X-ray transmission topography studies [22]. According to [23] a 1 μm particle between two 8-inch wafers (700 μm thickness) would cause a void of approximately 1 cm diameter.

It is apparent from the preceding considerations that extreme cleanness of the surfaces to be joined as well as extremely low particle count in the chemicals and

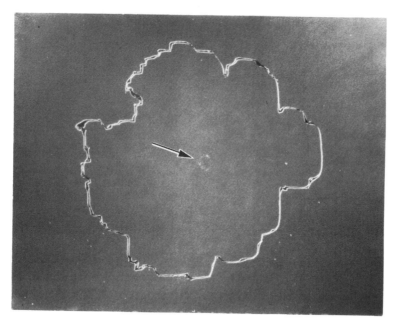

Figure 5.6 Void-producing particles on a (100) mm silicon wafer observed after grinding away most of the top wafer.

Figure 5.7 Examples of voids caused by dirt particles.

water rinses are a must if one wants to achieve void-free bonding. Moreover the ambient in which the two surfaces are contacted needs to meet the same stringent cleanness requirements. However, it has been reported [24] that even for wafers contacted in a class 1 clean room environment, about 70% of the wafers still contained one or more voids due to dust particles <1 μm in diameter. To circumvent the stringent requirement of a clean room environment, the same authors suggest a simple fixture to achieve void-free bond. The wafers to be bonded should be stacked horizontally in a Teflon rack with the polished sides facing each other. The wafers should be kept separated by a Teflon spacer, flushed with a stream of deionized water, and then spinned dry. After that the two surfaces are allowed to come in contact by removing the Teflon spacer. The void content of wafers bonded this way has been found to be influenced by the velocity of the water stream in the gap between the wafers and by the spin dryer speed. For a sufficiently high water velocity and spin dryer speed, void-free bonding could be achieved outside a clean room environment.

It should be pointed out that, during a high temperature bonding, some of the voids tend to shrink due to vacuum formation between the wafers. This in turn tends to pull the two surfaces together, thus reducing the unbonded areas. The vacuum formation is caused by the oxidation of Si in the cavity leading to oxygen depletion. This, of course, is not possible in the case of particulate-induced voids.

The study of voids dealing with such variables as bonding temperature and time, as well as surface condition (oxidized versus nonoxidized) has shown that the void content is different when two oxidized surfaces are joined, compared to two nonoxidized surfaces. These will be discussed in detail in the next two subsections.

5.3.1.1 Voids in $SiO_2//SiO_2$ Bonded Wafers

As has been pointed out there is a difference in bonding mechanism depending upon the presence or absence of SiO_2 between the surfaces. In this section we will consider the voids in $SiO_2//SiO_2$ bonded wafers. The void content of $SiO_2//SiO_2$ bonded wafers can be investigated as function of three variables: bonding temperature, bonding time, and oxide thickness. For illustration we will use an example where the bonding temperature was in the range between 900° and 1,200°C, the bonding time from 0.5 to 2 hours, and the oxide thicknesses in the range of 0.5 and 1 μm. Wafers of the same oxide thickness were mated. All wafers were bonded under wet oxidizing ambient. Wafers with 1 μm oxide were also mated to nonoxidized wafers. Prior to joining, the wafers were immersed in a solution designed to increase the hydroxyl content on the surface to be mated. All wafers were of standard semiconductor grade, and no care was taken to sort them by flatness or surface microroughness. Wafers were joined in a non-clean room environment thus the presence of some voids could not be avoided. The void content was analyzed by the SAT technique.

The results are summarized in Figure 5.8 for the case of 1 μm oxide bonded to 1 μm oxide. No difference in void content and general appearance of the voids was found when the oxide thickness was 0.5 μm or when nonoxidized wafers were bonded to oxidized ones. It is evident from the SAT images that the void content of oxidized silicon wafers is independent of the annealing temperature as well as of the annealing time. Except for a few particulate- and scratch-induced macrovoids, the bonded wafers are void free. There is an exception for the area 4 to 8 mm wide at the wafer periphery. These circular or semicircular areas contain a high density of small voids. The concentration of these voids at the wafer periphery seems to suggest a localized source of contamination. In a different study, where the bonding

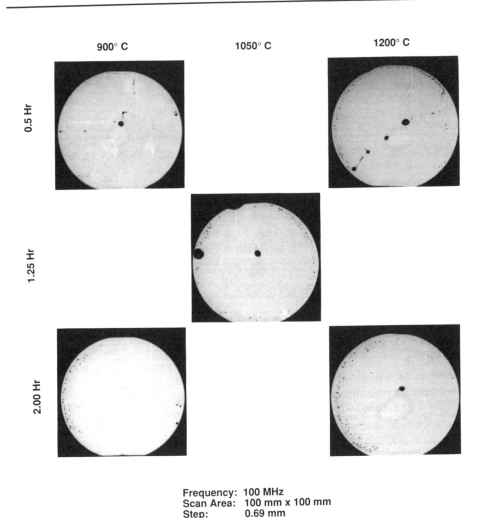

Figure 5.8 SAT images of the bonding interface between oxidized silicon wafers.

temperature was lowered, similar results were obtained, indicating the $SiO_2//SiO_2$ bond is stable at as low as 400°C.

K. Mitani et al. [25] have investigated the effect of organic contamination emanating from the boxes used to store wafers. They analyzed the void content after bonding and found that wafers which bond without voids at room temperature will produce voids after bonding and annealing at 200–800°C. Such behavior was attributed to the deposition of organics, presumably hydrocarbons, on the wafer surface and their subsequent evaporation at high temperature accompanied by void formation. The phenomenon was found to depend on the wafer manufacturer as well as on the storage conditions. The same authors [26] go so far as to recommend room temperature bonding as a way of storing wafers to protect them against surface contamination. Because room temperature bonding is reversible, the wafers can be separated for processing at a later date. It should be also pointed out that prolonged wafer storage can contribute to void formation. Figure 5.9 shows the SAT image of two bonded wafers that were oxidized with 1 μm of grown oxide and stored for a period of over a month in a plastic carrier. Prior to joining the wafers, the pair on the left side of Figure 5.9 was subjected to a hydroxyl enhancement treatment followed by scrubbing to remove particulates. The wafer pair on the right was treated with a cleaning solution known to remove organic contaminants and then received the same treatment as the wafers on the left. Both pairs were then annealed for 1 hour in an oxidizing ambient at a temperature of 1,200°C. As can be seen the wafer pair on the right side of Figure 5.9 is free of bonding voids with the exception of peripheral voids, whereas the wafer pair on the left side exhibit several large voids. This is attributed to organic contaminants.

5.3.1.2 Voids in Si//Si Bonded Wafers

The void content of Si//Si bonded wafers is different from that of SiO_2//SiO_2 bonded wafers described in the previous section. The reason for this is that, in addition to the causes previously outlined, another factor seems to play an important role in void formation: the wafer surface texture or microroughness. Abe, Nakano, and Itoh [27] have analyzed the effect of surface microroughness and correlated it to the void content. The authors list three conditions to achieve void-free bonding: cleaning, dust particles, and surface roughness. The surface roughness (Ra), which is usually measured by an optical interference method, is defined as the arithmetic average deviation of the centerline or peak to valley distance. The authors have analyzed four different levels of surface roughness 0.73 nm, 0.57 nm, 0.45 nm, and 0.37 nm. The values of 0.57 nm and 0.45 nm correspond to the roughness of commercially available wafers for VLSI. The 0.37 nm roughness was developed for wafer bonding. According to the analysis void-free bonding can be achieved using the 0.37 nm roughness, and the smaller is the roughness the fewer voids are present.

In another experiment the void content of Si//Si bonded wafers was measured utilizing the SAT technique. Commercially available VLSI-grade wafers were used for the study. The average roughness (Ra) values of the wafers was on the order 0.7 to 0.4 nm. The void content of Si//Si bonded wafers was investigated as a function of two variables: bonding temperature and bonding time. The investigated bonding temperature was in the range from 900° to 1,200°C. The bonding time was from 0.5 to 2 hours. The wafer preparation, prior to joining them together, was the same as in the case of SiO_2//SiO_2 bonding. The wafers were bonded as received from the manufacturer, and no attempt was made to remove the native oxide. The results are quite different from the case of oxidized wafers. As in the case of oxidized wafers,

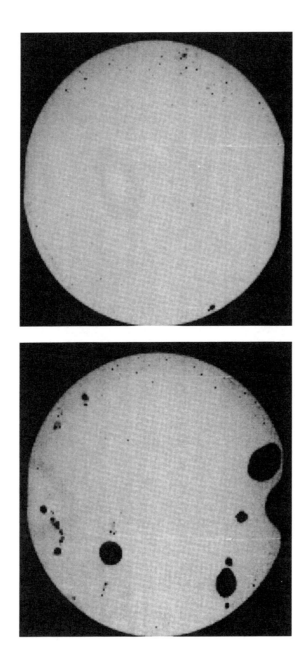

Figure 5.9 SAT images showing the effect of wafer contamination on the $SiO_2//SiO_2$ bond. The wafers were stored in a box for more than one month after (a) was boiled and scrubbed and (b) was piranha cleaned, boiled, and scrubbed.

the void content of Si/Si bonded wafers seems to be independent of the bonding time. However, there is a strong dependence on the bonding temperature. As can be seen in Figure 5.10, a high density of homogeneously distributed voids is revealed by the SAT technique in Si//Si bonded wafers. The void density seems to be independent of the annealing time. Annealing the same wafer at successive higher temperatures showed that the density of voids gradually decreases as a function of the temperature (Figure 5.11(a)). Figure 5.11(b), shows microvoid disappearance as a function of temperature for Si//Si bonded wafers.

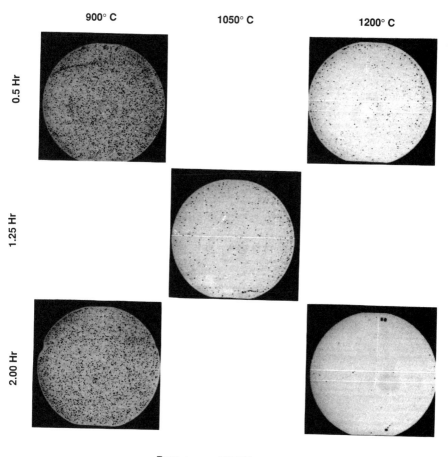

Frequency: 100 MHz
Scan Area: 100 mm x 100 mm
Step: 0.69 mm

Figure 5.10 SAT images of the Si//Si bond showing the dependence of the void content on the bonding temperature.

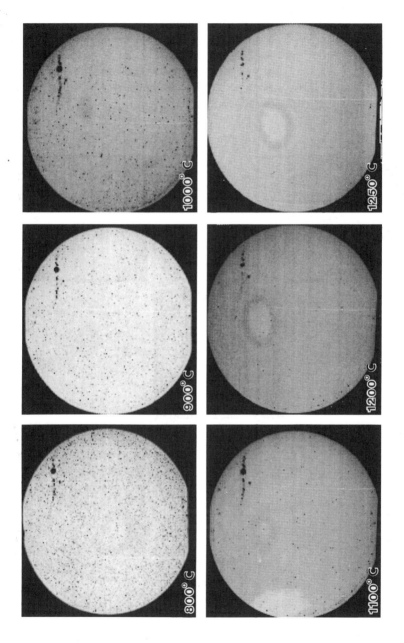

Figure 5.11(a) SAT images showing the disappearance of the microvoids in Si//Si (native oxide//native oxide) bonded wafers as function of the temperature.

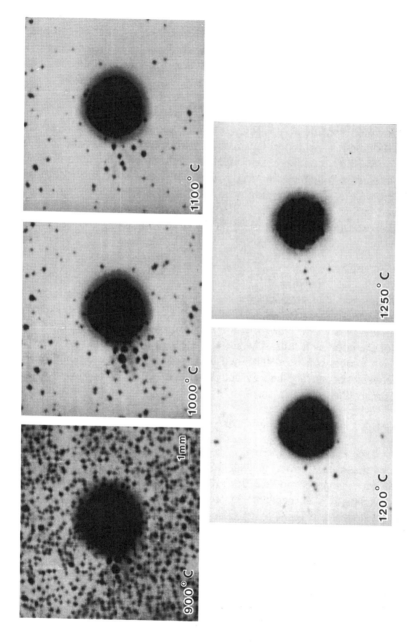

Figure 5.11(b) Higher magnification of the voids. The large void is dirt-induced.

The void annihilation as a function of temperature for Si//Si bonded wafers is interpreted by the closing of the gap between by two microrough surfaces due to the increased mobility of Si atoms. The movement of silicon atoms during bonding has been reported by Abe et al. [27]. The authors have observed that silicon bulges to fill an empty space leading to planarization of the surface during the bonding process. At higher temperatures (1,250°C) increased Si atoms mobility should be sufficient to completely fill the gap produced by mating two silicon surfaces even with a *Ra* microroughness of up to 0.7 nm.

It should be pointed out that particulate- or scratch-induced voids do not close even after extended anneals at 1,250°C (see, for example, Figure 5.11(a)). Si//Si direct wafer bonding is no more sensitive to particulate-induced void than either SiO$_2$//SiO$_2$ or Si//SiO$_2$ bonding. Also after the high temperature annealing, which eliminates the surface roughness-induced voids, peripheral voids are still present in Si//Si bonded wafers. These are similar in their spatial distribution and density to those observed in SiO$_2$//SiO$_2$ bonded wafers. The peripheral voids do not disappear upon annealing and are attributed to particulate contamination.

5.3.2 Bond Strength

One of the parameters that describes the bonding process is the bond strength. Maszara et al. [16] have introduced a quantitative method for measuring the strength of the bond between two wafers. The method is based on the measurement of the surface energy of the bond as it fails. The surface energy (or energy per area) necessary to separate two wafers can be determined by measuring the crack length. The relationship between the surface energy and the crack length, which was derived by Gillis and Gilman [28], is given as

$$g = 3Et^3y^2/8L^4 \qquad (5.3)$$

where g = surface energy (ergs/cm^2), E = Young's modulus of Si (dynes/cm^2), t = thickness of the wafer (cm), y = half thickness of a razor blade (cm), and L = crack length (cm). Young's modulus for (100) silicon is 1.66×10^{12} dynes/cm^2.

The crack length L is measured by inserting a blade of thickness $2y$ into the edge of the bonded pair (Figure 5.12). The separation between the two wafers appears under IR light as a series of interference fringes parallel to the blade edge. The distance between the blade edge and the tip of the crack is the crack length. Using this method one can estimate the bond strength for both Si//Si bonding and SiO$_2$//SiO$_2$ bonding. Figure 5.13 shows the experimental results. The surface energy of Si//Si bonded wafers at a temperature of 500°C is as high as that of oxide bond at 1,200°C. Above 600°C the surface energy of the Si//Si bond becomes too high to be measured by the blade method. Above about 800°C the surface energy of the Si//SiO$_2$ bond also becomes too high to be measured by the blade method. The

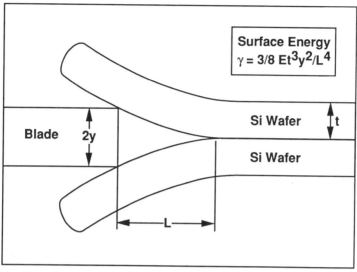

Figure 5.12 Measurement of the bond strength by the blade method.

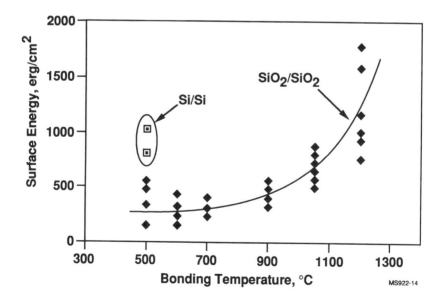

Figure 5.13 Bond strength of Si//Si and SiO_2//SiO_2 bonded wafers as a function of the bonding temperature.

results for surface energy of $SiO_2//SiO_2$ bonded wafers could be summarized as follows. The surface energy: (1) increases with increasing temperature; (2) does not depend on the oxide thickness or the annealing time for temperatures below 1,200°C; (3) is function of the annealing time for temperatures $\geq$ 1,200°C; and (4) is too large to be measured by the blade method for temperatures > 1,200°C.

5.3.3 Shaping Bonded Wafers

In addition to the bond itself an important aspect of the silicon direct wafer bonding technology is a thinning of wafers. Thinning is usually done on the wafer that is called the *active wafer* (the one used for fabrication of active devices). The second wafer of the bonded pair is called the *handle wafer*. Both wafers play important roles in shaping the bonded wafers. A key factor of the thinning process will be discussed in this section.

The thinning usually starts with a grinding operation in which the thickness of the active wafer is mechanically reduced within several micrometers of the final required thickness. After that, the wafers are chemically etched to remove the damage induced by the grinding operation. Finally, the wafers are polished down to the final thickness specification. For reasons that will soon become apparent the thinning has to be accomplished without significant change of flatness of the wafer pair. Purely mechanical polishing of silicon, such as the old diamond polishing, can maintain a high degree of wafer flatness and parallelism. However, it does introduce considerable subsurface damage, which is intolerable in silicon technology. This is the reason why diamond polishing has been replaced by tribochemical polishing or "syton" polishing, which utilizes a slurry of colloidal silica suspended in a sodium hydroxide–water solution. Syton polishing provides the kind of damage-free silicon surface necessary for device fabrication.

To understand the problems one faces during thinning it is necessary to deal with basic parameters used to describe the flatness. Therefore, we will first describe flatness parameters and then discuss the thinning process itself.

5.3.3.1 Wafer Flatness

The flatness of silicon wafers is usually defined by four parameters (Figure 5.14): total thickness variation (TTV), focal plane deviation (FPD), total indicated reading (TIR), and taper. The total thickness variation is the difference between the highest and the lowest elevation of the top surface of the wafer. As we will see later, this is the most important parameter to consider in the tribomechanical polishing of silicon wafers. The focal plane deviation is the greatest distance above or below a chosen reference plane (focal plane) in either a positive or negative direction. FPD is the parameter of great importance for VLSI IC processing, due to the limited depth

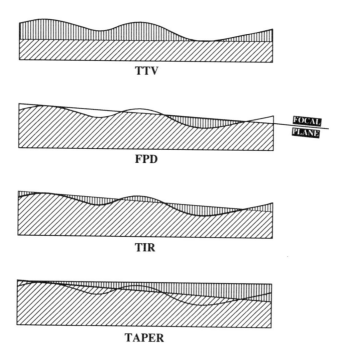

Figure 5.14 Parameters used to describe the flatness of silicon wafers.

of focus of advanced scanning and stepping projection aligners. The total indicated reading is the difference between the highest point above the focal plane and the lowest point below the focal plane. Finally, the taper is the difference between TTV and TIR and is an indication of the lack of parallelism between the back surface of the wafer and the selected focal plane. In wafer manufacturing the process capabilities for TTV, TIR, and FPD are closely monitored at each shaping step from cutting to the polishing operation. Each of the shaping steps can modify the wafer shape, either increasing or decreasing the flatness, depending on process variables.

It should be mentioned that flatness and thickness measurements are usually taken with the wafer clamped to a reference flat vacuum chuck. The chuck is calibrated to a horizontal plane and provides the back reference plane for TTV and taper. The back reference plane is then retilted (using software) by the value of taper to provide a horizontal "front referenced" plane used for TIR and FPD measurement. In the mid 1970s, TTV and taper were the only flatness parameters measured. The equipment used was based on contact-type gauges such as bench dial gauges and micrometers. Measurements were typically taken at five points with the maximum to minimum value being described as TTV. The contact gauges are now obsolete, as new and more accurate equipment using laser interferometry or noncontact ca-pacitance sensors have been developed in the 1980s. The new equipment allows a full surface scan of the wafer providing a true TTV, sometimes referred to as an *all point TTV*. Both phase measuring interferometer systems and capacitance gauges provide a "picture" of a wafer surface in three-dimensional topographic form. The ability to see the actual surface contour makes it possible to determine unusual effects and specific process details at each processing step.

5.3.3.2 Thinning Bonded Wafers

In shaping bonded wafers one must be concerned with the flatness parameters just described to produce bonded wafers that will be suitable for sensor and VLSI pro-cessing. Study has shown that the flatness of the starting wafers is crucial in deter-mining the thickness uniformity of the bonded wafers.

Let us repeat here that we make a distinction between the wafers in the bonded pair: one is called the *handle wafer*; the other is called the *active wafer*. The active wafer is the one that gets thinned. Its mechanical proprieties, primarily flatness, are significantly affected by the proprieties of the handle wafer. It should be emphasized that, to achieve a uniform thickness of the active wafer by mechanical and chemical thinning, it is necessary to have good parallelism of the handle wafer or TTV. The reason for this can be explained by the help of Figure 5.15. Let us suppose that a silicon thickness of 2 μm is required for the active wafer. It is evident from the figure that if the handle's TTV is also 2 μm (Figure 5.15(a)) the final thickness of active wafer is unachievable. The reason for this is that during the grinding and polishing operations the original shape of the handle wafer remains unchanged, and the active wafer layer will, on average, have the same thickness variation as the TTV of the handle wafer (Figure 5.15(b)). If the final shaping of the active wafer tends to match the contour of the handle wafer, the thickness uniformity could im-prove (Figure 5.15(c)), but when there is a mismatch, the thickness variation it will be even greater than the TTV of handle wafer. (Figure 5.15(d)).

The current typical manufacturing of IC MOS devices uses 100 mm diameter polished wafers with an upper control limit (UCL) TTV of about 3 μm. The IC manufacturing is often concerned more with the FPD parameter, which for the same wafers and process will result in a UCL of about 1 μm. It should also be noted that the control of flatness parameters is roughly proportional to wafer surface area, thus, a 150 mm diameter product of equivalent process would have a UCL of about 6 μm; that is, the best "ultraflat" wafer product now commercially available and is not adequate for wafer bonding.

The processing equipment that seems to have the greatest potential for im-proving the manufacture of ultraflat wafers is the automatic in-feed surface grinder. The machine allows a rotating vacuum chuck to be ground flat in-situ with the same

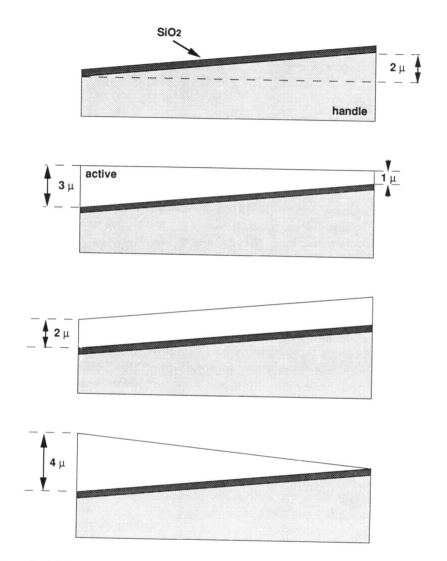

Figure 5.15 Effect of TTV on the final thickness of the active wafer.

wheel that is used for grinding. This provides excellent parallelism between the back reference surface of the wafer and the front surface being thinned. The grind wheel can have a very fine abrasive grit that produces less surface damage than the usual double-side lap process. This reduces the amount of polish removal required for a damage-free surface and is an aid to final flatness control, as no currently available polish process will produce submicrometer TTV flatness. A combination of the in-feed grinder with automated individual wafer polishing has a best chance for obtaining sub micrometer uniformity.

5.3.3.3 Polishing Bonded Wafers

Because of the limitations of purely tribochemical thinning significant effort is required to achieve thickness of submicrometer layers utilizing techniques that do not rely on the flatness and parallelism of the handle wafer. An excellent overview of the various techniques is given in [29]. These techniques can be divided into three broad categories: polishing stops, chemical etch stops, and electrochemical etch stops. We will briefly describe polishing stops techniques here. The other two approaches have been already covered in Chapter 3.

Polishing stops rely on the presence of layers resistant to syton polishing to slow the silicon polishing action. Silicon dioxide and silicon nitride layers buried in the active wafer are polished at a different rate than silicon and can therefore be used to achieve planarization of the silicon layer independent of the shape of the handle wafer. For example MOSFETs fabricated in SOI layers having a thickness of 0.1 μm and produced via the polishing stop technique have been reported [30]. An example of a polish-stop process is given in Figure 5.16. First, using standard techniques, grooves are etched into the active wafer (Step a). The depth of the grooves will determine the final thickness of the SOI layer. A polish-stop layer such as SiO_2 or LPCDV Si_3N_4 or both is then grown on the wafer (Step b). Polycrystalline silicon or thick glass such as TEOS is then deposited to fill the grooves (Step c). The poly-silicon layer is then polished down to the oxide layer (Step d). An alternate method is to polish the polysilicon layer to a mirror finish without reaching the oxide layer. The active wafer is then bonded to a handle wafer (Step e). The bonding operation is followed by grinding the active wafer in the proximity of the oxide layer and finally by polishing it down to the oxide layer (Step f). Even though this application is mask specific and more complex than simple polishing, it can give a better control of the SOI thickness while achieving a fully isolated SOI layer. An added advantage of this technique is that, in principle, the reach of the oxide layer can be monitored. This is because a small drop in the temperature of the polishing pad can be noticed when the oxide layer is reached. If the temperature of the wafer can be accurately and constantly monitored during polishing, the drop in temperature will signal that the oxide layer has been reached thus establishing an endpoint. The experimental results suggest that the time of temperature decline and rate of decline depend on the thickness of the SOI layer and its uniformity (TTV).

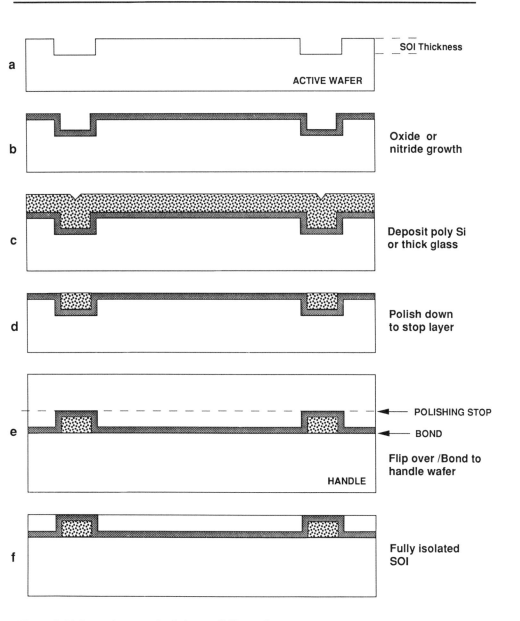

Figure 5.16 Processing steps (a–f) for a polishing etch stop.

5.3.4 Microdefects in Bonded Wafers

Another important factor of the silicon direct wafer bonding technology is the presence of microdefects. They are crucial for integrated sensors because active devices are an integral part of the solution and their functions can be affected adversely by the presence of microdefects. In this section we will outline briefly some key elements pertinent to microdefects in bonded wafers.

Our attention will be directed to the active wafer because it acts as a substrate for active devices. It should be pointed out that interstitial oxygen concentration $[O_i]$ and oxygen precipitation in the active wafer of the bonded pair play a crucial role in determining the microdefect content of bonded wafers: The high temperatures necessary for bonding also lead to oxygen precipitation and out diffusion. At the same time it is known that the oxygen precipitation phenomenon improves the wafer quality by gettering fast diffusing metal impurities. This could be used to control the defects in bonded wafers.

Since the pioneering work of Tan, Gardner, and Tice [31], a significant amount of research has been done to understand the relationships between $[O_i]$ and oxygen precipitation and between bulk and surface microdefects. In addition, there have been studies of implementations of the annealing steps necessary to achieve a defect-free wafer surface. Process-induced surface defects, such as oxidation-induced stacking faults (OISF) and smaller defects like metal-associated "haze" (or saucer pits), are found to adversely affect device performance when situated in the proximity of the device's active area. All the findings are fully applicable to bonded wafers.

Generally speaking the defect proprieties of bonded wafers appear to be dictated primarily by the thermal history of the individual wafers. For instance the oxygen precipitation and the thickness of the oxygen denuded zone (DZ) will be directly affected by the temperature used for the wafer bonding as well as the initial interstitial oxygen concentration. Bonding two silicon wafers with the same oxygen concentration and the same thermal history will lead to a symmetrical situation with respect to the bonding plane. This is shown in Figure 5.17(a), for two $SiO_2//SiO_2$ bonded wafers. It can be seen that no defects intersect the DZ on either side of the bond. Only when a bonding defect is present, such as a void probably induced by a small particle, will defects be present in the form of dislocations propagating along the slip planes on opposite directions of the bond plane (Figure 5.17(b)).

We should emphasize that the interaction between surface and bulk defects, DZ, and final thickness of the active wafer is important in wafer bonding technology. An example of this interaction is depicted in Figure 5.18, which shows a cross-section diagram of two bonded wafers. As mentioned already the thickness of the DZ will depend on the initial interstitial oxygen concentration as well as on the wafer thermal history and bonding temperature. If, for instance, the thickness of the DZ is 16 μm, shaping the active wafer down below 16 μm (polishing line *B*) will lead to a defect-free surface; if, on the other hand, the active wafer is shaped above the

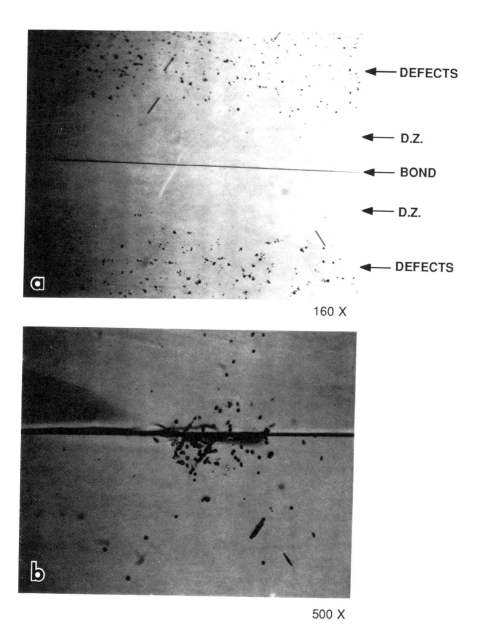

160 X

500 X

Figure 5.17 (a) Wafer cross section showing the symmetrical nature of microdefects on either side of the bond between two oxidized wafers. (b) Dislocations emanating from a void at the bonding interface.

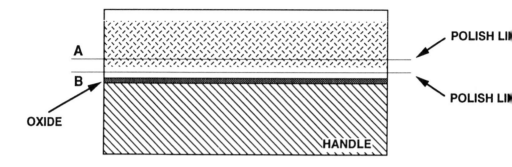

Figure 5.18 Cross-sectional diagram showing the effect of oxygen precipitation, denuded zone thickness, and polishing thickness on the microdefect content of bonded wafers.

16 μm line (polishing line A), the oxygen-induced bulk defects will be exposed. In Figure 5.19(a), the active wafer was polished down to ≈ 30 μm, well above the thickness of the DZ estimated to be about 15 μm. As a consequence the bulk defects will be on the wafer surface. Figure 5.19(b) shows a high density of OISF of different length. On the contrary in Figure 5.20, the wafer was polished down below the DZ, and the surface is clear of defects.

These examples clearly show that SOI layers with thicknesses thinner than the DZ are free of defect. However, one should be also aware that oxygen-depleted layers lack intrinsic gettering capability, and they are susceptible to defect formation during further processing. For example, the effect of processing a 2-μm thick SOI layer can be seen in Figure 5.21, which shows a 100 mm wafer and an enlarged view of the surface following thermal oxidation. A high density ($\approx 10^4/cm^2$) of microdefects can be seen after the removal of the oxide. The defects are similar to the hillocklike defects seen on epitaxial layers lacking intrinsic gettering, such as those grown on substrates heavily doped with antimony [32,33]. The hillocklike microdefects are probably small metal-decorated stacking faults. They are known to adversely affect the minority carrier lifetime in n/n^+ epitaxial wafers [34], and they can be eliminated by subjecting the n^+ substrates to extended preepitaxial annealings to induce oxygen precipitation. This is not possible for thin SOI layers, hence other techniques will have to be developed to enhance the gettering characteristics of SOI layers.

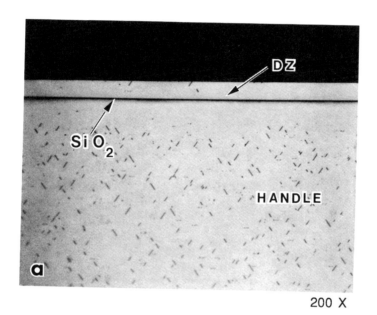

200 X

500 X

Figure 5.19 (a) Cross section of a SOI wafer showing a narrow denuded zone and bulk stacking faults. (b) Corresponding wafer surface showing high density of bulk stacking faults.

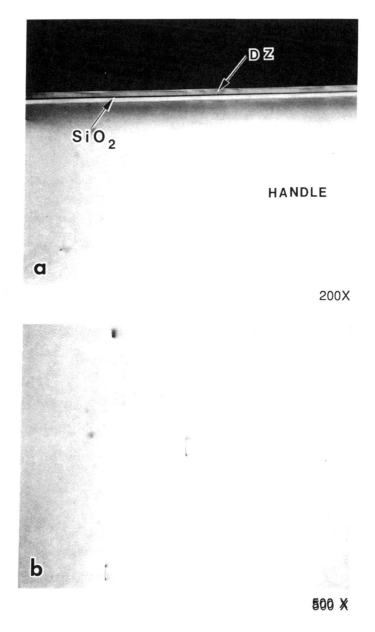

Figure 5.20 (a) Cross section of a SOI wafer showing a denuded zone clear of defects. (b) Corresponding wafer surface showing only a few oxidation induced stacking faults.

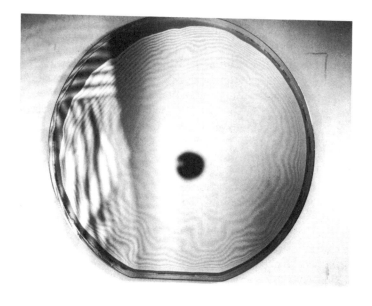

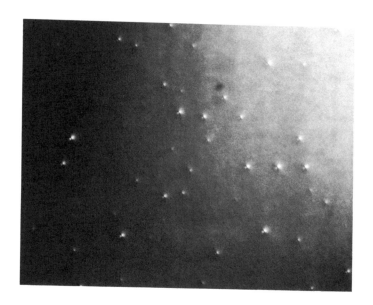

160X

Figure 5.21 SOI wafer following oxidation. Microdefects are present due to the lack of intrinsic gettering. (a) 2 μm SOI over 1 μm oxide. (b) "Hillocks" on the surface of the SOI wafer following oxidation and preferential etching.

5.4 APPLICATION OF DIRECT WAFER BONDING TO SENSORS AND ACTUATORS

Sensors and actuators that have been manufactured using direct wafer bonding technology are numerous. In this section we will describe only a few examples. The examples are adequate and they illustrate the basic capabilities of direct wafer bonding technology. The devices in this section include a threshold pressure switch, a pressure sensor, and a peristaltic membrane pump.

5.4.1 Threshold Pressure Switch

Numerous applications in industrial control systems and consumer products require a pressure switch that closes or opens at a preset pressure. Also, the pressure switch usually has to show a hysteresis in switching from the open position to the closed one, and vice versa, to prevent undesirable switching due to a small variation of pressure around the preset value. These requirements can be met by a threshold pressure switch such as the one shown in Figure 5.22 [35]. The pressure switch is made of spherically shaped and plastically deformed upper silicon electrode separated by insulating layer from the lower silicon electrode. The upper electrode will buckle down making the electrical contact with the lower electrode when exposed

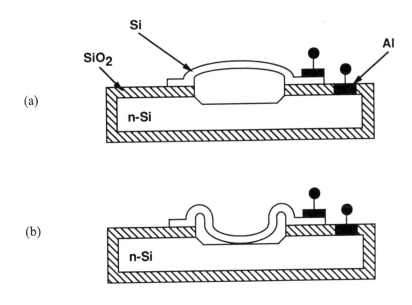

Figure 5.22 Threshold pressure switch: (a) switch open; (b) switch closed.

to a pressure above the buckling threshold. If the pressure is reduced one can find that the upper electrode does not "pop up" at the same buckling pressure, but rather at a lower pressure. The hysteresis can be explained by the plastic deformation of upper electrode, and it can be tailored by variations in the size and thickness of the electrode.

The fabrication process of the threshold pressure switch is shown in Figure 5.23. It starts with creation of the cavity in a handle wafer. After the oxidation and patterning, the handle wafer is bonded with active wafer, followed by a thinning of the active wafer. Then a low temperature oxide (LTO) is deposited, patterned, and etched, defining the upper electrode. This is followed by the plasma etching of the upper electrode and patterning and etching of LTO to enable the contact to the bottom electrode. After that the device is exposed to high temperature (1,050°C) for 1 hour in nitrogen. This is one of the crucial processing steps. During the high-temperature processing the upper electrode is plastically deformed due to expansion of the trapped gas in the sealed cavity. That is, the trapped gas loads the thin silicon layer (upper electrode) beyond the yield point, causing plastic deformation of silicon. After cooling, the upper electrode retains its spherical shape. It should be emphasized that the maximum yield stress of silicon decreases with temperature increase above the 600°C and the extent of the plastic deformation depends on temperature and the size of the

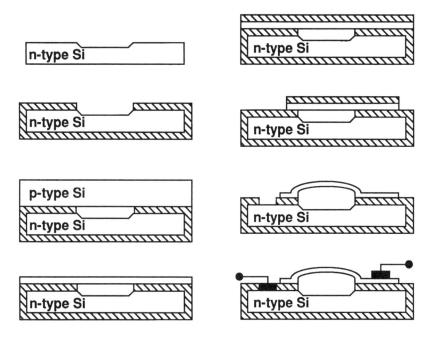

Figure 5.23 Fabrication process of the threshold pressure switch.

cavity as well as the size of the upper electrode. After the high-temperature pre cessing aluminum is deposited, patterned, and sintered to form the contacts to bot electrodes.

The devices are tested in a vacuum using the setup shown in Figure 5.24(a The test circuit consists of 20V battery in series with 1 kΩ resistor. Voltage acro the resistor is measured to get an indication of whether the switch is open or close (zero voltage across resistor indicates an open switch). Typical test results are show in Figure 5.24(b). The buckling of the upper electrode occurs at a repeatable valu

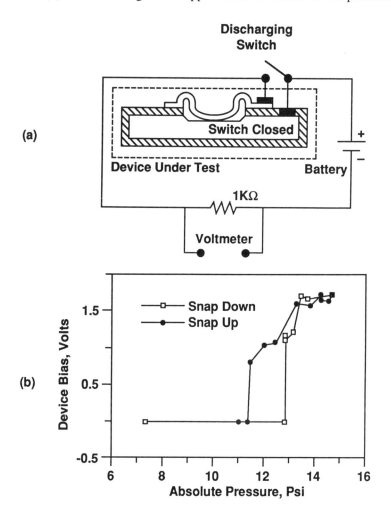

Figure 5.24 Testing of threshold pressure switch: (a) test setup, (b) voltage measured by voltmeter a a function of pressure.

of pressure. The snapping up of the upper electrode is also repeatable and occurs at a lower pressure, indicating the existence of hysteresis. The buckling behavior of the switch exhibits about 2 psi of mechanical hysteresis.

5.4.2 Pressure Sensor

The piezoresistive silicon pressure sensor is currently one of the most frequently used sensors. It is usually fabricated as a diaphragm with piezoresistors using a bulk micromachining technology. It shows high sensitivity, good linearity, and good stability. The main drawback of piezoresistive silicon pressure sensors is a temperature sensitivity that limits their application at higher temperatures. However, this drawback can be overcome if the pressure sensor is fabricated using DWB technology. The use of DWB technology enables dielectric isolation of piezoresistor and operating temperatures up to 350°C [36].

The cross section of a pressure sensor fabricated by the means of DWB is shown in Figure 5.25 [36]. This device utilizes a double SOI structure and the dielectric isolation of the piezoresistor. In this device an Al_2O_3 film, grown epitaxially, is used as dielectric isolation. The first SOI layer is used for precise control of the thickness of the diaphragm, and the second SOI layer is used for the sensing resistor itself, which is designed as a single-element four-terminal piezoresistor. The piezoresistor is located in the center of the diaphragm and aligned to ⟨100⟩ direction to optimize the piezoelectric effect.

The fabrication process of pressure sensor is shown in Figure 5.26. The process starts with two 4-in. p-type wafers with (100) orientation being oxidized (1) and bonded. Then, the active wafer is thinned down to 5 μm by both conventional lapping and polishing (2 and 3). This is followed by the epitaxial growth of 80 nm of Al_2O_3 using the LP-CVD method (4). The next step involves epitaxial growth of 1.3 μm of Si-layer on Al_2O_3 film (5). After that the structure is oxidized and the handle wafer micromachined from the back side to create the diaphragm (6). During the anisotropic etching of silicon the SiO_2 layer, formed by bonding of two wafers, acts as an etch stop and determines the thickness of the diaphragm. The diaphragm formation is followed by the patterning of the front side and the etching of piezoresistor. This is followed by the passivation (8) and metalization (9). The advantages of this type of pressure sensor are excellent control of the thickness of the diaphragm, which in turn determines the sensitivity, and a dielectrically isolated piezoresistor, which reduces the leakage and enables high-temperature operation. The gauge factors obtained by this device correspond to 80% to 90% of the gauge factor in bulk Si. This is an excellent result because the silicon on the top of Al_2O_3 is a thin film. The temperature characteristics of the sensitivity and the offset voltage of this device are shown in Figure 5.27 over the temperature range from -20°C to 400°C. As can be seen, the shift of the sensitivity and the offset voltage is less than $\pm0.3\%$ over the entire temperature range.

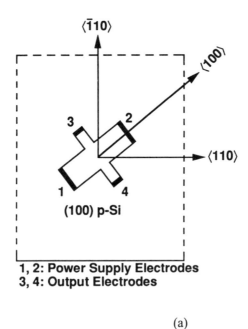

(a)

1, 2: Power Supply Electrodes
3, 4: Output Electrodes

(b)

Figure 5.25 Top view (a) and cross section (b) of pressure sensor fabricated by wafer bonding technology.

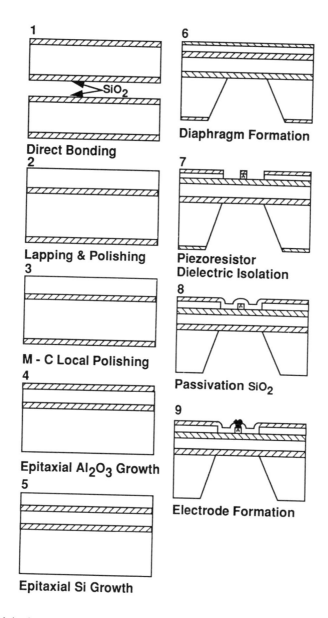

Figure 5.26 Fabrication process (1–9) of pressure sensor.

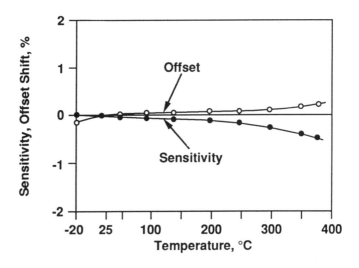

Figure 5.27 Change of sensitivity and offset as a function of temperature.

5.4.3 Peristaltic Membrane Pump

Miniaturized flow systems are becoming increasingly important in chemical analysis and clinical diagnosis because of their capability to handle and analyze very small quantities of specimen. An example of such a system is a thermopneumatically actuated peristaltic membrane pump schematically shown in Figure 5.28 [37]. The pump consists of three sequentially bulging membranes and can be manufactured by DWB using three wafers: a flow channel wafer, a membrane wafer, and a heater wafer. All three wafers are micromachined before the bonding. The pump has three independent thermopneumatic chambers and a channel with an inlet and an outlet. The basic principle of operation can be described as follows. When the leftmost heater is on it will heat the fluid within the left chamber causing expansion and bulging of the left membrane so that the bulging membrane can seal off the channel. After this, the center heater is turned on and it bulges the center membrane downward inducing a flow along the channel. When the right heater is on the left one is off and the fluid in the channel is pushed and drawn into the pump simultaneously. The repetition of the cycle creates a peristaltic pumping action. The advantage of this pump is the ability to provide precise and repeatable volumes of liquids. The thermal analysis of the heating and cooling cycles predicts that the pump should operate at 20 Hz and provide a flow of 7 mL/min for an inlet pressure of 15 psi. Proof-of-concept experiments, with an external pneumatic source to bulge the membranes, have shown that the flow of nitrogen with an inlet pressure of 5 psi can be reduced

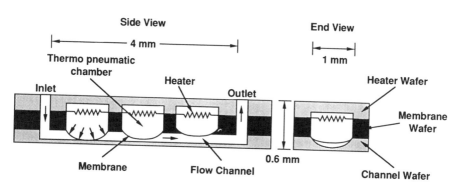

Figure 5.28 Cross section of the peristaltic membrane pump.

to 20% of its original flow if a pressure of 20 psi is applied to a bulging membrane. Also, the same pressure applied to a bulging membrane can reduce the water flow through the pump (with 5 psi of pressure at the inlet) to 8% of its original rate. These are encouraging results that call for further research and improvement of the device.

5.5 SUMMARY

The topic of this chapter was a silicon direct wafer bonding technology. The first part of the chapter described a mechanism of direct wafer bonding, emphasizing two special cases, $SiO_2//SiO_2$ and $Si//Si$ bonding. The central part of the chapter was devoted to processing considerations, discussing key elements of the wafer bonding technology: interface integrity and voids in bonded wafers, bond strength, shaping bonded wafers, and microdefects. The last part of the chapter included several applications, demonstrating unique capabilities of wafer bonding technology.

REFERENCES

[1] T. Nakamura, "Method of Making a Semiconductor Device," U.S. Patent No. 3,239,908, 1962.
[2] E. F. Cave, "Method of Making a Composite Insulator Semiconductor Wafer," U.S. Patent No. 3,290,760, 1966.
[3] J. P. White, R. Amantea, and H. W. Becke, "Current Limiting Integrated Circuit," U.S. Patent No. 3,769,561, 1973.
[4] R. Dahlberg, "Negative Semiconductor Resistance," U.S. Patent No. 4,317,091, 1982.
[5] R. Dahlberg, "Thyristor Having a Center *pn* Junction Formed by Plastic Deformation of the Crystal Lattice," U.S. Patent No. 4,441,115, 1984.
[6] I. M. Shimbo, K. Furukawa, K. Fukuda, and K. Tanzawa, "Silicon-to-Silicon Direct Bonding Method," *J. Appl. Phys.*, Vol. 60, 1986, p. 2987.
[7] A. Nakagawa, K. Watanabe, Y. Yamaguchi, H. Ohashi, and K. Furukawa, "1800 V Bipolar-Mode MOSFETs: A First Application of Silicon Wafer Direct Bonding (SDB) Technique to a Power Device," *Tech. Digest*, IEDM 86, 1986, p. 5.6.

[8] M. Shimbo, K. Fukuda, and Y. Ohwada, "Method of Manufacturing Semiconductor Substrate," U.S. Patent No. 4,638,552, 1987.

[9] M. Shimbo and K. Fukuda, "Method of Bonding Crystalline Silicon Bodies," U.S. Patent No. 4,671,846, 1987.

[10] A. Nakagawa, H. Ohashi, T. Ogura, and M. Shimbo, "Method of Manufacturing Semiconductor Device Wherein Silicon Substrates Are Bonded Together," U.S. Patent No. 4,700,466, 1987.

[11] W. A. Krull, J. F. Buller, G. V. Rouse, and R. D. Cherne, "Electrical and Radiation Characterization of Three SOI Material Techniques," *IEEE Circuits and Devices Mag.*, 1987, p. 20.

[12] R. B. Doremus, *Glass Science*, John Wiley & Sons, New York, 1973.

[13] G. A. C. M. Spierings and J. Haisma, "Diversity and Interfacial Phenomena in Direct Bonding," in *Semiconductor Wafer Bonding: Science, Technology and Applications*, ed. U. Gosele, T. Abe, J. Haisma, and M.A. Schmidt, The Electrochem. Soc. Press, Vol. 92-7, 1991, p. 18.

[14] J. B. Lasky, "Wafer Bonding for Silicon-on-Insulator Technologies," *Appl. Phys. Lett.*, Vol. 48, 1986, p. 78.

[15] J. B. Lasky, S. R. Stiffler, F. R. White, and J. R. Abernathey, "Silicon-on-Insulator (SOI) by Bonding and Etch Back," *Tech. Digest*. IEDM 85, 1985, p. 684.

[16] W. P. Maszara, G. Goetz, A. Caviglia, and J. B. McKitterick, "Bonding of Silicon Wafers for Silicon-on-Insulator," *J. Appl. Phys.*, Vol. 64, 1988, p. 4943.

[17] S. Wolf and R. N. Tauber, *Silicon Processing for the VLSI Era*, Vol. 1. *Process Technology*, Lattice Press, 1986, p. 215.

[18] W. Kern and D. A. Puotinen, "Cleaning Solutions Based on Hydrogen Peroxide for Use in Silicon Semiconductor Technology," *RCA Rev.* Vol. 31, 1970, p. 187.

[19] M. Grundner and H. Jacob, "Investigation on Hydrophilic and Hydrophobic Silicon (100) Wafer Surfaces by X-Ray Photoelectron and High-Resolution Electron Energy Loss-Spectroscopy," *Appl. Phys.*, Vol. A39, 1986, p. 73.

[20] Y. Backlund, K. Ljungberg, and A. Soderbarg, "A Suggested Mechanism for Silicon Direct Bonding from Studying Hydrophilic and Hydrophobic Surfaces," *J. Micromech. Microeng.*, Vol. 2, 1992, p. 158.

[21] K. Mitani, V. Lehmann, and U. Gosele, "Bubble Formation During Silicon Wafer Bonding: Causes and Remedies," *Tech. Digest*, IEEE Sensors and Actuators Workshop, Hilton Head Island, SC, 1990, p. 74.

[22] A. Yamada, Bai-Lin Jiang, G. Rozgonyi, H. Shirotori, O. Okabayashi, and T. Iizuka, "Structural Evaluation of SOI Fabricated by a Direct Wafer Bonding and Numerically Controlled Polishing Technique," in *Proc. of the Fourth Int. Symp. on Silicon-on-Insulator Technology and Devices*, ed. D.N. Schmidt, Montreal, The Electrochem. Society, 1990, p. 225.

[23] V. Lehmann, K. Mitani, I. W. K. Ong, R. Stengl, and U. Gosele, "Semiconductor Wafer Bonding and Thinning Techniques," in *Proc. of the Fourth Int. Symp. on Silicon-on-Insulator Technology and Devices*, ed. D.N. Schmidt, Montreal, The Electroch. Society, 1990, p. 213.

[24] R. Stengl, K.-Y. Ahn, and U. Gosele, "Bubble-Free Silicon Wafer Bonding in a Non-Clean Room Environment," *Jpn. J. Appl. Phys.*, Vol. 27, 1988, p. L 2364.

[25] K. Mitani, V. Lehman, R. Stengl, D. Feijoo, U. Gosele, and H. Z. Massoud, "Causes and Prevention of Temperature-Dependent Bubbles in Silicon Wafer Bonding," *Jpn. J. Appl. Phys.*, Vol. 30, 1991, p. 615.

[26] V. Lehmann, U. Gosele, and K. Mitani, "Contamination Protection of Semiconductor Surfaces by Wafer Bonding," *Solid State Technology*, 1990, p. 91.

[27] T. Abe, M. Nakano, and T. Itoh, "Silicon Wafer-Bonding Process Technology for SOI Structures," in *Proc. of the Fourth Int. Symp. on Silicon-on-Insulator Technology and Devices*, ed. D.N. Schmidt, Montreal, The Electroch. Society, 1990, p. 61.

[28] P. P. Gillis and J. J. Gilman, "Double-Cantilever Cleavage Mode of Crack Propagation," *J. Appl. Phys.*, Vol. 35, 1964, p. 647.

[29] J. Haisma, G. A. Spierings, U. K. P. Biermann, and J. A. Pals, "Silicon-on-Insulator Wafer Bonding-Wafer Thinning Technological Evaluations," *Japn. J. Appl. Phys.*, Vol. 28, 1989, p. 1426.

[30] M. Hashimoto, A. Ogasawara, M. Shimanoe, A. Nieda, H. Satoh, A. Yagi, and T. Matsushita, "Low Leakage SOIMOSFETs Fabricated Using a Wafer Bonding Method," *Proc. 21st Conf. on Solid State Devices and Materials*, Tokyo, 1989, p. 89.

[31] T. Y. Tan, E. E. Gardner, and W. K. Tice, "Intrinsic Gettering by Oxide Precipitate Induced Dislocations in Czochralski Si," *Appl. Phys. Lett.*, Vol. 30, 1977, p. 175.

[32] C. W. Pearce and G. A. Rozgonyi, "Intrinsic Gettering in Heavily Doped Si Substrates for Epitaxial Devices," in *VLSI Science and Technology*, ed. C. J. Dell' Oca and W. M. Bullis, Montreal, The Electroch. Society, Vol. PV 82-7, 1982, p. 53.

[33] F. Secco d'Aragona, J. W. Rose, and P. L. Fejes, "Outdiffusion, Defects and Gettering Behaviour of Epitaxial n/n^+ and p/p^+ Wafers Used for CMOS Technology," in *Proc. of the Third Int. Symp. on Very Large Scale Integration Science and Technology*, ed. W. M. Bullis and S. Broydo, Montreal, The Electroch. Society, 1985, p. 106.

[34] P. L. Fejes, F. Secco d'Aragona, and J. W. Rose, "Electrical Characterization of Epitaxial Wafers for Use in CMOS," in *Proc. of the Third Int. Symp. on Very Large Scale Integration Science and Technology*, ed. W. M. Bullis and S. Broydo, Montreal, The Electroch. Society, 1985, p. 118.

[35] M. A. Hauff, A. D. Nikolich, and M. A. Schmidt, "A Threshold Pressure Switch Utilizing Plastic Deformation of Silicon," *Tech. Digest*, Transdicers '91, San Francisco, 1991, p. 177.

[36] G. S. Chung, S. Kawahito, M. Ashiki, M. Ishida, and T. Nakamura, "Novel High-Performance Pressure Sensors Using Double SOI Structure," *Tech. Digest*, Transducers '91, San Francisco, 1991, p. 676.

[37] J. A. Folta, N. F. Raley, and E. W. Hee, "Design, Fabrication and Testing of Miniature Peristaltic Membrane Pump," *Tech. Digest*, IEEE Sensor and Actuator Workshop, Hilton Head Island, SC, 1992, p. 186.

Chapter 6
Packaging for Sensors

R. Frank, M. L. Kniffin, and Lj. Ristic
Motorola

6.1 INTRODUCTION

Sensor packaging techniques have been derived from standard semiconductor and hybrid electronic packaging technology [1–10]. However, there are some significant differences between packages for standard semiconductor devices and packages for silicon sensors. These differences arise from the fact that sensors must frequently interact with the environment and also must often be exposed to harsher operating conditions than standard integrated circuits or semiconductor components. Designing a package to meet these requirements represents a unique and difficult challenge to the packaging engineer.

It should also be pointed out that cost often introduces additional restrictions. Sensor packaging can represent a substantial portion of the total sensor cost. The packaging and assembly costs can range from as low as the cost of the silicon sensing element for simple applications to several orders of magnitude greater than the cost of the silicon for extreme sensing ranges. The challenge to the sensor manufacturer is to develop packaging technologies that meet all of the necessary performance and reliability criteria, while keeping the costs of assembly at a minimum.

Many of the requirements regarding sensor packaging are uniquely related to the sensor type and its intended application. As semiconductor sensors have proliferated, the number of applications and thus the number of package designs has become virtually limitless. Indeed, in the area of sensor packaging, custom package design is the rule rather than the exception. It would be impossible to review all of the currently available approaches to sensor packaging in the space of a single chapter. Instead we will discuss some of the common aspects of semiconductor packaging technology and their application to the assembly of sensors, along with a brief overview of sensor-specific technologies, such as wafer-level packaging. This will be followed by a detailed description of a few specific examples of custom sensor packages, as an illustration of how package design can be modified to fulfill application-specific requirements.

6.2 BASIC CONSIDERATIONS FOR SENSOR PACKAGING

One has to take into consideration several factors when designing packages for sensor devices. These include the general requirements common to semiconductor packaging technology [11–13], as well as a number of special requirements related to a given sensor and its intended application [1–10]. General requirements include consideration of the electrical, thermal, and mechanical performance of the packaged part. Special considerations include issues such as the requirement that the active area of the device interface with the external surroundings and the need for the sensor to function in a corrosive or high-temperature environment. The final design of both the sensor device and the package must take all of these factors into consideration.

Electrical considerations come from the need to ensure stable operation of the sensor device and the desire to eliminate electrical interference and parasitic signals to ensure accuracy and reliability. These requirements include parameters such as minimum power supply voltage drop (provision has to be made for low resistance), minimum self-inductance (requires short signal leads), minimum cross talk (if any), and minimum capacitive loading. All of these factors must be taken into account when designing the layout of the sensor package. The configuration of the sensor package will in turn determine many of the design parameters used in the layout of the sensor chip(s); for example, the size and placement of the bond pads and the distance between adjacent metal lines.

The thermal design considerations arise from the need to keep the temperature of the sensor and its support circuitry low enough to avoid either short-term catastrophic or long-term degradation failures. This can be particularly challenging for sensors that draw large amounts of power or are required to operate with the active element at temperatures well above the standard operating conditions for semiconductor devices. The maximum operating temperature of a packaged device is strongly dependent upon the heat conductivity of the materials used in the assembly of the package, the package layout, and the operating conditions of the device. Simplistic models for heat transfer are usually a good start in estimating thermal resistance of the package, but the best way to obtain accurate data is to use experimental methods.

Mechanical considerations fall into one of three categories: minimization of stresses induced by mismatches in the thermal expansion coefficients of the materials used to fabricate the package, minimization of stresses induced by external loading of the package, and prevention of mechanical failure of the package during service. The last issue is of concern for all varieties of semiconductor devices, and considerable effort has been devoted to the optimization of package designs and packaging materials to alleviate these problems. However for the assembly of stress-sensitive devices, such as pressure sensors and piezoresistive force sensors, the first and second considerations frequently drive the design of the sensor package. The reduction of package-induced stresses to levels that will not affect the output of a micromechanical sensor is a difficult and challenging task.

The differential thermal expansion between various elements of the assembly (for example, the chip and the package substrate or lead frame) during thermal cycling is the primary source of package-induced stress on the sensor die. Complete elimination of thermal stresses requires the use of materials with comparable thermal coefficients of expansion (TCEs). However, as one can see from the information presented in Table 6.1, it is usually necessary to use materials that have a wide variety of thermomechanical properties to assemble the package. Consequently the design of the package must be optimized to minimize the effects of thermal mismatch between the package components on performance of the sensor chip, especially for applications that require a wide range of operating temperatures. Prediction of the

Table 6.1
Thermal Expansion Coefficients of Materials
Used to Assemble Semiconductor Packages

Application	Material	Coefficient of Thermal Expansion ($°C^{-1}$)
Die	Silicon	2.6×10^{-6}
	Gallium Arsenide	5.7×10^{-6}
Lead frames	Copper	17.0×10^{-6}
	Alloy 42	$4.3–6.0 \times 10^{-6}$
	Kovar	4.9×10^{-6}
	Invar	1.5×10^{-6}
Subtrates	Alumina (99%)	6.7×10^{-6}
	AlN	4.1×10^{-6}
	Beryllia (99.5%)	6.7×10^{-6}
Adhesives	Lead glass	10×10^{-6}
	Au–Si eutectic	14.2×10^{-6}
	Pb–Sn solder	24.7×10^{-6}
	Ag-filled epoxy	32×10^{-6}*
	RTV silicone	$300–800 \times 10^{-6}$†
Conformal coating	Polyimide	$40–50 \times 10^{-6}$
	RTV silicone	$300–800 \times 10^{-6}$†
	Silicon gel	$\sim300 \times 10^{-6}$†
Epoxy molding compounds	Unfilled epoxy	$60–80 \times 10^{-6}$*
	Silica-filled epoxy	$14–24 \times 10^{-6}$*

Tg: glass transition temperature.   *Below Tg.   †Above Tg.

sign and magnitude of stresses that will be generated in a package fabricated from several components and many different materials is a rather complex problem that requires a significant effort to be analyzed. Modern packaging technology relies heavily on tools such as the finite element analysis (FEA) for efficient solutions. FEA can be an extremely effective tool for design optimization.

A number of additional factors somewhat unique to the assembly of sensor devices must be considered during the design of the package. Many of these present far greater challenges to the package designer than those faced in the design of standard semiconductor packages. We will briefly discuss some examples of application specific packaging requirements in the paragraphs that follow.

In designing packaging for magnetic field sensors based on galvanomagnetic effects one has to pay attention to the magnetic properties of the materials from which the package is fabricated. In this case all of the materials used in the assembly of the package must be nonmagnetic, as use of a magnetic material in external package would distort the magnetic field being measured. Since these devices are usually noncontact, low-cost sensors, the preferred package is a plastic package with a nonmagnetic lead frame.

Optical emitters and detectors must use a combination of materials that are optically transparent to the wavelengths of light emitted or detected by the active sensing element and materials that will absorb this electromagnetic radiation. For certain applications a window or a lens that transmits and focuses light in a certain direction is also required. The package and the materials from which it is fabricated must also maintain their optical properties over the lifetime of the device.

Other examples of application-specific package design are seen in the design of housings for pressure sensors. In the case of differential pressure sensors, provision for interfacing with two different external pressures has to be assured. This is typically done by providing two separate inlets in the package. For pressure sensors that must have some interface with a liquid environment the package must be designed to protect the device from the corrosive effects of the liquid, while providing accurate transmission of the fluid pressure to silicon diaphragm. One approach is to use a stainless-steel diaphragm on the top of micromachined diaphragm. In this case the stainless-steel diaphragm serves as the interface with the liquid under measurement, thus isolating the internal wiring and active circuit elements from the deleterious effects of exposure to fluids.

Chemical sensor packages require even more critical materials selection and design optimization than pressure sensor packages. Isolation of the sensing portion of the device is not a design option in most chemical sensors because the chemicals to be detected must be able to directly interact with the active surfaces. Approaches to the packaging of these devices have included partitioning of the electrical circuit elements and chemical sensing elements and the development of special media compatible packaging materials.

6.3 WAFER-LEVEL PACKAGING

Protection of the sensor from environmental factors that can affect the sensor's durability and performance must be taken into account at the earliest stages of packaging design. To meet this challenge, many sensor manufacturers have developed wafer-level packaging schemes that provide a low-cost, protective environment for the sensor device during all stages of assembly and testing [1,3,7,9,10,14]. The prevalent use of wafer-level packaging is unique to the manufacture of sensor components.

The most common wafer-level assembly technique involves bonding a micromachined support wafer [3] (for pressure sensors) or cover wafer (for accelerometers [1] or chemical sensors [10]) to the substrate carrying the sensor devices, using either an adhesive layer or an anodic bonding process [7]. As an illustration, a cross section of a silicon differential pressure sensor that has been attached to a silicon constraint wafer is shown in Figure 6.1. In this application the constraint wafer provides me-

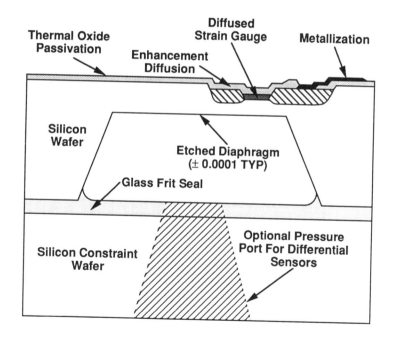

Figure 6.1 Schematic cross section of the wafer-level packaging scheme for a differential pressure sensor.

chanical support and protects the delicate micromachined membrane from physical damage during subsequent handling. It also creates sealed reference cavity for calibration of the device. The use of silicon for both the top and bottom of the sensor structure minimizes differences in thermal coefficient of expansion that can affect the offset and span of the assembled sensor device.

Several methods have been developed to create the hermetic seal between the cover wafer and the underlying substrate. In the following sections we will briefly review the two techniques most frequently used in the fabrication of sensor components: glass sealing and anodic bonding.

6.3.1 Glass-Sealed Technique

One approach to wafer-level packaging uses a frit glass seal to attach the silicon constraint wafer to the wafer that carries the sensing element [1,3,7]. Generally, a low-melting-point inorganic oxide glass such as a lead oxide or boric oxide is used to form the bond between the two wafers. These materials are similar to the glass used to attach the hermetic seal in CERDIP semiconductor packages (see Section 6.4.4) [15].

The sealing glass is typically applied to the constraint or cover wafer in the form of a paste, which consists of a mixture of glass frit and an organic binder. This glass paste is usually deposited through a standard silkscreen process. Following the silkscreening process the coated wafer is allowed to dry, then fired at high temperatures to burn off the organic binder and sinter the powdered glass. The wafers are then ready to be bonded to the sensor device wafer to form the wafer-level package.

To join the sensor wafer and the glass-coated constraint wafer, the two are aligned and placed in intimate contact with one another. The assembly is then heated to a temperature exceeding the softening point of the glass and thermocompression bonded together to form the hermetic seal. Because a low enameling powdered glass is used, the process temperature can be held below 450°C to avoid damaging the electronic circuitry on the sensor wafer and still achieve excellent sealing characteristics. The quality of the seal can be verified by using a helium bomb leak test.

Figure 6.2 shows an SEM cross section of a wafer-level accelerometer package that has been fabricated in this manner. The two silicon wafers and the glass seal are clearly visible, as is the cavity in which the surface-micromachined element resides. The gas sealed in this cavity can be used to modify the mechanical damping of the accelerometer through the squeeze film effect.

An alternative glass-sealing process that has been used in assembly of pressure sensors employs Pyrex glass (7740) as the bonding media. The Pyrex glass is sputtered onto one of the wafers in a layer about 4-μm thick [16]. The wafers are then placed together and a voltage is applied across the interface. Approximately 50V is sufficient to bond the two wafers.

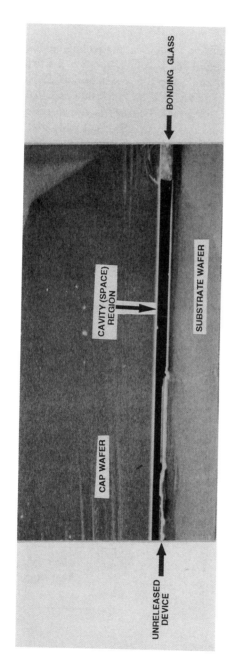

Figure 6.2 SEM cross section of the wafer-level package for a surface-micromachined accelerometer. The cover wafer has been bonded to the accelerometer substrate using a low-melting-point glass.

6.3.2 Anodic Bonding

Anodic bonding is another wafer-to-wafer bonding technique used in the fabrication of silicon sensors [17]. This technique involves bonding a silicon wafer to a glass wafer at an elevated temperature, using the assistance of electrostatic field. Hermetically sealed sensors and actuators can be fabricated using this method [18].

A typical apparatus used for anodic bonding is shown schematically in Figure 6.3. It consists of a chamber and two electrodes: a hot plate that serves as an anode and aluminum block as a cathode. The silicon and glass wafers to be bonded are placed between the two electrodes in such a way that silicon wafer is electrically connected to the anode and the glass wafer to the cathode. The critical process parameters are the magnitude and duration of the applied potential, the temperature of the wafers during the bonding cycle, and the area to be bonded. The required voltages range from 500V to 1000V, depending upon the materials to be bonded and the process temperature. The required process times are short (on the order of 10 min) and typical temperatures used in anodic bonding are between 200°C and 500°C, so the thermal budget of this process is relatively small. Consequently, the sensor wafer can be bonded to the glass substrate without risk of damage to the signal conditioning circuitry. It should also be noted that the ambient in the chamber during

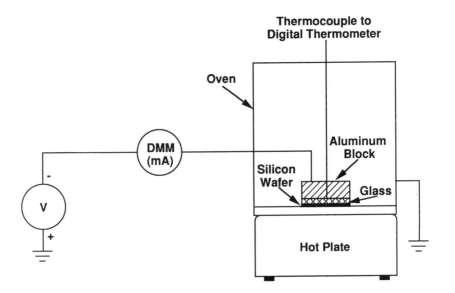

Figure 6.3 Schematic drawing of the typical apparatus used in the anodic bonding of wafers.

bonding can be varied from atmosphere to vacuum, so the ambient in the sealed wafer-level package can be tailored to fit the specific application of a particular sensor or actuator [19].

The bonding mechanism is considered to be an electrochemical process. At elevated temperatures mobility of the positive sodium ions present in the glass substrate is fairly high, and the presence of an electric field causes them to migrate to the negatively charged cathode at the back of the glass wafer. As the Na+ ions migrate toward the cathode, they leave behind permanently bound negative ions. These negative ions form a depletion region (space charge region) adjacent to the silicon surface, which gives rise to a large electric field at the wafer surface. As a result of this electric field, the glass wafer and silicon wafer are pulled into a contact with one another. At the same time, the extremely high electric field transports oxygen from the glass to the silicon–glass interface. Here the oxidation of silicon takes place, and the thin layer of SiO_2 that is formed creates a bond between the silicon and glass wafer. Measuring the current during the bonding process gives important information about the process. The current at the very beginning of the bonding process is high and starts to decay rapidly after a few seconds, as indicated in Figure

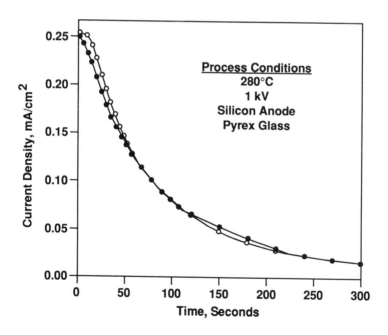

Figure 6.4 Typical variation in current during the anodic bonding process. Experimental curves for two different wafers.

6.4 [17]. The initial current peak corresponds to the initial transport of sodium ions from the glass to the cathode. Once the equilibrium is established the current drops rapidly.

Several factors need to be considered when designing an anodic bonding process for wafer-level packaging. First of all, the surfaces to be bonded have to be clean, free of organic residues and particulates, and smooth. Cleanliness can be achieved by using a combination of sulfuric–nitric acid cleanings and acetone rinsing. Smoothness is assured by using polished wafers and precludes the use of wafers that have patterned films. Furthermore, the thermal expansion coefficient of the glass must be close to that of silicon. Differences in the thermal expansion coefficients of the two wafers can lead to thermal mismatch stresses during cooling from the bonding temperature. These stresses can cause device offsets or, in severe cases, wafer breakage. A suitable glass is Pyrex glass (7740), whose temperature coefficient is close to silicon.

6.4 ASSEMBLY TECHNIQUES

In this section we will describe the basic steps involved in the assembly of semiconductor packages. Issues commonly associated with each stage of the assembly will be discussed, as well as requirements specifically related to the assembly of sensors. A description of the most common types of semiconductor packages will also be included.

6.4.1 Die Bonding

A critical step in the assembly of discrete sensor components is the attachment of the sensor die to the package substrate [15,19–21]. The basic structure of a chip bonded to a package substrate is shown in Figure 6.5. Three basic parts are brought together in this process: the sensor device with its backside metallization, the die attachment material, and the package substrate with its metallization. The attachment of the die to the substrate provides both mechanical support for the chip and a path for the conduction of heat and current away from the sensor die. The unique packaging demands for sensor devices can place stringent constraints on the thermal, mechanical, and electrical performance required of the die bonding material. As a consequence, considerable thought must be placed into selection of the materials and processes to be used to attach sensor chips to the package substrate.

Considerable stresses can be generated during the die attachment process due to differences in thermal expansion between the silicon chip and the package substrate [22,23]. These stresses are frequently of sufficient magnitude to cause mea-

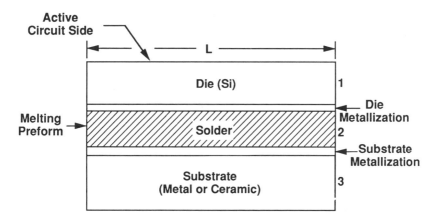

Figure 6.5 Schematic representation of the die bonding process.

surable parametric shifts in strain sensitive devices, such as pressure sensors or Hall detectors. The dimensions of the chip and package, the thermal mechanical properties of the die and substrate, the temperature of the die attachment process, and the compliance of the die bond material all have a significant impact on the magnitude of these stresses. All of these parameters must be considered and carefully optimized when designing the sensor package.

Flow anemometers, temperature sensors, and certain chemical sensors are often required to operate with the sensing element in contact with ambients at temperatures well above those commonly seen by conventional integrated circuits, although the signal conditioning circuitry must be kept at normal operating temperatures. The sensor and the package must be designed to handle these temperature and heat dissipation requirements. Because the thermal conductivity of the die bonding material can have a substantial impact on heat dissipation through the package, it must be selected with these specific requirements in mind.

Automotive, chemical and biological sensors are frequently exposed to environments substantially more corrosive than those commonly encountered by standard integrated circuits. Depending upon the design of the device and its exterior package, the die bond material may be required to encounter corrosive media. Again these requirements place considerable constraints on the materials that can be selected to affix the die to its package.

Tolerances on die placement are also critical for sensors designed to sense directional forces (such as gyroscopes and accelerometers) or electromagnetic fields (such as Hall sensors). Optimization of the die attach process to minimize die shift and careful selection of automated assembly equipment to ensure accurate die placement are needed for these applications. In some instances alternatives to conventional die attachment and wire bonding techniques have been employed to improve die placement accuracy [2].

Selection of the appropriate die bonding material(s) for sensor packaging must take into consideration all of the factors mentioned in the preceding section. In addition the issues of cost, manufacturability, and reliability must be factored into the decision. Fortunately, a wide variety of alternatives exist, so a solution is realizable for most applications. We will briefly describe some of the more common die attachment materials in the paragraphs that follow.

In the integrated circuit industry a variety of bonding media have been developed to attach individual die to package substrates. The available adhesives cover a wide range of different material classes: metal alloys (eutectic die attaches, solders), Ag-filled glasses, organic adhesives (polyimides, silicones, epoxies), and others. Each of these systems has a unique set of advantages, as well as an associated set of issues and drawbacks. In this section we will provide an overview of some of the more commonly used technologies [15,20,21].

The most common technique for bonding silicon die to ceramic packages or metal cans is a Au–Si eutectic bond with the silicon die. In this technique, the die is placed into a heated package (~425°C) that contains either a gold plated substrate or a Au–Si preform. When the die contacts the metallization, silicon diffuses into the gold until the Au–Si eutectic composition is reached. At this point the material at the die-substrate interface melts and forms an intimate, alloyed bond between the die and package. The resultant Au–Si bond is an excellent thermal and electrical conductor, hence is a good choice for applications in which heat or power dissipation are an issue. However, the compliance of the bond material is relatively low, so stresses imparted to the chip as a result of this die attachment process can be considerable. For this reason the Au–Si eutectic system is unsuitable for many sensor applications.

In many applications soft solders, such as 95Pb–5Sn, have replaced the Au–Si system as the material of choice. The bonding technique is similar to that used to form a gold–silicon bond, although the required temperatures are significantly lower as the melting temperatures of these materials are on the order of 300°C. The electrical and thermal performance of these materials is nearly comparable to that of Au–Si and the low yield strength of these materials limits the amount of shear stress that can be supported by the bond, thus decreasing the amount of stress exerted on the die as a result of the attachment process. However, these materials are also more susceptible to fatigue cracking and creep. Fatigue cracking can lead to eventual failure of the bond, and creep can lead to long-term drift in the output of strain sensitive devices. The low melting temperatures of these alloys also limits the useful operating range of these materials, so they may not be suitable for applications involving severe environments or for packages that require a high-temperature sealing step.

Silver-filled inorganic glasses have also been used as die attach materials for hermetic packages and are steadily replacing the gold–silicon eutectic in many VLSI applications [24]. This is due to the lower cost and higher yield of this die attachment process. The unprocessed adhesive is a paste consisting of silver flake and a low-melting-point lead borate glass dispersed in an organic binder. The die attachment process involves dispensing a controlled volume of the adhesive paste onto the package substrate, placing the die on top of the dispensed paste, and firing the parts at temperatures in excess of 400°C. The furnace annealing burns out the organic binder, sinters the silver–glass material, and promotes reaction bonding between the silicon substrate and the silver–glass. The bond quality is quite good, and the electrical thermal conductivity of the bond is generally comparable to that of a eutectic bond. However, the bond temperature is also quite high, and the silver–glasses tend to have elastic moduli similar to Au–Si, consequently stress is still a significant concern.

In many applications polymeric adhesives have replaced glasses and metal alloys as bonding media. The most common polymer die attachment materials are polyimide- or epoxy-based adhesives, although silicone elastomers have recently gained popularity for use in ultra-low-stress applications. Typically these are filled with 70 to 80 wt% silver to provide electrical and thermal conductivity, as the polymers themselves are poor conductors of heat and current. Alumina-filled versions are also available for applications that require high heat conduction in combination with electrical insulation.

Polymer adhesives are the most cost effective of the various die attachment methods because of the high throughput attainable with automated die bonding equipment. Most can be cured at temperatures below 150°C, which is particularly attractive for the assembly of chemical or biological sensors that might have heat-sensitive active layers. The low process temperatures and the compliant nature of these materials also reduce the stresses that arise from differential thermal contraction of the die and substrate as the piece is cooled from the die bonding temperature, making them attractive for the assembly of strain-sensitive devices.

Because of their comparatively low elastic modulus [25], silicone elastomers have found widespread application in the assembly of stress-sensitive sensor devices. Figure 6.6 shows a pressure sensor that has been mounted using a nonconductive silicone RTV adhesive to minimize stresses due to thermal mismatch between the pressure sensor and package substrate. Parts assembled in this manner exhibit excellent stability over time and temperature.

There are some significant drawbacks to polymer adhesives. These materials are still about an order of magnitude less conductive than eutectic or solder die attachments and so may not be suitable for the assembly of parts that have stringent heat or power dissipation requirements. They also tend to be less resistant to attack by solvents and other corrosive media and less able to withstand prolonged exposure to elevated temperatures than inorganic die bonding materials, hence their use in the assembly of sensors that must encounter harsh environments may be limited.

Figure 6.6 Silicon pressure sensor mounted to a plastic chip carrier, using a nonconductive RTV silicone adhesive. As it is quite compliant, the silicone provides mechanical isolation from the package substrate.

6.4.2 Wire Bonding

The wire bonding process provides the electrical connection between the sensor pads and the external packaging terminals [15,20,26]. The result of wire bonding is a wire loop with two bonds on the each side of the wire. Wire bonding is essentially a welding process, and it requires a special piece of equipment known as a *wire bonder*. There are two types of wire bonders: thermocompression and ultrasonic wire bonders (also known as *thermosonic*). The difference between these two is in the type of energy used for wire bonding. Thermocompression bonders rely on heating to create a bond between the wire and the bond pad, while ultrasonic bonders utilize ultrasonic vibration. Which machine is used for particular device depends, in addition to other factors, on the die-attachment process. Thermocompression bonding is not a viable technique at lower temperatures because it requires high thermal energy. On the other hand, ultrasonic bonding is a viable process at lower temperature because the needed high thermal energy is replaced by ultrasonic energy. Modern wire bonders are automated machines that use pattern recognition and operate at speeds of six wires per second.

Typically the wire is ball-wedge bonded: the ball bond is formed on the die side of the wire whereas the wedge bond is formed on the terminal side. The possibility to dress the wire in any direction from the ball-bonded side is what makes ball-wedge bonding so attractive. The best way to describe the wire bonding is to look at the capillary, which is one of the most important parts of the wire bonder. This part feeds the wire and also welds the wire to the bond pads and the terminals of the package. Figure 6.7 shows a capillary in different positions during the wire-bonding process [26]. The bonding process starts with the capillary targeted on the bond pad, Figure 6.7(a), followed by a capillary landing that brings the ball in contact with the pad. Using ultrasonic energy the ball bond is formed, Figure 6.7(b). Then the capillary rises to the loop's highest position, Figure 6.7(c). The next step is targeting the terminal, Figure 6.7(d). In the example shown in Figure 6.7 the lead frame is used as a representative for the package. The bonding of the wire to the lead frame (edge bond) is shown in Figure 6.7(e). Then the capillary rises, leaving the stitch bond. The wire is broken at the thinnest cross section of the wedge bond, Figure 6.7(f). Finally, a hydrogen flame or electric spark is used to form the ball on the wire ready for next bond, Figure 6.7(g).

It should be pointed out that a number of requirements must be met for proper wire bonding. For example, on the bonding sites the metal has to be clean, with no contamination or oxide coatings. The work piece has to be rigidly clamped during the wire-bonding process. The temperature has to be elevated to a proper level when the thermocompressive bonding is used, or ultrasonic power has to be sufficient and the ultrasonic frequency tuned if ultrasonic bonding is used. Further, the choice of capillary depends on the design of the pads, because the capillary face diameter is restricted by the center-to-center spacing of the bonding pads. The unique package

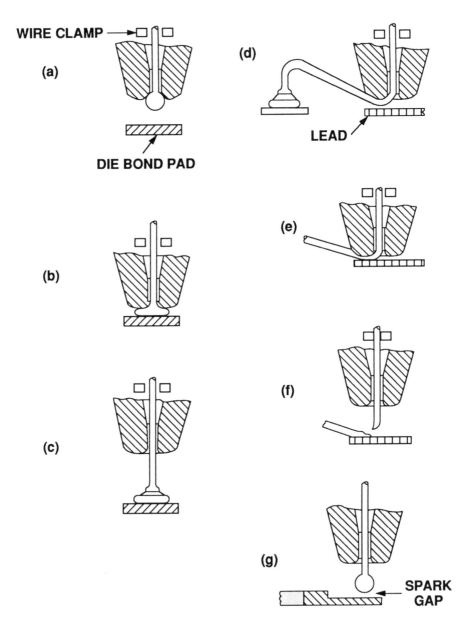

Figure 6.7 The steps involved in the wire bonding process: (a) capillary targeted on the bond pad; (b) capillary lands, bringing the ball in contact with the pad; (c) capillary rises to the loop's highest position; (d) targeting the terminal; (e) wire bonds to the lead frame (edge bond); (f) broken wire at the thinnest cross section of the wedge bond; (g) hydrogen flame or electric spark forms the ball on the wire ready for the next bond.

demands for a given sensor can place constraints on wire size and the spacing between them. Long wire loops are vulnerable to sagging, and the use of larger diameter wire can help solve the sagging problem. On the other hand, larger wires can increase parasitic capacitances, which could be very critical for some sensors. Short wire loops also have drawbacks. The short wire could be tight and break when subjected to mechanical shock. This, again, could be of great importance for some sensors. In conclusion, one can say that the wire-bonding process has to be optimized and tailored to the specific performance of the given sensor.

6.4.3 Chip Coatings

The most common use of chip coatings in the packaging of integrated circuits is in the assembly of overmolded plastic packages [20]. The thermal mismatch between the epoxy mold compound and the silicon die give rise to large stresses at the chip surface [27]. These stresses can cause passivation cracking and smearing of metal lines, both of which compromise the reliability of the packaged part [28]. Chip coatings are used to isolate the die or die surface from the stresses generated by shrinkage of the epoxy mold compound [29]. The most commonly used die coating materials are polyimides and silicone rubbers. These coatings have been shown to reduce the level of stress at the surface of the packaged die, reduce metal line smearing, and minimize passivation cracking. However, some other reliability issues are associated with the use of die coatings, such as wire bond failure [30] and the formation of voids that can trap moisture near the die surface [20].

Chip coatings are also finding application in the assembly of sensors which must face relatively harsh environments. For standard integrated circuit components, the package is designed to provide protection from environments that can include moisture and gaseous or liquid chemicals. In certain applications, these packages are further protected from the environment by additional epoxy potting compounds or conformal coatings that cover the printed circuit board to which the component is mounted. However, semiconductor sensors are frequently required to form an interface with this ambient, thus precluding the use of packaging schemes that completely isolate the device from the environment. The use of chip coatings is one means that has been developed to protect the active circuit components and wire bonds from the harmful effects of the environment, while allowing the measurand of interest to interact with the active sensor element.

An example of this approach is the media-compatible pressure sensor package described in detail in Section 6.5.1.2. In this design the chip is coated with a compliant coating such as paralyene or hydrostatic methyl–silicone gel. The coating allows pressure to be applied to the sensor diaphragm with minimal attenuation, while protecting the die surface and the wire bonding connections from exposure to corrosive chemicals. The applications for which these parts may be used are limited primarily by the media compatibility of the protective coating.

6.4.4 Package Types

The most commonly used packages for sensors are usually based on derivatives of conventional semiconductor packaging techniques. Ceramic, metal can, and plastic packages have all been used extensively in the integrated circuit industry. Each of these has been adapted in one form or another for the packaging of silicon sensors. In this section we will give a brief description of the package types and discuss their application to the packaging of sensor devices.

6.4.4.1 Ceramic Packaging

Ceramic packages are used extensively in the semiconductor industry for high-reliability applications [15]. Ceramic packages that have been used for the assembly of sensors include fired ceramic DIPs and ceramic flat packs. The CERDIP, or ceramic DIP, utilizes a lead frame that is attached to the ceramic (usually aluminum oxide) base through a sintered glass layer. A schematic cross section of a standard CERDIP is shown in Figure 6.8. Figure 6.9 shows a two-chip accelerometer that has been mounted in a CERDIP package. In the assembled part, the sensor resides in the cavity that is recessed into the body of the package. The die are placed into this cavity and bonded to the substrate using a gold eutectic bond or a polymeric adhesive. Connections to the outer leads are then made by gold or aluminum wire bonds from the pads on the chip to the inner lead tabs. Finally, the package is hermetically sealed, using a metal lid attached with solder or a ceramic lid attached with a lead oxide glass. The same assembly techniques are used for other types of ceramic packages including the ceramic flat pack.

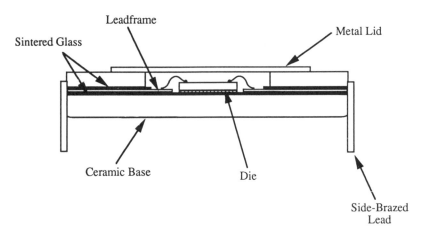

Figure 6.8 Schematic cross section of a standard CERDIP package.

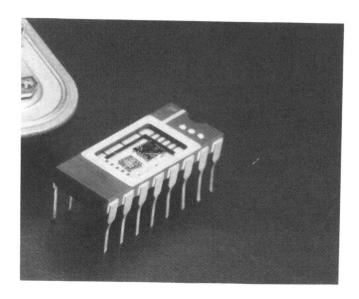

Figure 6.9 Two-chip accelerometer assembled in a 16-pin CERDIP package.

Ceramic packages are much more expensive than many other types of packages; however, they provide a reliable, hermetically sealed environment in which the sensor can be housed. This is particularly critical for the assembly of micromechanical devices that have not been packaged at the wafer level. They are also very useful in the development phase of a sensor because the silicon die does not have to be encapsulated. This allows various test points on the die to be easily probed and measured in packaged form.

Ceramic technology is also involved in hybrid assembly techniques for pressure sensors (see, for example, the package shown in Figure 6.10). The ceramic substrate provides a firm mounting platform for the sensor die. Stress isolation can be obtained by utilizing a compliant silicone for the die attachment. In addition, the ceramic substrate allows laser trimmable thick-film resistors to be packaged with the sensor component. These provide calibration for a signal-conditioned device. The metal ring around the sensor die in Figure 6.10 provides additional isolation for electromagnetic interference, which can be a problem in automotive applications.

6.4.4.2 Metal Can Packaging

Metal cans are cylindrically shaped packages that have an array of leads extending through the base of the package [15]. The leads are attached to the header via glass

Figure 6.10 Pressure sensor mounted in a hybrid ceramic package.

mounting seals. The die is mounted on the header using a eutectic die attachment, and connections to the outer leads are made by gold wire bonds to the inner posts. A metal lid is brazed or welded to the base, forming the completed package. A schematic cross section of this type of package is shown in Figure 6.11(a).

The metal can package may be used to provide a reliable hermetic environment for sensors that require no external physical interface. Surface-micromachined accelerometers and other micromechanical components that do not utilize wafer-level sealing techniques are frequently mounted in metal can packages to ensure a controlled, particle-free operating environment for the device. However, the cost of these packages can represent a significant portion of the total cost of the device, hence they are not desirable for use in high-volume, low-cost applications.

Pressure sensors that must face with severe environments are also commonly assembled in metal cans. A schematic cross section of a metal can modified to house a differential pressure sensor is shown in Figure 6.11(b). For this application the lid and header of the package usually have a hole to permit the pressure sensor die an interface with the environment and to provide an access port for supplying the reference pressure. The package cavity is also typically filled with an incompressible material to provide environmental protection for the signal conditioning circuitry, while allowing effective transfer of the external pressure to the sensor diaphragm. In terms of media compatibility, this type of assembly is one of the most robust pressure sensor packages available.

Body

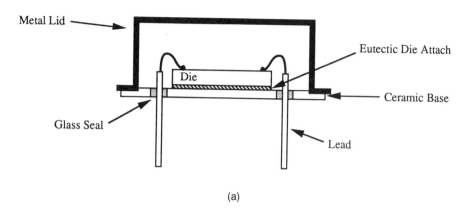

(a)

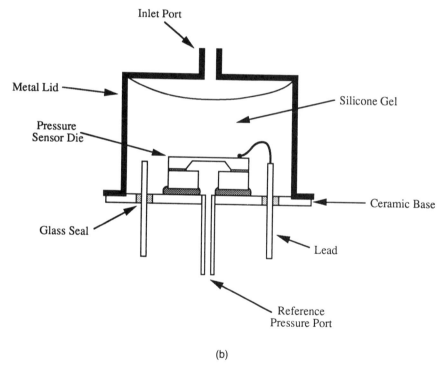

(b)

Figure 6.11 Schematic cross sections of two types of metal can packages: (a) standard package; (b) package modified for differential pressure sensing applications.

6.4.4.3 Plastic Packaging

The newest semiconductor packaging techniques utilize molded plastic packages with types that range from SIP (single in-line packages), DIP (dual in-line packages), and quad flat packages to surface mount devices (SMD) [20,31]. The lead frame and molding techniques used in plastic packages provide the lowest cost semiconductor packages. Though these packages are not hermetically sealed, improvements in mold compounds in recent years include enhanced moisture resistance. The development of improved low-stress mold compounds has also improved the long-term reliability of plastic encapsulated devices.

Plastic packages may be premolded packages or postmolded. The latter is the most cost effective package and is also the most commonly used form of plastic packaging for standard semiconductor components. Both versions have been used in the assembly of sensor devices. For pressure sensors and other sensor components that must come in direct contact with the environment, the use of premolded plastic chip carriers is the most common method of assembly.

Premolded plastic packages or chip carriers are similar to metal and ceramic packages in concept; however, they are not considered to be hermetically sealed because of the moisture permeability of epoxy molding compounds. A cross section of an individual plastic chip carrier is shown in Figure 6.12. The thermoplastic material is General Electric's Valox, molded around the lead frame. Assembly of this package is similar to the assembly of a CERDIP, although an encapsulating material such as a silicon gel is sometimes used to fill the open cavity for additional environmental protection. The properties of the die coating allow the pressure to be transmitted uniformly to the silicon diaphragm, while isolating the active elements on the die surface and the wire bonds from the external environment. The package is sealed using a metal or premolded plastic lid, which is attached with an adhesive.

A strip of premolded plastic chip carrier packages designed to house pressure sensors is shown in Figure 6.13 along with a number of external casings designed to house the chip carrier for a variety of different pressure sensing applications. A single molding operation forms both the body and back of the chip carrier. This unibody package provides lower cost, fewer process steps, higher pressure range capability, and greater media compatibility when compared to earlier versions that were made of separate body and metal back plate. The lead frame assembly technique allows the automation of assembly operations such as die bonding, wire bonding, and gel filling. Automation allows tight process controls to be implemented and provides high throughput.

Overmolded plastic packages are used in applications in which the silicon sensor is not required to come in direct contact with the external environment. Examples of these types of sensors are accelerometers, gyroscopes, and magnetic field sensors. This method of assembly is the most desirable form of packaging for low-cost, high-volume applications.

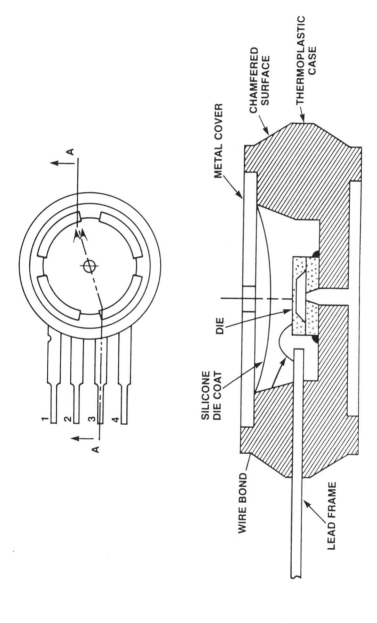

Figure 6.12 Schematic cross section of an individual plastic chip carrier designed for the assembly of pressure sensors.

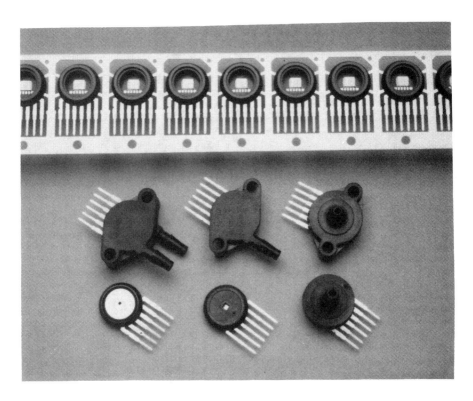

Figure 6.13 Premolded plastic chip carriers for pressure sensor applications: a strip of premolded chip carriers a variety of external housings designed to encase the plastic chip carrier alongside.

A schematic cross section of a postmolded plastic package is shown in Figure 6.14. To assemble this type of package, the die are first attached to a metal lead frame, using a gold–silicon eutectic or a silver-filled polymer adhesive. The most common lead frame materials are alloy 42 and copper. Connections from the bond pads to the outer leads are then made by gold or aluminum wire bonds. The lead frames are then placed in a molding tool and encapsulated with an epoxy molding compound, using a conventional transfer molding process.

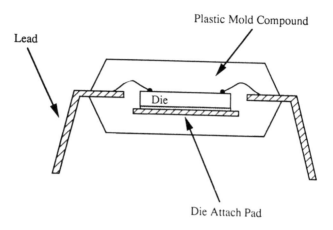

Figure 6.14 Schematic cross section of a postmolded plastic package.

6.5 PACKAGING FOR SPECIFIC APPLICATIONS

In the preceding sections we outlined the basic steps involved in the assembly of semiconductor packages and described the distinguishing features of some of the most common types of packages. In this section we will discuss the application of these packages to the fabrication of semiconductor sensors, with many of the specific examples drawn from the assembly of pressure sensors, because they represent the most frequently used type of silicon sensor. The packaging of accelerometers will also be discussed because they are projected to be manufactured in high volume in the very near future.

6.5.1 Pressure Sensor Packaging

Semiconductor pressure sensors are routinely used in a variety of applications, each of which has a unique set of performance and reliability requirements that must be met by the sensor component. Pressure sensor manufacturers have taken diversified approaches to packaging these devices in order to satisfy the needs of these varied applications. In this section, we will discuss some of these approaches in order to demonstrate how sensor packages can meet a wide range of performance criteria.

6.5.1.1 Media-Compatible Packaging

For standard integrated circuits and discrete components, the package provides protection from the environment, which can include moisture and gaseous or liquid chemicals. In certain applications (for example, automotive underhood mounted modules), these packages are often further protected by additional epoxy and silica potting compounds or conformal coatings such as acrylics, polyurethane, silicon, and ultraviolet curing compounds. The additional protection usually covers the printed circuit to which the component is mounted. However, semiconductor sensors frequently are in contact with the environment, and therefore complete isolation from the ambient atmosphere is not always feasible. For these applications, semiconductor sensors require alternative forms of media-compatible packaging.

The approach to media compatibility begins at the wafer level, where the active area of semiconductor devices is protected by a passivation layer. This passivation layer is deposited at the end of the wafer fabrication process. Silicon nitride and silicon dioxide doped with boron, phosphorus, or both are two materials most frequently used for passivation. The mechanical properties of these layers must be taken into account, as they can have an impact on the functionality of the sensor. For example, in silicon pressure sensors, circuit elements are protected by a passivation layer but the diaphragm must remain clear to avoid offsets due to stress arising from differences in thermal expansion between the passivation film and the silicon diaphragm.

Media-compatibility requirements usually drive the design of the external sensor package as well. For pressure sensors a number of packaging schemes that allow the pressure signal to be transmitted to diaphragm with minimal damping and distortion, while providing adequate environmental protection of the signal conditioning circuitry, have been developed.

The most robust media-compatible packages for pressure sensors are stainless steel housings filled with silicone oil used to transmit the pressure signal to the silicon sensor. In this type of package a thin, 300-series stainless steel diaphragm is in contact with the fluid or gas, and the ambient pressure is transmitted to the silicon diaphragm through the silicon oil. Care must be taken to minimize the volume of oil so that temperature effects are minimized. These packages are costly and thus can represent a significant portion of the total price of the device.

Plastic chip carriers molded from media-resistant materials have also been used to assemble parts for certain pressure sensing applications. In this case a material such as a compliant thin conformal parylene coating or hydrostatic methyl–silicone gel is used to transmit pressure to the active sensor element, while providing protection for the circuitry and external wire bonds. Parylene deposition is a vacuum process, where the reactive vapor is passed over a room temperature sensor and coats it with the polymer [32]. The equipment used to perform this process is quite sophisticated, especially relative to gel coatings. The kind of media to which parylene- and gel-protected sensors can be faced is somewhat limited. Consequently, for gauge pressure measurements in particularly harsh environments the pressure is frequently applied to the back of the sensor diaphragm. This avoids the deleterious effects of a corrosive environment on the protective coatings and ultimately the active surface of the sensor.

6.5.1.2 Industrial Applications

Industrial applications are one of the most common uses for silicon pressure sensors. These applications are frequently unique enough to require special packaging. The chip carrier package discussed in Section 6.4.4.3 has the flexibility needed to meet these requirements, because a custom housing that matches the packaging and assembly requirements of the system can readily be fabricated. This reduces system costs by eliminating the need for adapters and optimizing both the available space in the final product and manufacturing techniques.

Metal housings can also be used for those applications that require a rugged exterior or a threaded package. For example, industrial applications frequently require an interface with harsh liquids, which dictates the stainless steel isolation technique. This type of package is the most media-resistant package style currently available; however, the cost of these packages is substantial.

Metal packages that have been placed inside threaded metal housings are also commonly used in industrial applications, especially when measurement of high pressures (>100 psi) is required. These external metal housings can be designed to interface with readily available O-rings. The chamfered surface located on both the top and bottom of the chip carrier package shown in Figure 6.15 provides an O-ring interface that has several distinct advantages [3]. During the manufacturing process it permits quick testing of guaranteed parameters on 100% of the assembled parts. It also enables the parts to be easily attached to durability test fixtures that are used to determine the lifetime of the device. Customer designed adaptors and other modifications can also be reached through these O-rings. For example, if complete media isolation is required, an O-ring that has a diaphragm molded across it can provide a very low-cost solution to this problem. Pressure sensors with this type of assembly have withstood a high temperature, 30 psi water hammer cycle tests for over 2000 hr.

Many industrial pressure measurements require that the sensor package have a pipe thread connection. Figure 6.16 shows a chip carrier package captured between two 3/4-in. OD-3/32-in. wall O-rings. The O-rings act as both a high-pressure seal and a stack-up tolerance buffer. The differential configuration shown can measure the pressure drop across an orifice to calculate flow in a line operating at common mode pressures in excess of 100 psi.

6.5.1.3 Automotive Applications

Automotive sensors and the extreme operating conditions that can be encountered in certain automotive applications are covered in Chapter 11. However the requirements that arise from these extreme operating conditions affect their packaging and thus warrant further discussion at this point. Two automotive pressure sensor applications

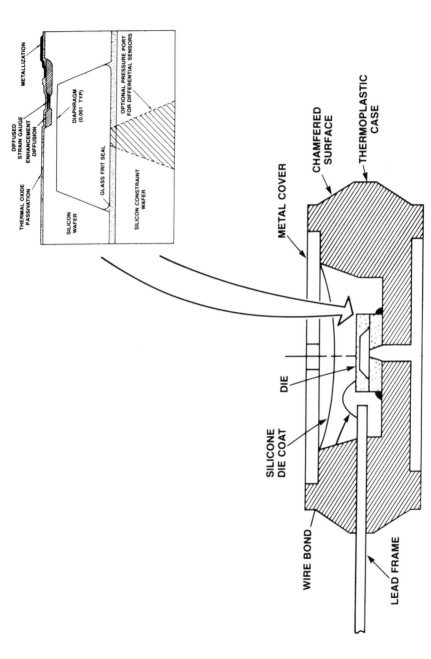

Figure 6.15 A pressure sensor chip carrier designed for industrial applications. The chamfered surface located at the top and bottom of the chip carrier package provides an O-ring interface that offers several distinct advantages.

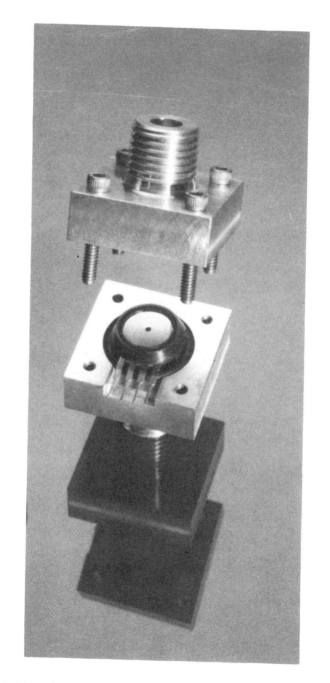

Figure 6.16 A chip carrier captured between two 3/4-in. OD-3/32-in. wall O-rings.

that have different packaging and reliability requirements will be covered in this section. These applications are intake manifold pressure measurement and oil pressure measurement.

Sensors for use in automotive applications such as manifold pressure measurement must be lightweight and small in size. These requirements have been the primary drivers behind package development for this particular application. In the past decade considerable progress has been made in the reduction of these sensors, as is graphically illustrated in Figure 6.17, where assembled part sizes are mapped as a function of time. Improvements in packaging technology and the development of single chip sensors have both played a crucial role in achieving these size reductions. With continued improvements in package design, the size of the silicon sensor is rapidly becoming the limiting factor in determining the size of the assembled part.

The measurement of liquid pressure (engine oil, transmission oil, fuel) or fluid level (oil, gas, coolant) requires the sensor package to be exposed to one or more fluids. Direct contact with these fluids is detrimental to the operation and reliability of the semiconductor circuit, yet cost requirements usually prohibit the use of stainless steel packages for media isolation. To resolve these issues, the approach has

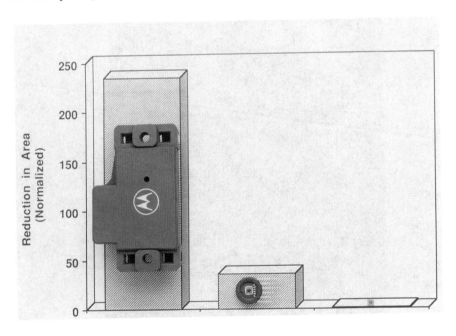

Figure 6.17 Graphical representation of the decrease in size of manifold pressure sensor packages in the past decade.

been to design packages fabricated from more cost effective protective polymers and chemically tolerant plastics.

A sensor package for oil pressure measurements is shown in Figure 6.18. This package was specially designed for easy assembly. It has an extremely reliable O-ring seal and it is able to withstand burst pressures over 400 psi [33]. Special materials were used for both the package and the protective gel that covers the sensor. These materials allow the final automotive assembly to survive qualification tests with over 1 million pressure cycles including portions conducted at high temperatures. The sensor has been designed for a 10 year/100,000 mile life, which exceeds the number of cycles that can be achieved by traditional diaphragm driven potentiometers.

6.5.1.4 Medical Applications

Silicon pressure sensors for medical applications are typically housed in the premolded plastic chip carriers described in Section 6.4.4.3. These chip carriers are easily fit with customer-designed polystyrene or polycarbonate external housings. The custom packaging personalizes the medical device manufacturer's product and provides differentiation in the hospital environment. It also addresses patient mounting and usage problems.

Many biomedical applications require the use of special materials that have passed U.S. Food and Drug Administration tests. The thermoplastic materials normally used to meet the durability and temperature requirements of the industrial and automotive markets cannot pass these tests. For biomedical applications, the sensor package must be molded utilizing a medically approved material such as white biomedical grade of Union Carbide's Udel polysulfone [3], which has passed extensive biological testing. The silicone gel for medical applications is a special grade dielectric gel, which has been used extensively in the fabrication of implants. This gel is a nontoxic, nonallergenic polymer system that passes pyrogen testing as well as meeting all U.S. Pharmacopia (U.S.P.) XX Biological Testing Class VI requirements. The aluminum cap is typically not used in this package due to biocompatibility requirements. Instead a biomedically approved, opaque white filler is mixed into the silicone gel to prevent bright operating room lights from affecting the performance of the light sensitive silicone chip.

Medical applications that do not involve invasive procedures, such as cuff blood pressure measurements (electronic sphygmomanometers) or respiratory pressure measurements, do not have the stringent biocompatibility requirements described in the preceding paragraph. Pressure sensors manufactured for these applications can utilize any of the previously discussed packaging techniques. Cost and performance will dictate the proper choice for the package.

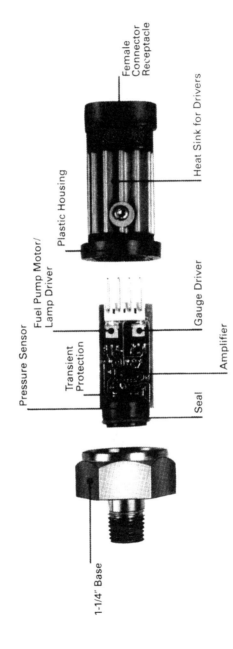

Figure 6.18 A sensor package designed for use in automotive oil pressure measurements.

6.5.2 Accelerometer Packaging

The packaging of bulk- and surface-micromachined accelerometers has to meet a number of requirements unique to the assembly of micromechanical devices. The most challenging of these issues is the need to provide an environment that protects the part from particles and physical damage, while providing a well-defined ambient that controls mechanical damping. To meet this challenge the producers of bulk- and surface-micromachined accelerometers have developed wafer-level packaging schemes that provide a low-cost, protective environment for the device during all stages of assembly and testing.

The most common wafer-level assembly scheme involves bonding a micromachined cover wafer to the device substrate using either a glass sealing or an anodic bonding process, as was discussed in Section 6.3 [1]. After bonding the two wafers, the parts can then be diced, probed, and sorted for final assembly and testing with minimal concern for mechanical damage to the micromechanical structure.

The final step in the assembly process involves mounting the sorted die in a conventional package, which can then be used to mount the device to a PC board. This package must be capable of protecting the sensor element from corrosive materials that tend to decrease the device lifetime. It should also be capable of buffering the device from externally applied stresses that can cause drifts in the accelerometer output. The external housing must also provide efficient transmission of the forces of acceleration to the active sensing element. The design of a package that fulfills all of these requirements represents a significant technological challenge.

The wafer-level packaging of Motorola's capacitive surface-micromachined accelerometer was specifically designed for compatibility with standard plastic packaging techniques. The accelerometer chip is hermetically sealed using the glass-sealing technique discussed in Section 6.3.1. The sensor die and the companion control die containing all of the signal conditioning circuitry can then be attached to a standard dual in-line lead frame, as shown in Figure 6.19 and encapsulated in an epoxy mold compound using conventional transfer molding techniques. This packaging approach offers the advantage of high-volume, low-cost assembly, which is a major concern for automotive applications.

Accelerometers that do not utilize wafer-level sealing techniques must be mounted in metal can or ceramic packages to ensure a controlled, particle-free operating environment for the device. A metal can may also protect the device from the spurious effects of electromagnetic fields that may be encountered in automotive applications. However, packages of this type are substantially more expensive than postmolded epoxy packages.

Figure 6.19 A two-chip accelerometer assembled on a PDIP lead frame. The entire assembly will be encased in an epoxy mold compound using conventional transfer molding techniques.

6.7 SUMMARY

In this chapter, packaging technology has been presented, with a focus on the performance and reliability requirements that are unique to the packaging of sensors. The discussion has included many of the common aspects of semiconductor packaging technology, along with techniques such as wafer-level packaging that are unique to the assembly of sensors. Several examples of sensor package designs have also been presented as an illustration of how packaging can be tailored to meet application-specific performance and reliability criteria.

REFERENCES

[1] V. Adams, R. Frank, and H. Hughes, "Low-Cost Accelerometer Packaging," unpublished manuscript.
[2] R. E. Bicking, L. E. Frazee, and J. J. Simonelic, "Sensor Packaging for High Volume Applications," *Tech. Digest*, Transducers '85, Philadelphia, 1985, p. 350.
[3] R. Frank and V. Adams, "Pressure Sensors Packaged in Plastic," *Electronic Packaging and Production*, 1986, p. 62.
[4] R. Frank and J. Staller, "The Merging of Microstructures and Microelectronics," *Proc. Third Int. Forum on ASIC and Transducers*, Banff, Canada, 1990, p. 53.
[5] F. Goodenough, "Sensor ICs: Processing, Materials Open Factory Doors," *Electronic Design*, 1985, p. 131.
[6] I. Igarashi, "New Technologies of Automotive Sensors," *Tech. Digest*, Transducers '85, Philadelphia, 1985, p. 246.
[7] W. H. Ko, J. T. Suminto, and G. J. Yeh. "Bonding Techniques for Microsensors," in *Micromachining and Micropackaging of Transducers*. Elsevier Science Publishers, 1985.
[8] J. Mallon, J. Bryzek, J. Ramsey, G. Tomblin, and F. Pourahmadi, "Low-Cost, High-Volume Packaging Techniques for Silicon Sensors and Actuators," *Tech. Digest*, IEEE Solid State Sensor and Actuator Workshop, Hilton Head Island, SC, 1988, p. 123.
[9] H. Reichl, "Packaging and Interconnection of Sensors," *Sensors and Actuators*, Vols. A25–A27, 1991, p. 63.
[10] S. D. Senturia and R. L. Smith, "Microsensor Packaging and System Partitioning," *Sensors and Actuators*, Vol. A15, 1988, p. 221.
[11] *Microelectronics Packaging Handbook*, ed. R. R. Tummala and E. J. Rymaszewski, New York, Van Nostrand Reinhold, 1989.
[12] Louis T. Manzione, *Plastic Packaging of Microelectronic Devices*, New York, Van Nostrand Reinhold, 1990.
[13] S. M. Sze, *VLSI Technology*, 2d ed., New York, McGraw-Hill, 1988.
[14] L. A. Field and R. S. Muller, "Fusing Silicon Wafers with Low Melting Temperature Glass," *Sensors and Actuators*, Vols. A21–A23, 1990, p. 935.
[15] *Microelectronics Packaging Handbook*, ed. R. R. Tummala and E. J. Rymaszewski, New York, Van Nostrand Reinhold, 1989.
[16] "Pyrex Bonding," *Design News*, 1984.
[17] K. B. Albaugh and P. E. Cade, "Mechanism of Anodic Bonding of Silicon to Pyrex Glass," *Tech. Digest*, IEEE Solid-State Sensor and Actuator Workshop, Hilton Head Island, SC, 1988, p. 109.
[18] P. R. Younger, "Hermetic Glass Sealing by Electrostatic Bonding," *J. Non-Crystalline Solids*, Vols. 38 and 39, 1980, p. 909.

[19] J. A. Minucci, A. R. Kirkpatric, and W. S. Kreisman, "Integral Glass Sheet Encapsulation for Terestrial Panel Applications," 12th IEEE Photovoltaic Specialist Conference, 1976.

[20] L. T. Manzione, *Plastic Packaging of Microelectronic Packages*, New York, Van Nostrand Reinhold, 1990.

[21] L. G. Feinstein, *"Die Attachment Methods,"* in *The Electronic Materials Handbook*, Vol. 1. *Packaging*, ed. Merril L. Minges, Materials Park, OH, ASM International Handbook Committee, 1989.

[22] E. Suhir, "Die Attachment Design and Its Influence on Thermal Stresses in the Die and the Attachment," *Proc. 37th Electronic Components Conf.*, 1987, p. 508.

[23] E. Suhir, "Calculated Thermally Induced Stresses in Adhesively Bonded and Soldered Assemblies," presented at ISHM International Symposium on Microelectronics, Atlanta, 1986.

[24] M. N. Nguyen, "Low Stress Silver–Glass Die Attach Material," *IEEE Trans. on Components, Hybrids, and Manufacturing Technology*, Vol. CHMT-13, 1990, p. 478.

[25] K. I. Loh, "Low Stress Adhesive," *Advanced Packaging*, 1991, p. 18.

[26] "Bonding Tools and Production Accessories," Bonding Handbook and General Catalog, Kulicke and Soffa Industries, 1980.

[27] D. S. Soane, "Stresses in Packaged Semiconductor Devices," *Solid State Technology*, 1989, p. 165.

[28] R. E. Thomas, "Stress-Induced Deformation of Aluminum Metallization in Plastic Molded Semiconductor Devices," *IEEE Trans. on Components, Hybrids and Manufacturing Technology*, Vol. CHMT-8, 1986, p. 427.

[29] D. B. Edwards, K. G. Heinen, J. E. Martinez, and S. Groothuis, "Shear Stress Evaluation of Plastic Packages," Proc. 37th Electronic Components Conference, 1987, p. 84.

[30] N. V. Chidambaram, "A Numerical and Experimental Study of Temperature Cycle Wire Bond Failure," presented at IEEE, 1991.

[31] C. A. Kovac. "Plastic Package Fabrication," in *The Electronic Materials Handbook*, Vol. 1. *Packaging*, ed. Merril L. Minges, Materials Park, OH, ASM International Handbook Committee, 1989.

[32] A. F. Benson, "Count on Conformal Coatings," *Assembly Engineering*, 1990, p. 20.

[33] "Acustar Electronic Oil Pressure Sensor," *Automotive Industries*, 1990, p. 44.

Chapter 7
Magnetic Field Sensors
Based on Lateral Magnetotransistors

Lj. Ristic
Motorola

7.1 INTRODUCTION

A magnetic field sensor can be described as a transducer that generates an electronic signal in the presence of a magnetic field, whereby the generated electronic signal is correlated to the intensity of the magnetic field and its direction. A variety of devices can generate the electronic signal in the presence of a magnetic field, ranging from a simple search coil to a superconducting quantum interference device. Applications of these devices are many and usually are classified into two groups [1,2]: direct measurement applications that include earth magnetic field measurements, reading of magnetic tapes and disks, magnetic apparatus control, and banknotes reading; and indirect applications that include contactless switching, displacement detection, current detection, watt meters, traffic detection, and biomagnetometry. The detection range of the magnetic field in all of these applications spans from 10^{-15}T (biomagnetometry) to 10^{6}T (magnetic apparatus control in high-energy physics).* The large segment of industrial applications is typically in the range of mT, and it can be covered by solid state sensors. These include Hall cells, magnetodiodes, MAGFETs, magnetotransistors, and carrier-domain magnetometers. With the exception of the Hall cell, which is commercially available (and probably the best understood device), all other solid state devices are still in the development phase. A lot of work remains for better understanding of the basic principles of operation and optimization of these devices.

It should be pointed out that the subject requires not one chapter but the book itself to deal with all of the different devices, materials and pertinent issues. How-

---

*The magnitude of magnetic induction is measured in Tesla (T), where 1 Gauss = 10^{-4} T; 1 gamma = 10^{-9} T. Note that, on earth at the equator, an average induction is about 35 μT.

ever, that would be impractical given the scope of this book and the size of thi. chapter. Therefore, this chapter will focus on one type of magnetic field sensors lateral magnetotransistors made in silicon. There are several reasons for this. First there is no need to repeat the well known facts about Hall devices and other magneti field sensors that have already been presented in several excellent review papers and books [2–6]. Second, lateral magnetotransistors have a high potential as magnetic field sensors because of their high sensitivity, linearity, and full compatibility with the existing silicon IC processing, and the possibility of their integration with signal processing circuitry. This may lead to a next generation of commercially available intelligent magnetic field sensors.

At the same time, it is my desire to shine the light on the lateral magneto transistor from different angles in a way that has not been done before and emphasize the high potential of this type of magnetic field sensor. To accomplish this, an effor has been made in this chapter to discuss universal issues such as sensitivity, offset noise, surface effects, and two- and three-dimensional sensing, all of these at the device level. It is my hope that the subject matter presented here will inspire the reader to the point that it could be applied in the search for additional and innovative solutions.

7.2 LATERAL MAGNETOTRANSISTORS

The first bipolar transistor to be sensitive to magnetic field was reported in 1949 [7], only two years after the discovery of the bipolar transistor. Later, when the bipolar transistor with two collectors was proposed as a device for magnetic field sensing [8], the foundation for use of bipolar transistor as a magnetic field sensor was laid. The expression *magnetotransistor* was coined for the first time for the lateral bipolar transistor with two collectors [9]. Today this name is commonly used for all bipolar transistors designed as magnetic field sensors.

Magnetotransistors can be divided into two major groups: vertical magneto-transistors [10,11] that depend on the vertically flowing carriers for their magnetic operation, and lateral magnetotransistors (LMTs) [12–16] that depend on the lateral flow of carriers for magnetic response. In both cases, when the magnetic field is applied to the magnetotransistor, the Lorentz force acts on flowing carriers, causing deflection. Depending on the magnetotransistor structure, the deflection of the carriers may further create a Hall voltage along the emitter-base junction and in turn, cause emitter injection modulation [17] or asymmetrical concentration of carriers within the device, causing a local modulation of conductivity called *magnetocon-centration* [13]. If the operation of the magnetotransistor is based solely on carrier deflection, the magnetotransistors exhibit linear response to the magnetic field [16,18–21]. On the contrary, magnetoconcentration and emitter injection modulation involve a nonlinear magnetic response. The exact analysis of the magnetotransistors is complex. Some of the examples of a simulation of magnetotransistors can be found in Chapter 2 of this book.

This chapter, as has already been pointed out, will focus on LMT. LMT can often be designed in such a way as to explore only carrier deflection or to have carrier deflection as the dominant mechanism of operation. This is the reason why the concise description of the basic quantities related to the deflection mechanism is included in this section.

7.2.1 Combined Action of an Electric Field and a Magnetic Field

The fundamental quantities used to describe a magnetic field are the magnetic field strength, $\mathbf{H}$, the magnetization, $\mathbf{M}$, and the magnetic induction (or magnetic flux density), $\mathbf{B}$. All of these are vector quantities determined by magnitude and direction. The relationship among these quantities is described by the equation

$$\mathbf{B} = \mu_0(\mathbf{H} + \mathbf{M}) \tag{7.1}$$

where μ_0 is the permeability of vacuum. In nonmagnetic material such as silicon the magnetization $\mathbf{M}$ is 0 and (7.1) is simplified to

$$\mathbf{B} = \mu_0\mathbf{H} \tag{7.2}$$

$\mathbf{B}$ is an important quantity when it comes to the combined action of an electric and magnetic field on moving carriers. When a charge carrier in a semiconductor moves with a velocity ϑ in the combined electric and magnetic field, it will be affected by both, which can be expressed as

$$m(d\vartheta/dt) + m(\vartheta/\tau) = q\mathbf{E} + q(\vartheta \times \mathbf{B}) \tag{7.3}$$

where m is the effective mass of the carrier, ϑ is the velocity vector of the carrier, τ is the mean collision time, q is the charge of the carrier, $\mathbf{E}$ is the electric field, and $\mathbf{B}$ as already pointed out is the magnetic induction. The first and second terms on the left side of (7.3) represent a rate of change of momentum and a damping force (the consequence of a collision of the carrier with the lattice), respectively. The first term on the right side of (7.3) represents a Coulomb force ($q\mathbf{E}$) and the second term is a Lorentz force ($q\vartheta \times \mathbf{B}$). The Lorentz force is dependent on $\mathbf{B}$ and ϑ vectors and causes deflection of the carriers from the direction of the electric field by the Hall angle, θ_H. The Hall angle can be expressed as

$$\tan \theta_{Hn} = \mu_H nB \text{ (electrons)} \tag{7.4a}$$

$$\tan \theta_{Hn} = \mu_H pB \text{ (holes)} \tag{7.4b}$$

where μ_{Hn} and μ_{Hp} represent Hall mobility for electrons and holes respectively. Equation (7.4) suggests that the deflection is proportional to the magnitude of **B**. Also, the deflection is proportional to the Hall mobility. Therefore, it is necessary to know the Hall mobility. The Hall mobility is directly proportional to the mobility of carriers and can be expressed as

$$\mu_{Hn} = r_n \mu_n \tag{7.5a}$$

$$\mu_{Hp} = r_p \mu_p \tag{7.5b}$$

where r_n and r_p are Hall factors for electrons and holes, respectively, and μ_n and μ_p are the mobility of electrons and holes, respectively. The values of r_n and r_p are dependent on the material but typically they are the same order of magnitude (close to 1). From this, we can draw another conclusion: electrons are more suitable for detection of a magnetic field because of higher mobility. For this reason, the majority of LMTs are designed as *npn* transistors.

7.2.2 Lateral Magnetotransistor in CMOS Technology

Today, CMOS technology has emerged as a main contender for IC manufacturing because of its low power consumption and its ability to meet the requirements of both digital and analog applications. Therefore, it is no surprise that CMOS is used extensively for the fabrication of LMTs. The basic structure of LMT is shown in Figure 7.1 [16]. The device is placed in a *p*-well, which serves as a base region. The base contacts, B_1 and B_2, are used to bias the device and form a lateral electric field. Two n^+ regions serve as the emitter and collector of LMT. These two regions

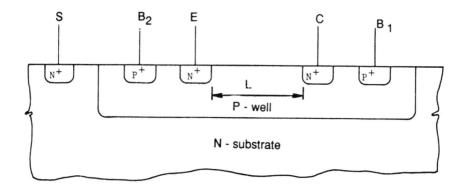

Figure 7.1 Lateral *n-p-n* magnetotransistor in CMOS technology, where L is the width of the lateral base.

are separated by the lateral base length L. The p-well–n-substrate junction, which is always reversed biased (to isolate the device), serves as a parasitic collector. Let us analyze the operation of the device in the forward active regime—this means that the emitter-base junction is forward biased and the collector-base junction is reverse biased. The electrons injected into the base travel in vertical and lateral directions, and they are collected by the collector or substrate. If the potential applied to B_1 is higher than the potential at B_2 the lateral electric field will help the drift of the injected electrons toward the collector. Let us assume that the collector current in the absence of a magnetic field is I_{CO}. If a magnetic field is applied perpendicular to the plane of Figure 7.1 (denoted **B**), it will cause a change of the collector current. That is, the velocity vector of carriers ϑ and magnetic induction **B** are perpendicular, which will deflect electrons toward the surface, increasing the collector current. When the direction of the magnetic induction is reversed, the electrons are deflected toward the substrate junction (parasitic collector). Consequently, fewer electrons contribute to collector current, which will decrease. The change of the collector current is directly related to the magnitude of **B** and its direction. A figure of merit used to describe the response of a device is called *relative sensitivity*. For the device shown in Figure 7.1 the relative sensitivity can be defined as

$$Sr = \frac{\Delta I_c}{I_{CO}} \times \frac{1}{B} \tag{7.6}$$

where $\Delta I_C = I_C - I_{CO}$, I_C is the collector current in the presence of B and I_{CO} is the collector current when $B = 0$. The experimental results show that relative sensitivity is dependent on biasing conditions and can be as high as 1.5T^{-1} or $150\%/\text{T}$. The results also show [16] that the relative sensitivity is directly proportional to the base length L.

7.2.3 Suppressed Sidewall Injection Magnetotransistor in CMOS Technology

It has already been mentioned that LMT can be designed as a device with two collectors. One version of such a device is a suppressed sidewall injection magnetotransistor (SSIMT) [18–20]. A number of important features of the SSIMT stand out: the linear response, high sensitivity, the possibility of working as a magnetic switch, and offset elimination. Analysis of this device can serve as good example of what is involved in the design of an LMT.

The SSIMT structure, shown in Figure 7.2, is fabricated in a standard CMOS process. The base region is formed by a p well. The emitter and both collectors are realized using the standard doping procedure for the source and drain of a n-channel MOS transistor. The p^+ stripes at the side of the emitter and the base contacts B_1

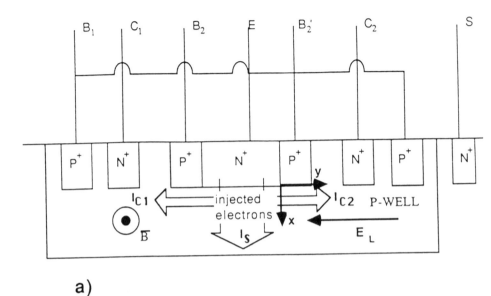

a)

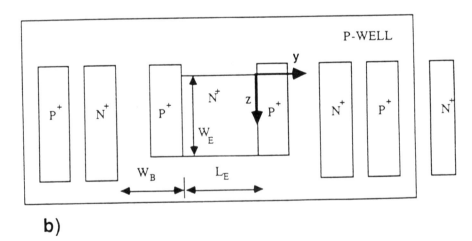

b)

Figure 7.2 SSIMT: (a) cross-section of the device: I_{C1} and I_{C2} are the collector currents, I_S is the substrate current; (b) top view, $L_E = 100~\mu m$, $W_E = 50~\mu m$, and $W_B = 8~\mu m$.

were formed using the standard doping procedure for a p-channel MOS transistor. The characteristic mask dimensions are also shown in Figure 7.2.

The specific feature of the SSIMT is the two p^+ stripes placed along the edges of the emitter parallel to the collectors. The stripes play a twofold role. First, they suppress carrier injection from the emitter in the lateral direction toward the collectors. If the p^+ stripes are biased with a voltage V_r and V_r is less than or equal to the potential of the emitter V_E; this effect is obvious. The suppression of laterally flowing electrons is a consequence of the reverse biasing of the junction between the p^+ stripe and the emitter; in fact, a small strip at the bottom of the emitter next to each stripe will also be reverse biased. The width of this reverse-biased strip will increase with increasingly negative V_r. The breakdown voltage of the n^+ emitter–p^+ stripe junction limits the magnitude of V_r. Carrier injection from the emitter is also confined to the vertical direction due to the formation of a potential hill (for electrons) around the p^+ stripes. The second effect of biasing the p^+ stripes is the creation of a lateral electric field E_l in the neutral base region. This field is due to the difference in potential between the base contacts B_1 and p^+ stripes. When the device is operated normally with $V_r \leq V_E < V_{b1}$ (V_{b1} is the potential of B_1), this field will be oriented in such a way as to sweep the injected electrons laterally toward the collectors (Figure 7.2(a)).

The biasing circuit of the SSIMT is shown in Figure 7.3. The values of I_b, V_s, V_C, V_{r1}, and V_{r2} determine the operating point of the device. In normal operation slightly different values for V_{r1} and V_{r2} are used for elimination of the offset of the collector current, and we define $V_r \equiv (V_{r1} + V_{r2})/2$ for convenience.

The device operation can be described in the following way. If the emitter-base junction is forward biased and electrons are injected from the emitter into the base region, the net effect of the p^+ stripes, when biased with a small negative voltage, is to force the minority electron current down toward the substrate. It is then split into three distinct current flows. Two of these flows, I_{C1} and I_{C2}, are collected by the collectors and the third, I_s, by the substrate. The application of a larger negative potential to the p^+ stripes amplifies the effect of the p^+ stripes, pushing the minority current further into the device and then establishing a lateral flow of current out to the collectors. In the absence of a magnetic field, the collector currents I_{C10} and I_{C20} are equal because of the device's symmetry, and the device is balanced.

The application of a magnetic field $B = B_z$ parallel to both the chip surface and the collectors will produce an imbalance in the two collector currents I_{C1} and I_{C2} due to the following "double-deflection" effect. The Lorentz force acts on all three current components: I_s, I_{C1}, and I_{C2}. The I_s component is deflected in the y direction, increasing I_{C2} and decreasing I_{C1}. Moreover, the current components I_{C1} and I_{C2} are deflected in the $+x$ and $-x$ directions, respectively, causing a further increase of I_{C2} and a decrease of I_{C1}. These two deflections will collaborate to cause one collector current to increase at the expense of the substrate current and the other to decrease

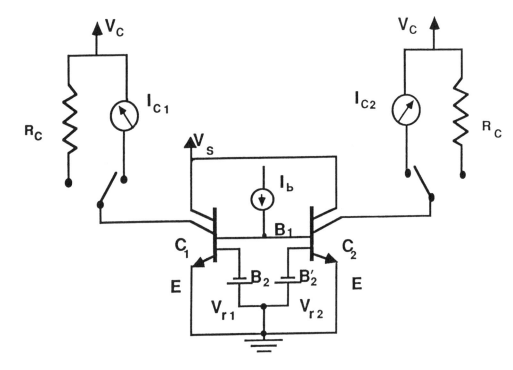

Figure 7.3 Biasing circuit for the SSIMT: I_B is the base current, I_{C1} and I_{C2} are the collector currents, V_S is the substrate potential, V_C is the collector supply voltage, V_{r1} and V_{r2} are p^+ stripe potentials.

with a corresponding gain in the substrate current. The net effect on the substrate current should be 0.

An analytical model of the SSIMT can be developed based on the assumption that a magnetic flux density, B, causes a linear displacement of the minority carriers in the neutral base region. This deflection of the minority carriers is due to the action of the Lorentz force. To obtain the magnetic response of the SSIMT one can consider the action of the magnetic field on only one-half of the device. The flow of the minority carriers within half of the device is shown in a cross section of right half of the structure; see Figure 7.4. Let us analyze the part of the electron flow that contributes to the collector current, see Figure 7.4(b). For simplicity we assume that this flow of electrons from the emitter to the collector consists of a current tube of constant cross-sectional area with a uniform current density and that only the carriers that are inside this region will contribute to the collector current. This current flow can be modeled as first a vertical flow and then a lateral flow, see Figure 7.4(c). The vertical flow is a consequence of the negative potential applied to the p^+ stripes, and the lateral flow is due to the strong lateral electric field in the neutral base region.

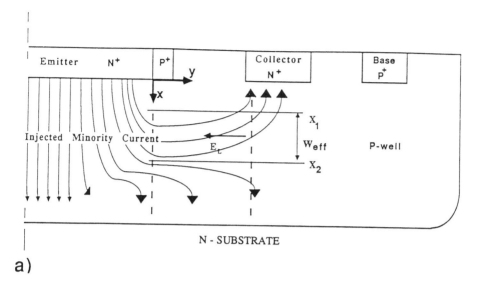

a)

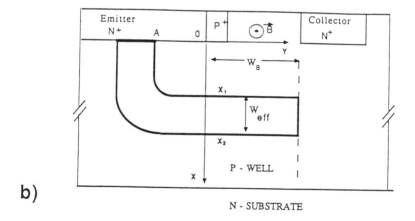

b)

Figure 7.4 Right-side cross section of the SSIMT: (a) minority current flow, (b) minority carriers that contribute to the collector current, (c) L-shaped model of the collector current, (d) vertical and horizontal box as equivalent for L-shaped region.

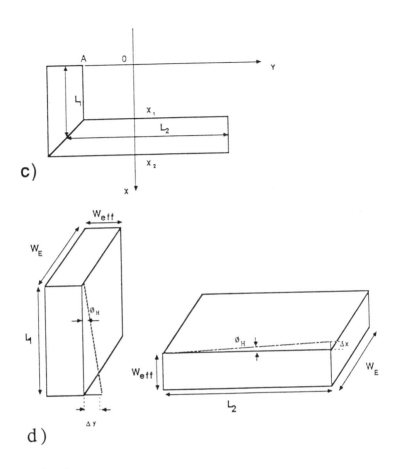

Figure 7.4 continued.

To facilitate the analysis of the deflection of the carriers, the L-shaped region can be further broken down into two boxes, one vertical and the other horizontal, see Figure 7.4(d). The corresponding current densities within these two boxes are J_{nx} and J_{ny} for the vertical and lateral directions, respectively. It remains to determine the boundaries of these boxes. The top and bottom edges of the horizontal box are divided by two planes, $x = X_1$ and $x = X_2$. The first plane, $x = X_1$, is defined to provide a means to analyze the effect of the negative potential V_r applied to the p^+ stripes. The negative potential applied to the stripes will block the injection of electrons from the emitter into the p^+ stripes and also the injection across a small region of the bottom of the emitter next to each stripe. This will prevent movement of the injected electrons laterally, forcing the minority current to flow down into the device. The plane $x = X_1$ is positioned so that a majority of the laterally flowing electron current will flow below this plane. One can also assume the $0A = X_1$ (penetration of the negative potential is equal in all directions). The plane $x = X_2$ is used to analyze the effect of the substrate potential V_S on the electron flow. The electron current above this plane is presumed to be collected by the collector; conversely, the electron current below this plane is collected by the substrate. According to this model all of the electron current flow contributing to the collector current will flow in the region

$$X_1 \leq x \leq X_2 \tag{7.7}$$

The distance between these two planes X_1 and X_2 is defined as $W_{eff} = X_2 - X_1$ and can be thought of as the effective width of the stream of laterally flowing electrons. We assume that all the electrons flowing in the region defined by (7.7) that reach the plane $y = W_B$ are collected by the collector and form the collector current.

The length of the vertical box L_1 can be expressed in terms of the two planes previously defined as

$$L_1 = X_1 + \frac{W_{eff}}{2} \tag{7.8}$$

With the assumption given above, L_2 can be expressed as

$$L_2 = W_B + X_{1+} \frac{W_{eff}}{2} \tag{7.9}$$

or

$$L_1 = \frac{X_1 + X_2}{2} \tag{7.10a}$$

$$L_2 = W_B + L_1 \tag{7.10b}$$

The total current collected by the right collector at a zero magnetic field I_{C20} can be expressed in terms of the current density of laterally flowing electrons J_{ny}. If we presume that a uniform current distribution exists in the region between the two planes X_1 and X_2, then we have

$$I_{C20} = J_{ny} W_{eff} W_E \tag{7.11}$$

with W_E denoting the emitter width.

To obtain the current change ΔI_{C2} due to a magnetic field B, we have to determine the effect of the magnetic field on both the vertical and lateral electron flows. Both flows will be deflected by Hall angle θ_H, which is defined by $\tan(\theta_H) = \mu_H B$, where μ_H is the electron Hall mobility. This is shown in Figure 7.4(d). The change in the collector current I_{C2} due to the deflection of the vertical component can be calculated by integrating the current density J_{nx} in the plane $x = L_1$ over the area defined by ΔY and the emitter width W_E. For the case of a uniform current distribution J_{nx}, we obtain

$$\Delta I_{C2V} = J_{nx} W_E \Delta Y \tag{7.12}$$

where

$$\Delta Y = L_1 \mu_H B \tag{7.13}$$

Using (7.12) and (7.13) we obtain

$$\Delta I_{C2V} = J_{nx} W_E L_1 \mu_H B \tag{7.14}$$

Here, ΔI_{C2V} is an increase in the collector current due to electrons being deflected that would otherwise be part of the substrate current.

The change in the collector current I_{C2} due to the deflection of the lateral component can be calculated by integrating the current density J_{ny} in the plane $y = W_B$ over the area defined by ΔX and the emitter length W_E. If we assume, as before, a uniform current distribution in the area of integration, we have

$$\Delta I_{C21} = J_{ny} W_E \Delta X \tag{7.15}$$

where

$$\Delta X = L_2 \mu_H B \tag{7.16}$$

Substituting (7.16) into (7.15) we obtain

$$\Delta I_{C21} = J_{ny} W_E L_2 \mu_H B \tag{7.17}$$

The electrons contributing to ΔI_{C21} would also, without the deflection, be a part of the substrate current.

As this analysis presumes a continuous current flow of constant cross-sectional area, and we have assumed uniform current density in both boxes; the current density in both boxes must be equal, $J_{nx} = J_{ny}$. Using (7.11) to express J_{ny} as a function of I_{C20}, we write (7.14) and (7.17) as

$$\Delta I_{C2V} = \mu_H \frac{L_1}{W_{eff}} I_{C20} B \qquad (7.18)$$

$$\Delta I_{C21} = \mu_H \frac{L_2}{W_{eff}} I_{C20} B \qquad (7.19)$$

The total current change is then the sum of ΔI_{C2V} and ΔI_{C21};

$$\Delta I_{C2} = \mu_H \frac{L_1 + L_2}{W_{eff}} I_{C20} B \qquad (7.20)$$

Defining the relative sensitivity as

$$S_r = \frac{1}{I_{C20}} \times \frac{\Delta I_{C2}}{B} \qquad (7.21)$$

one can get

$$S_r = \mu_H \frac{L_1 + L_2}{W_{eff}} \qquad (W_{eff} > 0) \qquad (7.22)$$

This result for the magnetic response of the collector currents and the relative sensitivity was derived for current flow of half of the device. In the actual device there are two collector currents, I_{C1} and I_{C2}, and the relative sensitivity is defined as

$$S_r = \frac{1}{I_{c0}} \times \frac{\Delta I_C}{B} \qquad (7.23)$$

where $\Delta I_C = \Delta I_{C2} - \Delta I_{C1}$ and $I_{C0} = I_{C10} + I_{C20}$. The structure is symmetrical, and we can assume $\Delta I_{C1} = -\Delta I_{C2}$ and $I_{C10} = I_{C20}$. The final expression for sensitivity is therefore unaffected as both the current change and the total current are increased by a factor of 2.

If we express the collector currents I_{C1} and I_{C2} as a function of the magnetic field in terms of the zero field current and ΔI_{C1} and ΔI_{C2}, we get

$$I_{C1} = I_{C10} + \Delta I_{C1} \qquad (7.24)$$

$$I_{C2} = I_{C20} + \Delta I_{C2} \qquad (7.25)$$

Using $\Delta I_{C1} = -\Delta I_{C2}$ we have

$$I_{C1} = I_{C10}\left(1 - \mu_H \frac{L_1 + L_2}{W_{\text{eff}}} B\right) \qquad (W_{\text{eff}} > 0) \qquad (7.26)$$

$$I_{C2} = I_{C20}\left(1 - \mu_H \frac{L_1 + L_2}{W_{\text{eff}}} B\right) \qquad (W_{\text{eff}} > 0) \qquad (7.27)$$

The last two equations suggest that both collector currents are linear functions of the magnetic induction and can be expressed as

$$I_{C1} = I_{C10}(1 - S_r B) \qquad (7.28)$$

$$I_{C2} = I_{C20}(1 + S_r B) \qquad (7.29)$$

where S_r is defined by (7.22). From (7.28) and (7.29) it is clear that the higher is the sensitivity, the higher is the change in the collector currents. It should also be noted that (7.22) can be used as a guideline for the design, as well as to set the operating point for the device. To obtain higher relative sensitivity L_2 should be designed larger and W_{eff} should be made smaller. Because W_{eff} is a function of V_r, V_s, and I_B [20], S_r could also be set by choosing biasing conditions of the device.

In addition to relative sensitivity sometimes the absolute sensitivity S_a is used to characterize the device. The absolute sensitivity is defined as

$$S_a = \frac{\Delta V}{B} \qquad (7.30)$$

where ΔV is expressed as

$$\Delta V = (I_{C2} - I_{C1})R_C \qquad (7.31)$$

and R_C is the load resistivity.

For characterization of the sensor one can use the circuit shown in Figure 7.3. The circuit can operate in two configurations. Either the collector currents I_{C1} and I_{C2} can be measured directly and ΔI_C calculated, or the collector load resistors can be placed between the supply voltage and the collectors and the potential difference between the two collectors measured, $\Delta V = V_{C1} - V_{C2}$. The two stripe potentials V_{r1} and V_{r2} are adjusted to make $I_{C10} \approx I_{C20}$. V_r is defined as $(V_{r1} + V_{r2})/2$ and used as a representative value for the stripe potential.

In Figure 7.5, the response of $\Delta I_C = I_{C2} - I_{C1}$ to a magnetic field is shown. This figure clearly shows the symmetrical nature of the SSIMT, with the measured

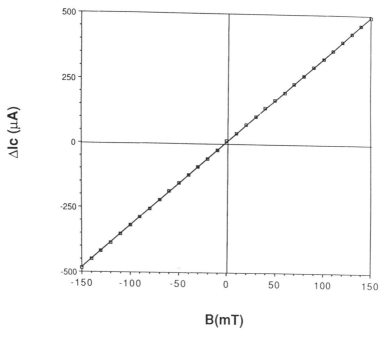

Figure 7.5 I_C as a function of $B(I_C = I_{C2} - I_{C1})$; $I_B = 8.5$ mA, $V_C = 5$ V, $V_S = 5$ V, $V_r = -250$ mV; $I_{C10} = 801$ μA, $I_{C20} = 795$ μA.

value of ΔI_C switching sign as the magnetic field is reversed from the positive z direction to the negative.

It has already been stated that S_r is a function of biasing conditions. Figure 7.6 shows dependence of S_r as a function of V_r. This result clearly shows the crucial role of V_r for attainment of high sensitivity. V_{rc} represents a critical value of V_r when W_{eff} is reduced to 0. Similar results are obtained for S_r as a function of V_s and I_B [20]. Overall, this device offers a great choice of relative sensitivities, from 50 to 3000%/T. The combination of a linear response and the great range of sensitivities makes attractive the magnetic switching applications of the SSIMT. That is, by setting appropriate operating conditions it is possible to choose at what critical magnetic field B_C one of the collector currents is driven to almost 0. The range of B_C can be set from 30 mT to over 1 T [20].

Very often the circuitry supporting an MFS uses a voltage signal produced on two load resistors rather than measure the current change directly. The absolute sensitivity S_a defined by (7.30) serves as a parameter to characterize the device for this situation. It is apparent from the definition of S_a that ΔV is proportional to ΔI_C and

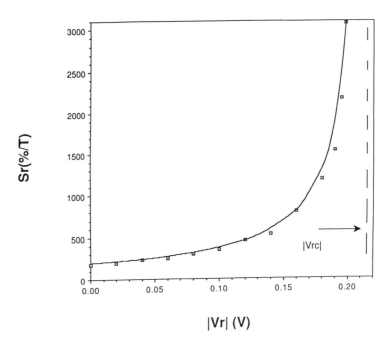

Figure 7.6 *Sr* as a function of *Vr*: $I_B = 7$ mA, $Vc = 5$ V, $Vs = 5$ V, $B = 30$ mT.

R_C. This, of course, implies that the absolute sensitivity can be chosen by an appropriate choice of load resistance. For magnetic sensors displaying small relative sensitivity, R_C is sometimes chosen to be on the order of megaohms. However, when a sensor provides a much larger current change, R_C can be much smaller. Figure 7.7 shows S_a as a function of *Vr*. It can be seen that the absolute sensitivity increases with an increase in the negative potential of V_r. This can easily be understood when S_a is expressed in terms of the relative sensitivity using (7.28) and (7.29)

$$S_a = S_r I_{C0} R_C \qquad (7.32)$$

With the SSIMT structure one can obtain very high absolute sensitivity. Values of up to 9 V/T, at fields of 30 mT and with $R_C = 500\Omega$ are reported. This is a very high absolute sensitivity for a magnetotransistor using load resistors this small.

Finally, it should also be pointed out that the SSIMT can operate in the floating mode (with p^+ stripes floating). The results show that even then SSIMT is more sensitive than the regular LMT [22,23].

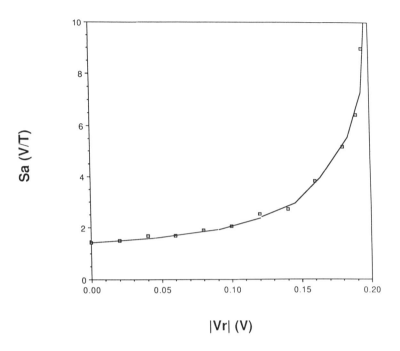

$|Vr|$ (V)

Figure 7.7 *Sa* as a function of *Vr*: $I_B = 7$ mA, $Vc = 5$ V, $Vs = 5$ V, $Rc = 500$ Ω, $B = 30$ mT.

7.2.4 Supressed Sidewall Injection Magnetotransistor in Bipolar Technology

Magnetotransistors are usually designed as *npn* devices because the higher mobility of electrons makes it easier to achieve higher sensitivity. However, some practical analog applications require complementary pairs of *pnp* and *npn* transistors. That is why it is worth having *pnp* LMT.

The *pnp* version of SSIMT can be fabricated in bipolar technology [24,25]. As shown in Figure 7.8, it is possible to design two slightly different structures. One is with an n^+ buried layer, which is a standard feature of bipolar technology, and the other is without the n^+ buried layer. The *n* epitaxial layer serves as the base region, and the *p* diffusion (which normally serves as the base region of a standard vertical *npn* transistor) is used to make the emitter (E) and the collectors (C_1 and C_2). An n^+ diffusion is used to make the base contacts (B_1), and n^+ stripes (B_2).

The device is composed of a differential lateral transistor (E, B, C_1 and C_2) and a parasitic vertical transistor (E, B, and substrate). In the forward active regime,

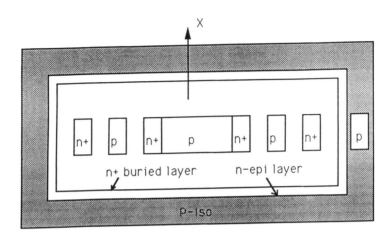

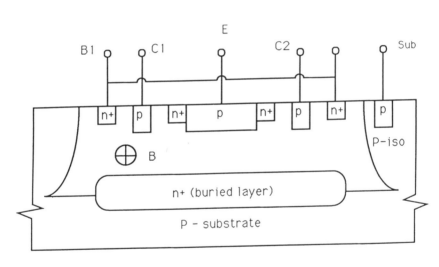

Figure 7.8 Lateral *pnp* magnetotransistor with n^+ buried layer (device *A*).

holes are injected from the emitter into the base, and then split into two lateral components collected by C_1 and C_2 and one vertical component collected by the substrate. A constant base current I_B is supplied to the base contact, B_1.

Experimental results show that the change of the collector currents with magnetic field is linear, Figure 7.9. This is in agreement with the results in [20]. Furthermore, the response of device *B* is much higher than that of device *A*. The maximum relative sensitivity of LMT, with the n^+ buried layer (device *A*) is

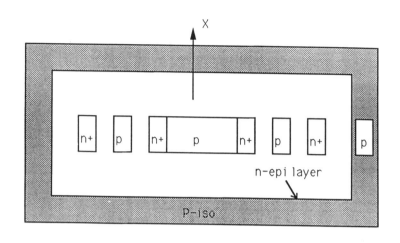

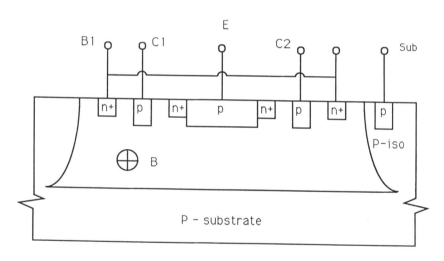

Figure 7.9 Lateral *pnp* magnetotransistor without n^+ buried layer (device *B*).

12%/T, and without n^+ buried layer (device *B*) is 135%/T. Although 135%/T is among the highest sensitivity achieved for *pnp* LMT, it is still smaller than the sensitivity of *npn* LMT [18–20]. The reason for this is a smaller mobility of holes compared to electrons.

As for the role of the n^+ buried layer, the absence of the buried layer is crucial to achieve high sensitivity—an increase of more than 10 times is obtained. The lateral transistor, which has less favorable standard electrical characteristics (device

B) [25], shows higher magnetic response. The reason for this can be explained as follows. "Shaping" the flow of carriers in the neutral base region is essential for high sensitivity, as already explained for *npn* SSIMT. This "shaping" is partially controlled by the substrate potential. The substrate potential for reverse biasing of *n*-epi–*p*-substrate junction increases the depletion region of this junction, which in turn reduces the thickness of the neutral base region and so increases the sensitivity. This behavior of the device without the buried layer is identical to the original structure in CMOS technology [20]. By contrast, the reverse biasing of the n^+-buried layer–*p*-substrate junction does not produce the same effect because of the high doping of the n^+ buried layer and negligible change of the depletion region on the side of n^+ buried layer.

7.2.5 Offset in Magnetotransistors

An important aspect of the performance of magnetic field sensors is resolution; that is, the minimum field intensity that can be detected. One cause of limited resolution is offset: the signal at the output of the sensor at zero B. This type of error is particularly critical when a static magnetic field is measured because the offset signal of the output is indistinguishable from the useful signal. Hence, consideration of the offset is often one of the crucial problems for designers of systems involving sensors.

The main causes of offset are imperfections in the process technology, including mask misalignment [26] and strain introduced by packaging and aging [27]. Hence, one of the approaches for offset reduction is based on the improvement of process technology, but because of technological limitations it is not possible to eliminate offset completely. Other offset-reduction methods are calibration [28], which demands the presence of a known value of the measurand; compensation [29], which demands the use of two sensors where one is used as an offset reference; and the sensitivity-variation offset-reduction method [30]. Each approach has advantages and disadvantages, and the method chosen depends on the type of magnetic field sensor and the application.

The SSIMT offers an elegant solution for the offset problem [31]. In view of the confinement of the carrier injection from the emitter to the bottom center of the emitter-base junction, there is the possibility of shifting the center of injection toward one of the collectors. If we focus on SSIMT in CMOS technology this can be done by applying different negative potentials to each p^+ stripe, B_2 and $B_{2'}$ (Figure 7.10). This procedure allows adjustment of both collector currents until they are equal and thus annuls the offset.

To check the behavior of the collector currents with respect to the negative potential applied to the p^+ stripes, an investigation can be made for a set of different values of the base current, I_B. The different values for V_{r1} and V_{r2} are used for offset

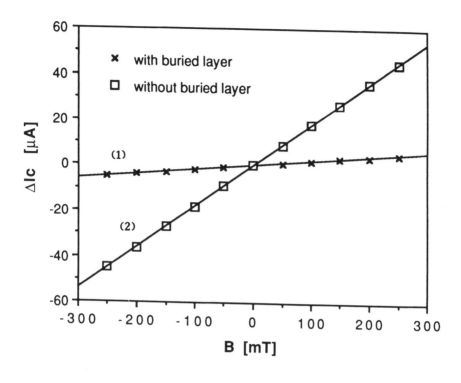

Figure 7.10 I_C as a function of B: $V_C = -5$ V, $V_S = -5$ V, $I_B = -5$ mA; (1) device A, (2) device B.

elimination, and we define $V_r = (V_{r1} + V_{r2})/2$ as a representative value for stripe potential. This is done because for the given I_B, each particular value of $I_{C10} = I_{C20}$ is determined by one unique set of V_{r1} and V_{r2}.

The offset investigation can be done with respect to the potential difference, $\Delta V_r = |V_{r2} - V_{r1}|$, used for offset elimination (V_{r1} is applied to one stripe and V_{r2} to the other in such a way that I_{C10} is adjusted to equal I_{C20}). The average value of V_{r1} and V_{r2} ($V_r = (V_{r1} + V_{r2})/2$) can be used as representative for both sources. The results, presented in Figure 7.11, show the required voltage difference decreasing with an increasing $|V_r|$, which suggests that it is easier to null the offset for smaller values of I_{C10} and I_{C20} than for larger currents. It can also be seen from Figure 7.11 that the smaller is the supplying base current, the smaller is the difference, ΔV_r, needed to null the offset. This is also related to the values of collector currents: the smaller is the base current, the smaller the collector currents I_{C10} and I_{C20} are, and the device is more sensitive to the difference between V_{r2} and V_{r1}.

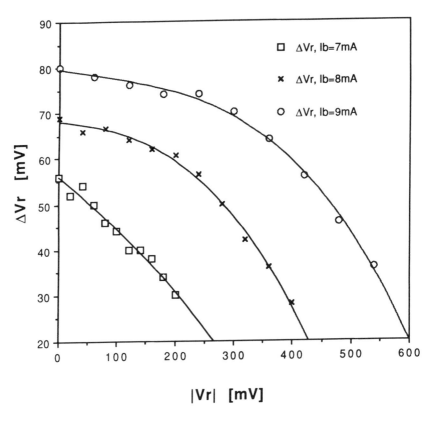

Figure 7.11 The potential difference $V_r = /V_{r2} - V_{r1}/$ (needed to achieve offset elimination, $I_{C10} = I_{C20}$) as a function of V_r: $V_C = 5$ V, $V_S = 5$ V.

7.2.6 Noise in Magnetotransistors

The characterization of noise in a device plays an important part in the optimization of sensors because noise severely limits sensor resolution. When it comes to LMT it is particularly interesting to compare $1/f$ noise in single and differential magnetotransistors and investigate the noise correlation. As a test vehicle for this investigation we use several LMTs, differing in the number of emitters and collectors [32,33]. The devices fabricated by using the CMOS process are shown in Figures 7.12–7.14. The basic idea behind these test structures is to determine the source of $1/f$ noise in LMT: Is it the emitter-base junction, the base region, or the collector-base junction? The single lateral magnetotransistor is shown in Figure 7.12, whereas the two differential magnetotransistors are shown in Figures 7.13 and 7.14. The differential device in Figure 7.13 has two emitters, and the one in Figure 7.14 has only one

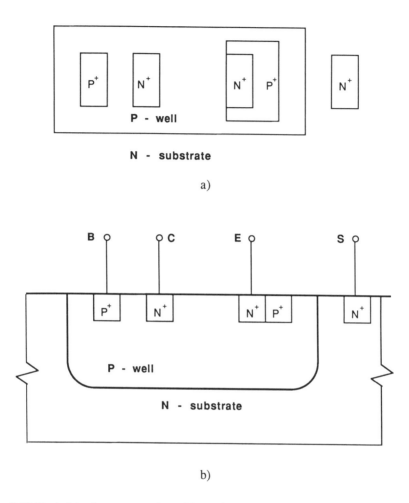

Figure 7.12 Single lateral magnetotransistor: (a) top view; (b) cross section.

emitter. All collectors and emitters are realized with n^+ regions. The p well serves as a base in all three cases. The p^+ regions are used to make the base contacts. The device shown in Figure 7.13 is just a differential version of the device shown in Figure 7.12. That is, two completely independent lateral magnetotransistors (two separate p wells) are placed next to each other to make a differential lateral magnetotransistor. The emitters in Figure 7.12 and 7.13 are surrounded by p^+ regions on three sides to create lateral flows of injected carriers toward the collectors only. The device shown in Figure 7.14 is a differential lateral magnetotransistor having only one common emitter.

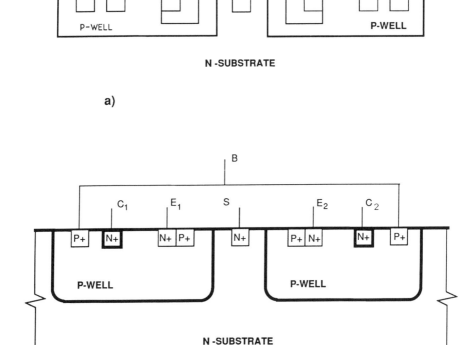

Figure 7.13 Two lateral magnetotransistors making a differential pair: (a) top view; (b) cross section.

The operation of all three devices is similar and based on the lateral flow of the injected carriers from the emitter, as already explained in Section 7.2.2. In differential structures, the magnetic field will change both collector currents at the same time in such a way that one increases and the other one decreases (because the velocity of the carriers in lateral directions is opposite). Thus, the difference between two collector currents gives information about the strength of the applied magnetic field.

To characterize the noise one can use power spectral density (PSD) of the voltage fluctuations of the collectors when LMT is biased. Let us designate the PSD

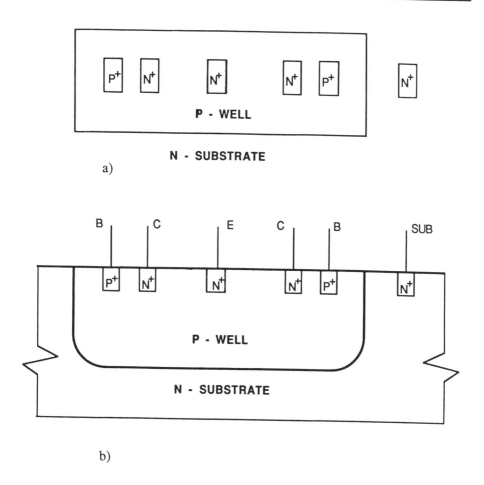

Figure 7.14 Differential magnetotransistor with a common emitter: (a) top view; (b) cross section.

of voltage fluctuations at a collector of a single lateral magnetotransistor as $S_{11}(w)$, and the PSDs of voltage fluctuations at the collectors of a differential lateral magnetotransistor as $S_{11}(w)$ and $S_{22}(w)$. Measurements by spectrum analyzer show that $S_{11}(w)$ and $S_{22}(w)$ are identical. Therefore, for convenience, $S_{11}(w)$ and $S_{22}(w)$ are denoted as $S_{XX}(w)$. We also designate the PSD of differential collector noise as $S_D(w)$. This value can be expressed as

$$S_D(w) = 2S_{XX}(w) - 2\,\text{Re}[S_{XY}(w)] \qquad (7.33)$$

where $\text{Re}[S_{XY}(w)]$ represents the real part of the cross-spectral density. Measuring the $S_{XX}(w)$ and $S_D(w)$, we can find the normalized cross-spectral density as

$$\rho = \frac{\text{Re}[S_{XX}(w)]}{S_{XX}(w)} = 1 - \frac{S_D(w)}{2S_{XX}(w)} \tag{7.34}$$

From this result, assuming the function is even and applying the inverse Fourier transformation, one can find the mutual time average correlation function between the two collector noise voltages.

The estimation of the correlation can be done also by approximating the rms noise voltages from an oscilloscope, using the approach described in [34]. In that case the correlation coefficient γ can be expressed as

$$\gamma = 1 - \frac{\langle v^2 nd \rangle}{2\langle v^2 n1 \rangle} \tag{7.35}$$

where $v_{nd}(t)$ is the differential noise voltage, and $v_{n1}(t)$ and $v_{n2}(t)$ are the corresponding noise voltages at each collector. Equation (7.35) is obtained by using $\langle v_{n1}^2 \rangle = \langle v_{n2}^2 \rangle$.

The comparative results are presented in Figures 7.15 and 7.16. Figure 7.15 shows the noise voltage power spectral density of the single ended and the differ-

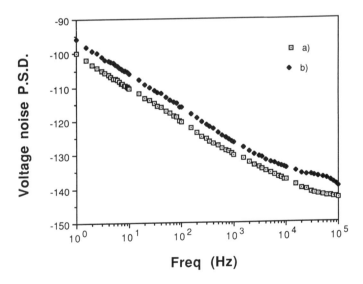

Figure 7.15 Noise voltage power spectral density for structure shown in Figure 7.13: (a) single lateral magnetotransistor; (b) differential approach (two collectors).

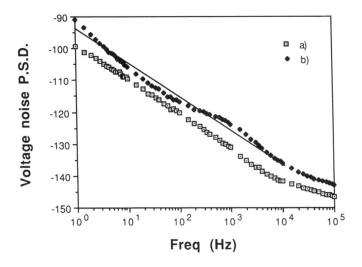

Figure 7.16 Noise voltage power spectral density for device shown in Figure 7.14: (a) single collector (one side of the device); (b) differential approach (both collectors).

ential lateral magnetotransistor shown in Figure 7.13. The graph of the noise voltage power spectral density versus frequency for the differential structure shown in Figure 7.14 is presented in Figure 7.16. In both cases the noise voltage power spectral density shows an approximate $1/f$ spectrum. Further, the differential noise voltage is higher than the noise from the single lateral magnetotransistor. The correlation coefficient is close to 0. In other words, there is no positive correlation between the noise voltages in differential lateral magnetotransistors. However, results have been published for vertical magnetotransistors that show quite the opposite [35,36]; that is, the differential noise voltage is an order of magnitude smaller than the noise of a single collector counterpart, and the correlation coefficient is close to 1. Why is there such a difference?

It is worth mentioning that there is no definite answer about the collector $1/f$ noise in silicon bipolar transistors. The few attempts to prove or disprove its existence have failed [37–40]. Also some results suggest that the collector noise is just the amplified base noise source [41,42]. The results published in [35,36] identify the emitter-base junction as a source for $1/f$ noise. If the speculations that $1/f$ noise is governed by events localized at the emitter-base junction vicinity are correct, then the differential noise for the structure shown in Figure 7.14 should be lower than the differential noise for structure shown in Figure 7.13, because only one emitter exists in Figure 7.14 compared to the two emitters in Figure 7.13. That, however, is not the case. The level of the differential noise for both structures is about the same. Therefore, the higher differential noise for LMT should rather be attributed to the base region and to the collector-base junctions. In both structures, Figures 7.13 and 7.14, there are two collector-base junctions as compared to the one collector-base junction of the single lateral magnetotransistor, Figure 7.12. And in both cases, the differential noise is higher. As for the correlation, it should be pointed out that the analyzed vertical magnetotransistor structures in [35,36] are not "real" differential structures, because there is only one collector-base junction and the collector current is simply split between the two collector contacts. The strong correlation found there represents the correlation of the collector signal with itself; therefore, the "differential" noise voltage is much smaller.

To check this assumption on LMT one can use another single lateral magnetotransistor structure (Figure 7.17). The only difference between this device and the

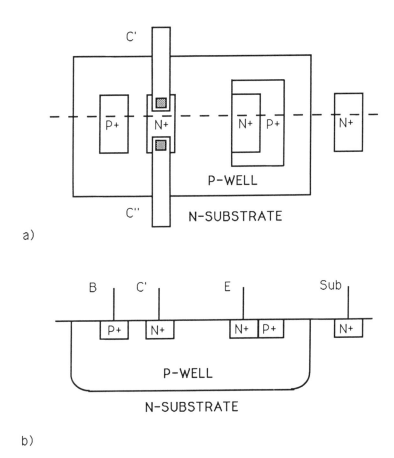

Figure 7.17 Single collector lateral magnetotransistor with split collector contact: (a) top view; (b) cross section.

one shown in Figure 7.12 is in the design of the collector contact. The device in Figure 7.17 is designed with split collector contact (C' and C'') [33]. This LMT can be tested as a single LMT (using either C' or C'') and as a "differential" LMT, as shown in Figure 7.18. It should be noted that two sets of measurement can be used for the "differential" LMT. One approach is to measure the noise signal between C' and C'', Figure 7.18(a), and the other approach is to measure noise signal between C_1 and C_2, Figure 7.18(b), in which case two external resistors, $R_C = 300\Omega$, are added. The reason for the addition of the external resistors is to investigate the role of resistance in the collector region. It is very important to emphasize that the resistance r_c on the chip (between C' and C'') is about 100Ω, which rules out a possible speculation that C' and C'' are shorted. The noise results are shown in Figure 7.19. As can be seen the "differential" noise in both cases is much lower than the noise of the single lateral magnetotransistor. It has also been found that the correlation

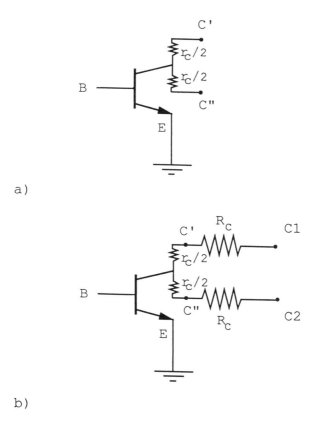

Figure 7.18 Equivalent circuit of LMT with a split collector contact for differential mode: (a) LMT as it is on the chip; (b) LMT with external resistors, $R_c = 300$ ohms.

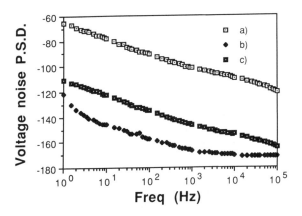

Figure 7.19 Noise voltage power spectral density for device shown in Figure 7.17: (a) single ended noise at C' or C''; (b) differential noise $C' - C''$; (c) differential noise $C_1 - C_2$.

coefficient in both cases is close to 1. Furthermore, the results in Figure 7.19 show that the noise level in "differential mode" depends on the resistivity of collector region—the higher the resistivity, the higher is the "differential" noise. This is the expected result.

In summary, the analysis of the $1/f$ noise in LMT suggests that the differential magnetotransistors should be designed in such a way to create a condition that would correlate the signals between the two terminals. This, in turn, will reduce the $1/f$ noise.

7.2.7 Surface Effects

The specific feature of LMT is a lateral flow of carriers parallel to the surface that is used for magnetic field sensing. This flow is in the vicinity of the Si/SiO_2 interface. Therefore, one might expect that the surface effects would have an influence on the sensitivity of LMT. The results presented in [43] show that this is indeed the case. The analysis of the role of surface effects in LMT was done on a simple LMT device, similar to the one in Figure 7.1. The fact that the device was manufactured in CMO technology made the investigation of surface effects possible. The inherent feature of CMOS technology is the gate, which exists on the top of the base region between the emitter and the collector, because an *npn* LMT essentially is an *n*-channel MOS transistor. The existence of the gate makes it easy to control the base surface potential, which in turn controls the rate of surface recombination. A biasing circuit for this characterization is shown in Figure 7.20. The change in the collector current is measured in the presence of a magnetic field for a whole range of V_G, from -10

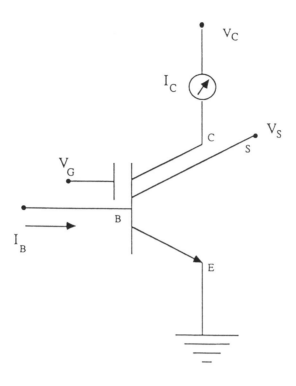

Figure 7.20 Biasing circuit of lateral magnetotransistor for the investigation of surface effects: $I_B = 0.5$ mA, $V_C = 5$ V, $V_S = 5$ V.

V to the threshold voltage and beyond. The relative sensitivity for the case when the gate is floating is designated as S_{rf}, which is used as a reference value. The relative sensitivity when the gate voltage is applied is designated as Sr_G. The relative change of sensitivity as a function of V_G is shown in Figure 7.21. It can be seen that the relative sensitivity for negative values of V_G is increased by 10% compared to the floating-gate case, and the relative sensitivity decreases dramatically for positive values of V_G. This can be explained as follows. The application of negative voltage to the gate enriches the surface of the base (accumulation condition) and pushes the electrons away from the Si/SiO_2 face. This means that the recombination-generation current is reduced, thus increasing the sensitivity. The saturation in sensitivity for higher negative values of V_G occurs probably because the recombination-generation current is small and after a few volts does not change anymore. The dramatic drop in sensitivity when going from a negative gate potential to a positive one is caused by the depletion of the surface and finally inversion (beyond the threshold voltage). The inversion shorts the base region between the emitter and the collector, which violates the basic principle of operation of the LMT.

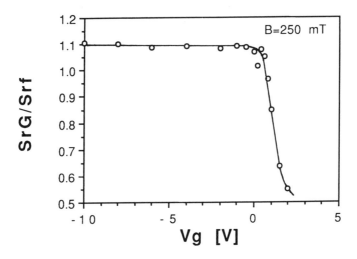

Figure 7.21 Relative change of sensitivity of lateral magnetotransistor as a function of V_G: S_{rf} is the relative sensitivity for the floating gate; S_{rG} is the relative sensitivity when V_G is applied to the gate.

Overall, the results suggest that for optimal operation of LMT one should set operating conditions to minimize recombination-generation effects at the Si/SiO$_2$ interface.

7.3 MULTIDIMENSIONAL SENSING

Magnetic field sensors are usually designed to be sensitive to the one component of a magnetic field which is either parallel to the chip surface or perpendicular to the surface of the device. However, very often there is a need to sense two components or all three components simultaneously. Several examples can be found in literature about two-dimensional sensing [11,44,45] or three-dimensional sensing [46–50]. These devices include vertical magnetotransistors, Hall devices and a combination of vertical DMOS and split-drain lateral DMOS transistors.

In this section we will focus only on the LMT structures designed for multi-dimensional sensing. The discussion will start with the simple structure that could sense components of the magnetic field either parallel or perpendicular to the chip surface, followed by some more complex structures for simultaneous detection of two and three components of a magnetic field.

7.3.1 Lateral Magnetotransistor Structure Sensitive to Magnetic Field Either Parallel or Perpendicular to the Chip Surface

The simple LMT structure of the device sensitive to the magnetic field either parallel or perpendicular to the chip surface is shown in Figure 7.22 [51,52]. It is an LMT fabricated in the CMOS process. The device is composed of a single emitter and collector situated in a p well. The emitter is surrounded by p^+ region on three sides of its perimeter to shape the lateral flow of carriers toward the collector. The collector is deliberately placed asymmetrically between the emitter and the base contacts to sense B applied either parallel or perpendicular to the chip surface. The characteristic dimensions are shown in Figure 7.22(a,b).

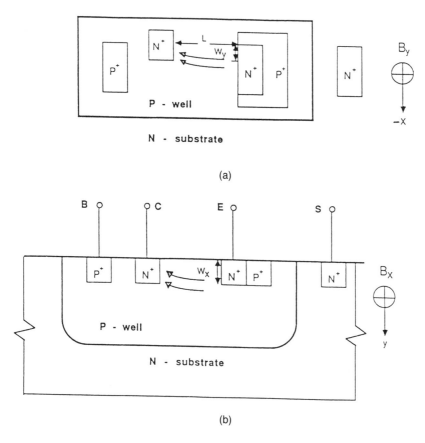

(a)

(b)

Figure 7.22 Lateral magnetotransistor sensitive to magnetic field in two different directions: (a) top view, deflection of carriers cased by B_y, $W_y = 12$ μm, L = 70 μm; (b) cross section, deflection of carriers cased by B_x, $W_x = 1.5$ μm.

The device operation is as follows. In the forward active regime the electrons are injected from the emitter into the base. The lateral component plays a crucial role in magnetotransistor action. One can distinguish two different situations: when B is applied parallel to the chip surface (Figure 7.22(a)), and when B is applied perpendicular to the chip surface (Figure 7.22(b)). When B_x, parallel to the chip surface, is applied the Lorentz force deflects the lateral carriers toward the chip surface in the $-y$ direction, thereby increasing the collector current by ΔI_{Cx}. If the direction of magnetic induction is opposite, $-B_x$, the deflection is in the opposite direction $+y$, and the collector current decreases. When B_y, perpendicular to the chip surface, is applied the Lorentz force will act again on the parallel flow of carriers, deflecting them toward the collector in the $+x$ direction, thus increasing the collector current by ΔI_{Cy}. If the direction of magnetic induction is reversed, $-B_y$, the carriers are deflected out of the collector (in the $-x$ direction) and the collector current decreases. As can be seen, the response of the device to the magnetic field perpendicular to the chip surface is possible because of the asymmetrical placement of the collector.

The change of the collector current as a function of B is shown in Figure 7.23 for both cases, when B is parallel to the chip surface (ΔI_{Cx}) and when B is perpendicular (ΔI_{Cy}). The response of the device in both cases is linear. One can estimate the sensitivity using the first-order approximation. The sensitivity of the device is directly proportional to the base length W_B and inversely proportional to the width

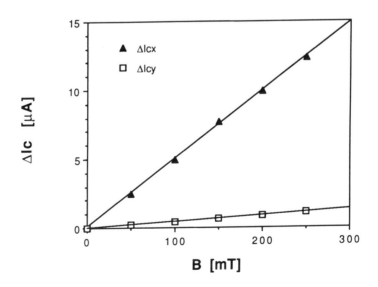

Figure 7.23 Change of the collector current as a function B: $I_{Cx} = I_C - I_{C0}$, when B_x is applied; $I_{Cy} = I_C - I_{C0}$, when B_y is applied; $V_C = 5$ V, $V_s = 5$ V, $I_B = 0.3$ mA.

of the lateral flow that creates collector current seen by the applied magnetic field [52]. The width of the lateral flow seen by B_x is determined by the depth of the n^+ region (w_x), and the width of the lateral flow seen by B_y is equal to w_y. Therefore for the given dimensions in Figure 7.22, ΔI_{Cx} should be approximately 10 times ΔI_{Cy}, which is in good agreement with experimental results shown in Figure 7.23. This structure plays an important role in the design of three-dimensional LMT, as will be shown in Section 7.3.3.

7.3.2 Two-Dimensional Sensing

The first two-dimensional magnetotransistor was designed as vertical magnetotransistor in bipolar technology having four segments of buried layer [11]. The device was sensitive to two-dimensional in-plane magnetic field. LMT can also be designed for 2-D sensing. An example of the 2-D LMT manufactured in CMOS technology [53] is shown in Figure 7.24. The structure is composed of a single emitter, a common base region, and four symmetrical collectors. The device has four base contacts on each of the emitter sides, designed to achieve the symmetry of the structure. These contacts are connected together by the metallization process. There are also four p^+ regions placed on the four corners of the emitter. These regions suppress the lateral emitter injection of electrons through the corner areas and shape the lateral flow of electrons toward the collectors, as shown in Figure 7.24(a).

The biasing circuit of the 2-D LMT device is shown in Figure 7.25. The device under analysis is operated in a common-emitter configuration. The base is driven with a constant current, the four collectors are biased with a voltage source V_C, and the substrate with V_S. The basic operation of the device is as follows. When the emitter-base junction is forward biased by the applied constant base current, the electrons injected into a neutral base region are shaped into four lateral components (in the x and y directions) and one vertical component (z direction). The vertical flow of carriers originating from the bottom of the emitter is collected by the substrate and do not contribute to device sensitivity. On the other hand, the four lateral components are essential to device sensitivity. The carriers of these four lateral flows are collected by four collectors (one collector for each lateral flow). In the absence of a magnetic field, these collector currents are equal because of the device symmetry. Upon application of a magnetic field consisting of two components, B_x and B_y, the collector currents will change due to the action of the Lorentz force. For example, the B_x component, which is parallel to the chip surface and oriented in the x direction, will cause a deflection of carriers flowing in the y direction, as is shown in Figure 7.24(c). As a result, I_{C3} will increase and I_{C1} will decrease. Therefore, the difference between I_{C3} and I_{C1} can be used to detect B_x. It should be pointed out that ideally B_x should not affect I_{C2} and I_{C4} because the flow of these components is parallel to B_x. However, these two collector currents will be affected by B_y, in a way

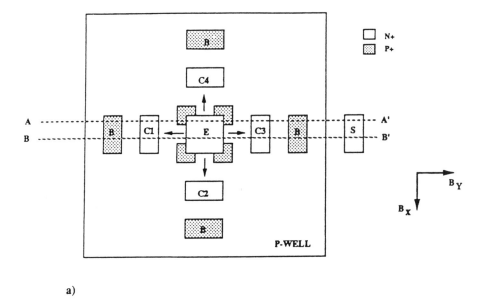

a)

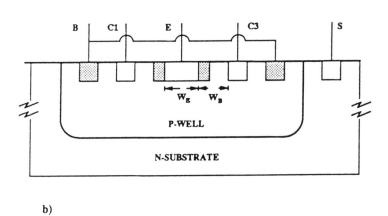

b)

Figure 7.24 2-D lateral magnetotransistor: (a) top view; (b) cross section A–A', suppressed lateral injection, $W_E = 12$ μm, $W_B = 12$ μm; (c) cross section B–B', L is lateral flow, V is vertical flow.

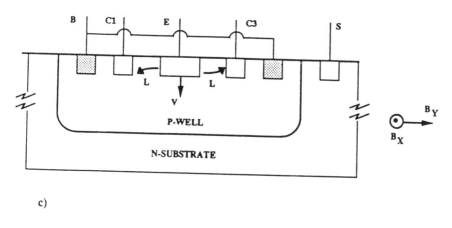

c)

Figure 7.24 continued.

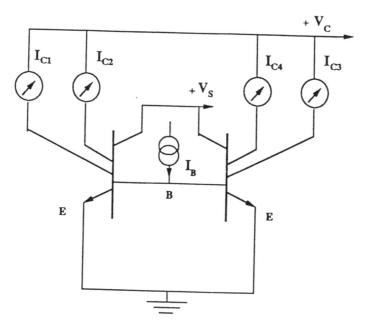

Figure 7.25 Biasing circuit of 2-D lateral magnetotransistor: I_{Ci} are collector currents, $i = 1-4$, I_B is the base current, V_C is the collector voltage, V_S is the substrate voltage.

similar to that of I_{C3} and I_{C1} affected by B_x. The difference between I_{C4} and I_{C2} can be used to detect B_y.

Figure 7.26 shows the change in current with magnetic field. Here ΔI_{Cx} is the difference between collectors C_3 and C_1, and ΔI_{Cy} is the collector current difference between collectors C_4 and C_2. The response of the sensor is linear with the magnetic field. At the same time, it should be mentioned that there is no cross sensitivity in this particular structure. More specifically, the x channel (C_1 and C_3) does not show sensitivity to B_y, and the y channel (C_2 and C_4) does not show sensitivity to B_x.

The relative sensitivity per channel is defined as

$$S_{rj} = \frac{\Delta I_{Cj}}{I_{Cj0}} \times \frac{1}{B} \tag{7.36}$$

where $j = x$; y: ΔI_{Cj} represents the difference between corresponding collector currents in the presence of a magnetic field; ΔI_{Cj0} is the sum of the corresponding collector currents at zero magnetic field; and B is the applied magnetic induction. Both channels show relative sensitivity of 45%/T.

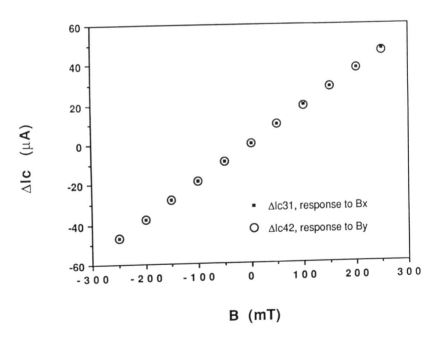

Figure 7.26 I_{Ci} as a function of B, $i = x$ or y: $I_B = 1$ mA, $V_C = 5$ V, $V_S = 5$ V, all collectors are drawing the current; $I_{Cx} = I_{C31}$, $I_{Cy} = I_{C42}$; $B_j = 200$ mT ($j = x, y$).

7.3.3 Three-Dimensional Sensing

As one might expect the structures for three-dimensional sensing are more complex structures than those for one-dimensional and two-dimensional devices. Indeed, to sense all three components of the magnetic field at least three pairs of collectors are needed. Kordic and Munter have designed devices that are a combination of vertical magnetoresistor and LMT [48]. The structure is similar to the vertical 2-D magnetotransistor device [11] with the addition of a third pair of collectors to sense the component perpendicular to the chip surface.

Another option for three dimensional sensing is the LMT structure with eight collectors [54,55]. This structure is basically the combination of devices described in Sections 7.3.1 and 7.3.2, see Figure 7.27. The device is fabricated using the standard CMOS process.The structure comprises a single emitter, a common base region, and eight symmetrical collectors.The device has four base contacts on each of the emitter sides, designed in such a way as to accomplish the symmetry of the structure. There are also four p^+ regions placed on the four corners of the emitter as in the case of 2-D LMT. These regions suppress the lateral emitter injection of electrons through the corner areas.

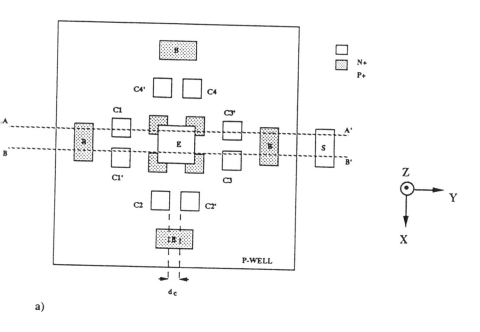

a)

Figure 7.27 A 3-D lateral magnetotransistor: (a) top view; (b) cross section A–A′; (c) cross section B–B′.

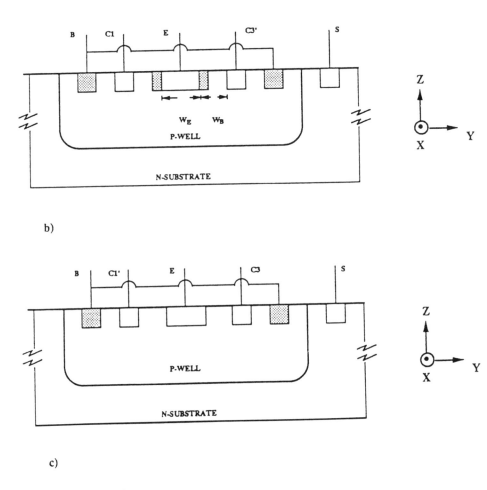

b)

c)

Figure 7.27 continued.

The biasing circuit of the 3-D LMT sensor is shown in Figure 7.28. The base is driven with a constant current while the eight collectors are biased with a voltage source V_c, and the substrate with V_s. When the emitter-base junction is forward biased by the applied constant base current, the electrons injected into a neutral base region are shaped into four lateral components (in the x and y directions) and one vertical component (z direction). The vertical flow of carriers does not contribute to device sensitivity. The four lateral components, however, are important. The carriers of four lateral flows are collected by eight collectors (two collectors for each lateral flow). In the absence of a magnetic field, these collector currents are equal because of device symmetry. When a magnetic field, consisting of all three components B_x

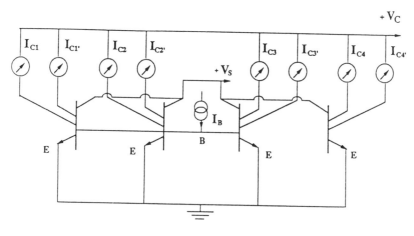

Figure 7.28 Biasing circuit of 3-D lateral magnetotransistor: I_B is the base current, V_C is the collector voltage, V_S is the substrate voltage.

B_y, and B_z, is applied the collector currents will change due to the action of the Lorentz force.

For example, the B_z component, which is perpendicular to the chip surface, will cause a deflection of the carriers as shown in Figure 7.29. As a result, the collector currents I_{ci} will increase and $I_{ci'}$ will decrease ($i = 1$–4). This means that the collector pairs on the same side of the emitter can be used to detect the z component of a magnetic field (the example in Figure 7.29 shows C_1 and $C_{1'}$). Note that, for this condition, the difference between I_{C3} and I_{C1}, or I_{C4} and I_{C2}, is 0 because of the symmetry of the device (the same is valid for $I_{C3'}$ and $I_{C1'}$, and so on).

In a similar manner, the deflection mechanism can be used to sense the B_x and B_y components. That is, the B_x component, which is parallel to the chip surface and oriented in the x direction, will cause a deflection of carriers flowing in the y direction, see Figure 7.30(a). As a result, I_{C3} and $I_{C3'}$ will increase and I_{C1} and $I_{C1'}$

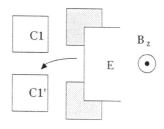

Figure 7.29 Deflection of carriers caused by B_z.

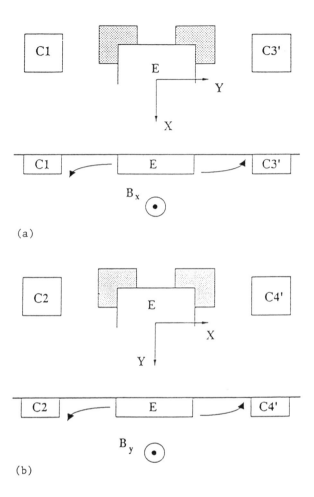

Figure 7.30 (a) Deflection of carriers caused by B_x. (b) Deflection of carriers caused by B_y.

will decrease. The difference between I_{C3} and I_{C1} (which was not affected by B_z) can be used to detect B_x. It is worth mentioning that ideally B_x should not affect I_{C2}, $I_{C2'}$, I_{C4}, and $I_{C4'}$, because the flow of these components is parallel to B_x. But these four collector currents will be affected by B_y similar to the way that I_{C3}, $I_{C3'}$, I_{C1}, and $I_{C'}$ were affected by B_x. The action of B_y is presented in Figure 7.30(b). In this case, I_{C4} and $I_{C4'}$ will increase and I_{C2} and $I_{C2'}$ will decrease. The difference either between I_{C4} and I_{C2} or $I_{C4'}$ and $I_{C2'}$ can be used to detect B_y.

Figure 7.31 shows experimental results of the change in current with the magnetic field. The collector current difference, ΔI_{Cx} is between collectors C_1 and C_3,

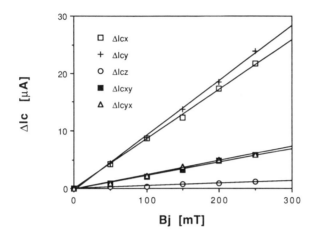

Figure 7.31 ΔI_c as a function of B_x, B_y, and B_z: $I_B = 1$ mA, $V_C = 5$ V, $V_S = 5$ V, all collectors are drawing the current, x is the channel I_{Cx}; $= I_{C31}$; y is the channel $I_{Cy} = I_{C42}$; z is the channel $I_{Cz} = I_{C11}$; cross sensitivity is $I_{Cxy} = I_{C13}$, $I_{Cyx} = I_{C24}$. $B_j = 200$ mT ($j = x, y, z$).

ΔI_{Cy} is the collector current difference of the collectors C_2 and C_4, and ΔI_{Cz} is the collector current difference of the collectors C_1 and $C_{1'}$. The response of the sensor is linear with the magnetic field. At the same time, note that there is a cross sensitivity in this particular structure. More specifically, the x channel (C_1 and C_3) shows sensitivity to B_y, which is represented by ΔI_{Cxy}, and the y channel (C_2 and C_4) shows sensitivity to B_x, which is represented by ΔI_{Cyx}. These can be explained with the aid of Figure 7.32.

It has already been stated that the four lateral carrier components in the x and y directions are essential for device operation. As can be seen, two collectors for

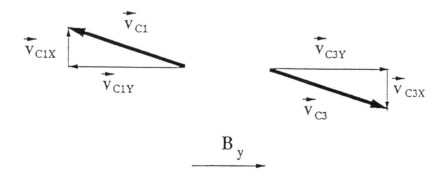

Figure 7.32 The velocity vectors of carriers collected by C_1 and C_3.

each of these lateral components are placed slightly off the x and y axes. This means that the resulting velocity vector for carriers collected by the collectors is also off the x and y axes. Figure 7.32 shows the velocity vectors for carriers collected by C_1 and C_3. Both of these have velocity components in the x and y directions. Because B_y has no effect on the y component of velocity, the resulting cross sensitivity $\Delta I_{C_{xy}}$ arises only from the x component of the velocity vectors. The same explanation holds true for the $\Delta I_{C_{yx}}$ cross sensitivity.

Each channel of the device can be described by relative sensitivity. The relative sensitivity per channel is defined as

$$S_{rj} = \frac{\Delta I_{Cj}}{I_{Cj0}} \times \frac{1}{B} \qquad (7.37)$$

where $j = x, y, z$, ΔI_{Cj} represents the difference between corresponding collector currents in the presence of a magnetic field, I_{Cj0} is the sum of the corresponding collector currents at zero magnetic field, and B is the applied magnetic induction. The experimental results show that the x and y channels have a relative sensitivity by one order of magnitude higher than the z channel and that cross sensitivity is also higher than the sensitivity for z channels. The increasing of the sensitivity of z channel remains one of the challenges for future designers.

7.4 SUMMARY

The focal point of this chapter is a lateral magnetotransistor. An effort has been made to describe the basic principle of operation of the LMT. Several devices are discussed for one-dimensional sensing, focusing on the very basic features such as sensitivity, noise, and offset. The discussion has included both CMOS and bipolar technology. Also, examples are discussed for two-dimensional and three-dimensional sensing. These structures represent the direct evolution of the devices for one-dimensional sensing. The whole matter was presented with three goals in mind: to discuss issues only at the device level, to keep it simple, and to try to capture the attention of the reader with the versatility of the LMT, which has a high potential as a magnetic field sensor.

REFERENCES

[1] V. Zieren, "Integrated Silicon Multicollector Magnetotransistors," Ph.D. thesis, Delft University of Technology, Delft, The Netherlands, 1983.

[2] H. P. Baltes and R. P. Popovic, "Integrated Semiconductor Magnetic Field Sensors," *Proc. IEEE*, Vol. 74, 1986, p. 1107.

[3] S. Kordic, "Integrated Silicon Magnetic Field Sensors," *Sensors and Actuators*, Vol. A10, 1986, p. 347.

[4] T. Nakamura and K. Maenaka, "Integrated Magnetic Sensors," *Sensors and Actuators*, Vol. A21–A23, 1990, p. 762.

[5] C. S. Roumenin, "Bipolar Magnetotransistor Sensors. An Invited Review," *Sensors and Actuators*, Vol. A24, 1990, p. 83.

[6] R. S. Popovic, *Hall Effect Devices. Magnetic Sensors and Characterization of Semiconductors*, Adam Hilger Series on Sensors, ed. B.E. Jones, Bristol, IOP Publishing, 1991.

[7] C. B. Brown, "Magnetically Biased Transistors," *Phys. Rev.*, Vol. 76, 1949, p. 1736.

[8] E. C. Hudson, Jr., "Semiconductive Magnetic Transducer," U.S. Patent, No. 3 389 230, 1968.

[9] L. W. Davies and M. S. Wells, "Magneto Transistor Incorporated in a Bipolar IC," *Proc. ICMCST*, Sydney, Australia, 1970, p. 34.

[10] J. B. Flynn, "Silicon Depletion Layer Magnetometer," *J. Appl. Phys.*, Vol. 41, 1970, p. 2750.

[11] V. Zieren and B. P. M. Duyndam, "Magnetic-Field-Sensitive Multicollector *n-p-n* Transistors," *IEEE Trans. Electron. Dev.*, Vol. ED-29, 1982, p. 83.

[12] L. W. Davies and M. S. Wells, "Magnetotransistor Incorporated in an Integrated Circuit," *Proc. IREE Australia*, 1971, p. 225.

[13] I. M. Mitnikova, T. V. Persiyanov, G. I. Rekalova, and G. Shtyubner, "Investigation of the Characteristics of Silicon Lateral Magnetotransistors with Two Measuring Collectors," *Sov. Phys. Semiconnd.*, Vol. 12, No. 1, 1978, p. 26.

[14] A. G. Andreou and C. R. Westgate, "The Magnetotransistor Effect," *Electronics Lett.*, Vol. 20, 1984, p. 699.

[15] R. S. Popovic and H. P. Baltes, "An Investigation of the Sensitivity of Lateral Magnetotransistor," *IEEE Electron. Dev. Lett.*, Vol. EDL-4, 1983, p. 51.

[16] R. S. Popovic and R. Widmer, "Magnetotransistor in CMOS Technology," *IEEE Trans. Electron. Dev.*, Vol. ED-33, 1986, p. 1334.

[17] A. W. Vinal and N. A. Masnari, "Magnetic Transistor Behavior Explained by Modulation of Emitter injection, not carrier deflection," *IEEE Electron. Dev. Lett.*, Vol. EDL-3, 1982, p. 203.

[18] Lj. Ristic, T. Smy, H. P. Baltes, and I. Filanovsky, "A Highly Sensitive Magnetic Field Sensor Based on Magnetotransistor Action with Suppressed Sidewall Injection," *Proc. Yugoslav Conf. Microelectronics*, Banja Luka, 1987, p. 25.

[19] Lj. Ristic, H. P. Baltes, T. Smy, and I. Filanovsky, "Suppressed Sidewall Injection Magnetotransistor with Focused Emitter Injection and Carrier Double Deflection," *IEEE Electron. Dev. Lett.*, Vol. EDL-8, 1987, p. 395.

[20] Lj. Ristic, T. Smy, and H. P. Baltes, "A lateral Magnetotransistor Structure with Linear Response to the Magnetic field," *IEEE Trans. on Electron. Dev.*, Vol. ED-36, 1989, p. 1076.

[21] T. Smy and Lj. Ristic, "On the Injection Modulation Effect," *Sensors and Materials*, Vol. 1, No. 4, 1988, p. 233.

[22] Lj. Ristic, T. Smy, H. P. Baltes, and I. Filanovsky, "A CMOS Bipolar Transistor with a Locally Doped Base in the Proximity of the Emitter as a Magnetic Field Sensor," *Proc. IEEE 1988 Bipolar Circ. and Tech. Meetings*, Minneapolis, 1988, p. 199.

[23] T. Smy and Lj. Ristic, "Optimization of Magnetotransistor Structure in CMOS Technology," *IEEE Trans. on Magnetics*, Vol. MAG-28, 1992.

[24] Lj. Ristic, K. Maenaka, T. Smy, T. Nakamura, and M. T. Doan, "*pnp* lateral Magnetotransistor and Influence of n^+ Buried Layer on Sensitivity," *Appl. Phys. Lett.*, Vol. 58, No. 2, 1991, p. 149.

[25] Lj. Ristic, M. T. Doan, K. Maenaka, and T. Nakamura, "*pnp* Bipolar Lateral Transistors as Magnetic Field Sensors," *Sensors and Materials*, Vol. 2, 1990, p. 163.

[26] G. Bjoklund, "Improved Design of Hall Plates for ICs," *IEEE Trans. on Electron. Dev.*, Vol. ED-25, 1978, p. 541.

[27] Y. Kanda and M. Migitoka, "Effect of the Mechanical Stress on the Offset Voltage of Hall Devices in Si," *Phys. Stat. Sol. (a)*, Vol. 35, 1981, p. 115.

[28] R. J. Braun et al., "FET Hall Transducers with Control Gates," *IBM Tech. Discl. Bull.*, Vol. 17, 1979, p. 7.

[29] J. T. Maupin and M. L. Gaske, "The Hall Effect in Silicon Circuits," in *The Hall Effect and Its Application*, ed. C. L. Chien and C. R. Westgate, New York, Plenum Press, 1980.

[30] S. Kordic and P. C. M. Van der Jagt, "Theory and Practice of Electronic Implementation of the Sensitivity-Variation Offset-Reduction Method," *Sensors and Actuators*, Vol. A8, 1985, p. 197.

[31] Lj. Ristic, T. Smy, and H. P. Baltes, "A Magnetotransistor Structure with Offset Elimination," *Sensors and Materials*, Vol. 2, 1988, p. 83.

[32] Lj. Ristic, M. T. Doan, and T. Q. Truong, "1/f Noise in Lateral Magnetotransistors," *Proc. Canadian Conf. Electrical and Comp. Eng.*, Vancouver, 1988, p. 772.

[33] Lj. Ristic and M. T. Doan, "On the 1/f Noise and Noise Correlation in Magnetotransistors," *Sensors and Materials*, Vol. 3, No. 5, 1992, p. 281.

[34] M. E. Gruchalla, "Measurements of Wide Band White Noise Using a Standard Oscilloscope," *Electrical Design News*, 1980, p. 157.

[35] H. P. Baltes, A. Nathan, and D. R. Briglio, "Noise Correlation in Magnetic Field Sensitive Transistors," *Tech. Digest*, IEEE Solid-St. Sensor and Actuator Workshop, Hilton Head Island, SC, 1988, p. 104.

[36] A. Nahtan, H. P. Baltes, D. Briglio, and M. T. Doan, "Noise Correlation in Dual Collector Magnetotransistors," *IEEE Trans. Electron. Dev.*, Vol. ED-36, 1989, p. 1073.

[37] X. C. Zhu and A. Van der Ziel, "The Hooge Parameters of n^+pn and p^+np Silicon Bipolar Transistors," *IEEE Trans. on Electron. Dev.*, Vol. ED-32, 1985, p. 658.

[38] A. Van der Ziel, P. H. Handel, X. C. Zhu, and K. H. Duh, "A Theory of the Hooge Parameters of Solid State Devices," *IEEE Trans. on Electron. Dev.*, Vol. ED-32, 1985, p. 667.

[39] X. N. Zhang, A. Van der Ziel, and H. Markoc, "Location of the 1/f Noise in Bipolar Transistors," in *Noise in Physical Systems and 1/f Noise*, ed. D'Amico and Mazzetti, Amsterdam, Elsevier, 1986, p. 397.

[40] X. N. Zhang and A. Van der Ziel, "Test for the Presence of Injection—Extraction and Umklapp 1/f noise in the Collector of Silicon Transistors," in *Noise in Physical Systems and 1/f Noise*, ed. D'Amico and Mazzetti, Amsterdam, Elsevier, 1986, p. 485.

[41] A. Van der Ziel, X. N. Zhang, and A. H. Pawlikiewicz, "Location of 1/f Noise Sources in BJT's and HBJT's—I. Theory," *IEEE Trans. on Electron. Dev.*, Vol. ED-33, 1986, p. 1371.

[42] A. H. Pawlikiewicz and A. Van der Ziel, "Location of 1/f Noise Sources in BJT's—II. Experiment," *IEEE Trans. on Electron. Dev.*, Vol. ED-33, 1987, p. 2009.

[43] Lj. Ristic, T. Q. Truong, M. Doan, D. Mladenovic, and H. P. Baltes, "Influence of Surface Effects on the Sensitivity of Magnetic Field Sensors," *Can. J. Phys.*, Vol. 67, 1989, p. 207.

[44] K. Maenaka, H. Fujiwara, T. Ohsakama, M. Ishida, T. Nakamura, A. Yoshida, and Y. Yoshida, "Integrated Magnetic Vector Sensor," *Proc. 5th Sensor Symp.*, Japan, 1985, p. 179.

[45] Lj. Ristic, M. Paranjape, and M. T. Doan, "2-D Magnetic Field Sensor Based on Vertical Hall Device," *Tech. Digest*, Solid-St. Sensor and Actuator Workshop, Hilton Head Island, 1990, p. 111.

[46] M. Paranjape and Lj. Ristic, "Micromachined Vertical Hall Magnetic Field Sensor in Standard Complementary Metal Oxide Semiconductor Technology," *Appl. Phys. Lett.*, Vol. 60, No. 25, 1992, p. 3188.

[47] K. Maenaka, T. Ohgusu, M. Ishida, and T. Nakamura, "Integrated Magnetic Sensor Detecting x, y, and z Components of the Magnetic Field," *Tech. Digest*, Transducers '87, Tokyo, 1987, p. 523.

[48] S. Kordic and P. J. A. Munter, "Three-Dimensional Magnetic Field Sensors," *IEEE Trans. on Electron Dev.*, Vol. ED-35, 1988, p. 771.

[49] L. Zongsheng, L. Yi, W. Guangli, and J. Hao, "A Novel Integrated 3-D Magnetic Vector Sensor Based on BCD Technology," *Sensors and Actuators*, Vol. A22, 1990, p. 786.

[50] M. Paranjape, Lj. Ristic, and I. Filanovsky, "A 3-D Vertical Hall Magnetic Field Sensor in CMOS Technology," *Tech. Digest*, Transducers '91, 1991, p. 1081.

[51] Lj. Ristic and M. Doan, "Lateral Transistor Structure Sensitive to Magnetic Field Applied Parallel or Perpendicular to the Chip Surface," *Proc. MIEL '90*, Ljubljana, 1990.

[52] M. Doan and Lj. Ristic, "Magnetotransistor Sensitive to Magnetic Field Applied in Two Different Directions," *Can. J. Phys.*, Vol. 69, 1991, p. 170.

[53] Lj. Ristic, M. Doan, and M. Paranjape, "2-D Integrated Magnetic Field Sensor in CMOS Technology," *Proc. 32nd Midwest Symp. Circuits and Systems*, Champaign-Urbana, IL, 1989, p. 701.

[54] Lj. Ristic, M. Doan, and M. Paranjape, "3-D Magnetic Field Sensor Realized as Lateral Magnetotransistor in CMOS Technology," *Sensors and Actuators*, Vols. A21–A23, 1990, p. 770.

[55] Lj. Ristic, M. T. Doan, and M. Paranjape, "Multi-Dimensional Sensing of Magnetic Field," *Proc. 33rd Midwest Symp. on Circuits and Systems*, Calgary, 1990, p. 858.

Chapter 8
Thermal Sensors

Wolfgang Rasmussen
University of Sherbrooke

8.1 INTRODUCTION

Electronic temperature sensing is a mature art, but there is still room for innovation. With temperature affecting most sensors, thermal sensors are being integrated along with other IC sensors for on-the-spot temperature compensation and correction. Silicon IC thermal sensors are restricted to an operating range between about $-55°C$ and $+150°C$, a temperature range that actually includes much human activity, from making ice cubes to boiling water. The advantage of IC sensors is that signal conditioning can be carried out on the same chip, so that weaker signals can be measured and the output used directly by another electronic system. Current trends toward all-digital control systems, including sensor data transmission, make CMOS silicon temperature sensors particularly attractive, because both digital data correction and communication can be included on the same chip. Digital data correction with an EPROM or an EEPROM may be easier and cheaper than analog data correction, which generally requires expensive thin-film resistors. Bipolar CMOS (BiCMOS) technology seems to be ideal for combining bipolar transistor temperature sensors with CMOS output processing.

Temperature measurement is less tangible than most other physical measurement: twice as hot is much more difficult to visualize than twice as long. Temperature is an intrinsic quantity that reflects the state of thermal agitation of an object. Temperature scales are defined on the basis of fundamental thermodynamic laws. Starting from the abstract thermodynamic definition, practical, internationally agreed upon temperature scales (such as IPTS-90 [1]) have been established by standards organizations; these scales are based on precisely defined fixed temperature points, such as the boiling or freezing points of various substances under precisely controlled conditions, and on the interpolation between these fixed reference points using specified, stable thermometric properties. Of primary concern in the choice of thermometric properties for interpolation, and for temperature sensor design in general, are the stability, repeatability, and adherence to a standard curve of the thermometric property, as well as the absence of cross sensitivities, a useable temperature range, and sensitivity to temperature. Over the temperature interval of interest here ($-55°C$ to $+150°C$), the platinum resistance thermal device (RTD), which is based on the variation of the resistance of ultrapure platinum with temperature, is the internationally agreed on practical reference.

High precision and accuracy are difficult to achieve in temperature measurement. The first problem is to accurately calibrate the IC sensor. To calibrate a sensor, the sensor must be compared to a reference sensor, usually a platinum RTD or a thermocouple, with which it is in thermal equilibrium. It is difficult to assure that the reference sensor and the sensor under test are at the same temperature, especially when electrical current flows that can cause self-heating. This is not as much a problem when an IC sensor is calibrated at the wafer level. Once in a package, heat transfer conditions may be more complicated. The package and the environment—whether the measurement is made in still air or while the sensor is immersed in a flowing liquid—determine the actual temperature difference between the chip and its surroundings, as well as the sensor's response time. In practice, the absolute accuracy of a temperature measurement is much lower than the precision with which temperature differences under similar environmental conditions can be determined.

Due to silicon's high thermal conductivity, a chip is nearly isothermal except for local hot spots where large currents flow. In designing high-precision thermal sensors, the small thermal inhomogeneities due to self-heating need to be taken into account by careful chip layout, separating the sensing elements from power elements, placing elements that must match close together, and generally designing for minimum power usage to avoid self-heating in the first place.

In this chapter, we focus on temperature sensors based on active devices such as diodes and transistors. We start with a discussion of IC temperature sensors made in bipolar technology. We then explore several CMOS solutions and consider some digitally compatible sensor designs. Finally, we discuss the properties of polysilicon resistors for temperature sensing.

8.2 BIPOLAR DEVICES AS TEMPERATURE SENSORS

Silicon diodes and bipolar transistors are regularly used for temperature sensing or compensation from cryogenic temperatures up to about 200°C; however, IC sensors based on bipolar devices are usually limited to temperatures between $-55°C$ and $+150°C$. These devices are based on the highly predictable, extensively studied, stable diode forward voltage-current or the diodelike base-emitter voltage-collector current characteristics. Our discussion starts with a brief look at diode sensors, followed by a discussion of the temperature dependence and the stability of bipolar transistors. A presentation of some circuit approaches and sensing ideas rounds out the bipolar technology section.

8.2.1 Diodes as Temperature Sensors

The simplest bipolar temperature sensing element is the *pn* diode. The classical diode equation for the current I is:

$$I = I_{os}(\exp [qV_F/kT] - 1) \tag{8.1}$$

It contains two strongly temperature-dependent factors, the reverse saturation current I_{os} and the Boltzmann factor $\exp [qV_F/kT]$, in which q is the electric charge, k is Boltzmann's constant, $(k/q = 86.1706 \ \mu V/K)$, V_F is the forward voltage, and T is the absolute temperature in degrees Kelvin ($0°C = 273.15K$). The reverse saturation current represents the drift current of thermally generated minority carriers from the neutral regions on either side of the depletion region into the junction depletion region and varies nearly exponentially with temperature, doubling approximately every 10°C around room temperature. The Boltzmann factor arises from the diffusion of charge carriers over the depletion region potential barrier.

For temperature sensing, single, calibrated *pn* junction sensors are normally forward biased by a constant current, with the forward voltage V_F serving as a measure of the temperature. The diode forward voltage V_F of OMEGA Engineering's temperature sensing diode [1] biased with an accurate 10 μA current is plotted versus the absolute temperature T in Figure 8.1. At very low temperatures, incomplete ionization of the dopant atoms, carrier freeze out, becomes important, leading to a steep increase in the forward voltage needed to conduct the 10 μA. Above 100K, V_F

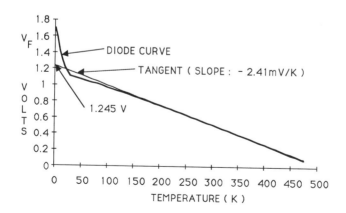

Figure 8.1 Diode forward voltage with tangent fit.

decreases nearly linearly with temperature as indicated by the tangent drawn to the curve at 320K:

$$V_F = V_R - C_R T = 1.245 - .00241(V/K)T \tag{8.2}$$

The tangent approximation can be used for simple temperature measurement. Its deviation from V_F expressed as a temperature difference is less than 0.5°C between 0°C and 100°C. In IC circuits, diodes are sometimes used for monitoring on-chip temperatures. In general, they can be used for accurate temperature measurements only after careful calibration and with sophisticated support electronics.

8.2.2 Bipolar Transistors as Temperature Sensors

The total diode current consists of not only a diffusion current but also generation-recombination currents coming from charge carriers that recombine or are generated in the depletion region and at the surface regions. These currents have a different temperature behavior than purely diffusion currents. Such unwanted currents can be extracted via the base in a bipolar transistor, so that the collector current of a forward biased transistor is a nearly pure diffusion current [2]. A bipolar transistor's base-emitter voltage-temperature characteristic for a constant collector current has the same basic form as the diode V_F curve (Figure 8.1). Furthermore, being three terminal devices, bipolar transistors are much more flexible than diodes for use in IC temperature sensors and their close cousins, the bipolar bandgap references. The ideal collector current I_c of a bipolar transistor in the forward active region, neglecting the Early effect and assuming low-level injection, is related to the base-emitter voltage V_{BE} [3] by

$$I_c = I_s \exp[qV_{BE}/kT] = I_s \exp[V_{BE}/V_T] \tag{8.3}$$

where $V_T = kT/q$ is the thermal voltage (≈ 26 mV at 300K) and I_s is the saturation current. I_s can be expressed by

$$I_s = BAT\bar{\mu}_n n_i^2 \tag{8.4}$$

where T is the temperature; $\bar{\mu}_n$ the average minority carrier mobility in the base, which depends on the base doping profile; A is the base-emitter area; n_i the intrinsic carrier concentration; and B is a temperature-independent constant. The temperature dependence of $\bar{\mu}_n$ and n_i are well represented by

$$\bar{\mu}_n \propto T^{-m} \tag{8.5}$$

and

$$n_i^2 \propto T^3 \exp[-V_{G0}/V_T] \tag{8.6}$$

where V_{G0} (≈ 1.205 V) is the bandgap voltage for silicon extrapolated to absolute 0, and m is an effective mobility exponent. Defining an exponent $\gamma = 4 - m$ and a constant C, I_c and V_{BE} become

$$I_c = CAT^{\gamma} \exp[(V_{BE} - V_{G0})/V_T]$$

$$V_{BE} = V_{G0} + V_T \ln\left[\frac{1}{C}\frac{I_c}{A}T^{-\gamma}\right] \tag{8.7}$$

In writing (8.6), the bandgap voltage was assumed to be a linear function of the temperature [4]. As Meijer [5] has pointed out, this is an excellent approximation if an empirical fit is used to determine γ and V_{G0}. The temperature dependence of I_c must still be specified. In practice, it is relatively simple to generate a collector current with a power-law temperature dependence

$$I_c = I_{c0}(T/T_0)^{\alpha} \tag{8.8}$$

where α is the temperature exponent and I_{c0} is the collector current at a reference temperature T_0. V_{BE} then becomes

$$V_{BE} = V_{G0} + V_T \ln\left[\frac{T_0^{\gamma}}{C}\left(\frac{T}{T_0}\right)^{-(\gamma-\alpha)}\frac{I_{c0}}{A}\right] \tag{8.9}$$

or, in terms of the base-emitter voltage V_{BE} at T_0, V_{BE0},

$$V_{BE} = V_{G0} - (V_{G0} - V_{BE0})\frac{T}{T_0} - (\gamma - \alpha)V_T \ln\left(\frac{T}{T_0}\right) \tag{8.10}$$

Expanding the logarithmic term around T_0 in a Taylor series in $(\Delta T/T_0) = (T - T_0)/(T_0)$, and regrouping the linear terms, we find the quadratic approximation

$$V_{BE} = (V_{G0} + (\gamma - \alpha)V_{T0}) - (V_{G0} + (\gamma - \alpha)V_{T0} - V_{BE0})(T/T_0)$$

$$- (\gamma - \alpha)(V_{T0}/2)[(\Delta T/T_0)^2] + \ldots \tag{8.11}$$

and the tangent approximation illustrated in Figure 8.1

292

$$V_{\mathrm{BE}} \approx V_R - (V_R - V_{\mathrm{BE0}})(T/T_0) = V_R - C_R T \qquad (8.12)$$

Here V_{T0} is the thermal voltage $(k/q)\, T_0$, V_R equals $V_{G0} + (\gamma - \alpha)\, V_{T0}$, and C_R equals $(V_R - V_{\mathrm{BE0}})/T_0$. On comparing (8.10) and (8.12), we can see that the contribution of the logarithmic term in (8.10) to the linear terms in (8.11) and (8.12) is simply equivalent to shifting V_{G0} to V_R by adding the factor $(\gamma - \alpha)\, V_{T0}$ to V_{G0} in the linear terms in (8.10).

The limits of equations (8.7) to (8.12) for IC temperature sensing are illustrated in Figure 8.2, where the deviation of the theoretical curves from measured values are plotted in terms of an equivalent temperature error for a diode [1] and a transistor [6] biased by a constant current. We used three data points in each case to determine the parameters V_{G0}, γ, and the constants in the ln term in (8.7). For the diode, $V_{G0} = 1.17$V and $\gamma = 2.61$, whereas for the transistor, $V_{G0} = 1.180$V and $\gamma = 3.27$. Figure 8.2 shows that equations (8.7) to (8.10) accurately describe both the $V_F - I$ and the $V_{\mathrm{BE}} - I_c$ characteristics to better than 0.05°C between −50°C and +125°C. The deviation of the first three terms in the Taylor series expansion of V_{BE} (8.11) from the transistor data is also plotted in Figure 8.2 as the "quadratic fit." With the inclusion of the quadratic "correction," the error falls below 0.25°C. Quadratic curvature correction is commonly used in temperature sensors and bandgap references to correct for the nonlinearity of V_{BE}.

V_{BE} can also be linearized by adjusting the temperature dependence of the collector current (8.10) so that $\alpha = \gamma$. This analog linearization is explored in Figure 8.3 for several values of the temperature exponent α (8.8) by plotting the deviation in degrees Celsius from the tangent fit and the theoretical expression (8.7). Ohte and Yamagata [6], for example, found that the theoretical curves (Figure 8.3) agreed, to better than 0.15°C, with experimental values between −50°C and +120°C. With α equal to γ, they found a deviation from linearity of less than 0.1°C.

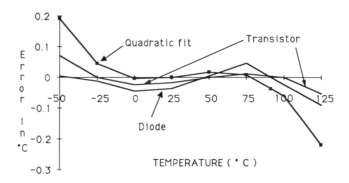

Figure 8.2 Error between theoretical fit and data for a diode and a transistor.

The constant V_R in the tangent approximation (8.12) depends only on the transistor's physical construction via the exponent γ, on the temperature T_0 and on the collector current's temperature exponent α, but not on the operating point at T_0, V_{BE0}. The slope C_R, however, does. C_R can thus be adjusted by modifying V_{BE0}, that is, by changing the current I_{c0} at T_0 or the emitter-base area. In Figure 8.4, V_{BE} curves are drawn with an exaggerated curvature to illustrate this point. The rotation of the tangents about $(0, V_R)$ corresponds to the slope adjustment as the current I_{c0} is changed. Adjusting V_{BE0} at one temperature, with α fixed, completely determines V_{BE} at any temperature, enabling a single point slope adjustment. This property is a reflection of the more general result (8.13) that the difference ΔV_{BE} (Figure 8.4) between the

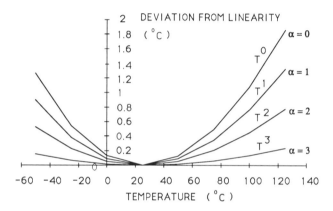

Figure 8.3 Deviation from linearity of the theoretical curves for several values of the collector current's temperature exponent.

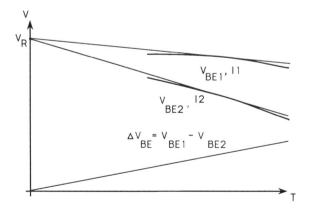

Figure 8.4 V_{BE} and ΔV_{BE} versus temperature for two different currents.

values of V_{BE} (8.7) of two transistors, identical except maybe for size, when operated at different current densities, depends only on the thermal voltage V_T and the ratio of the current densities and is independent of the fabrication details:

$$\Delta V_{BE} = V_{BE1} - V_{BE2} = V_T \ln\left[\frac{I_{c1}}{I_{c2}}\frac{A_2}{A_1}\right] \qquad (8.13)$$

Remarkably, ΔV_{BE} is strictly proportional to the absolute temperature (PTAT) if the ratio of the current densities is fixed. Equation 8.13 forms the basis for the classical PTAT temperature sensors. Real values of ΔV_{BE} are much smaller, typically only a few degrees of V_T, than indicated in Figure 8.4.

8.2.3 Stability

The repeatability and long-term stability of V_{BE} and ΔV_{BE} have been extensively tested and reported on by Meijer [5]. The repeatability of V_{BE} after thermal cycling depends on the transistor manufacturing process. For one process, on multiple thermal cycling between 25°C and 110°C, transistors showed an initial shift in V_{BE} equivalent to about 0.15°C, followed by decreasing shifts later. Transistors from a second manufacturer showed no shift at all. After a period of aging by continuous cycling, the stability of all transistors improved to better than .015°C. Interrupting the thermal cycling and resuming a month later with the same transistors, those from the first manufacturer showed a renewed initial shift of about half that of new transistors. In a remarkable series of once a day thermal cycling between room temperature and eight hours at 125°C over a period of about three years by Ohte and Yamagata [6], the values of V_{BE} showed an instability of less than .02°C. The main accuracy problem in the use of V_{BE} as a thermometric property is thus the stability of the transistors on thermal cycling and any calibration error. The PTAT voltage ΔV_{BE} offers neither any higher linearity nor stability than is achievable by using V_{BE} [5]. One reason is that ΔV_{BE} is a very small voltage, so that small, nonideal factors in V_{BE} can lead to significant error.

8.3 INTEGRATED TEMPERATURE SENSORS IN BIPOLAR TECHNOLOGY

Various bipolar temperature sensing schemes can be imagined by looking at Figure 8.4. The linearly decreasing values of V_{BE} and the linearly increasing PTAT voltage ΔV_{BE} can be used separately or combined, as illustrated in Figure 8.5, in four useful ways that have all seen application.

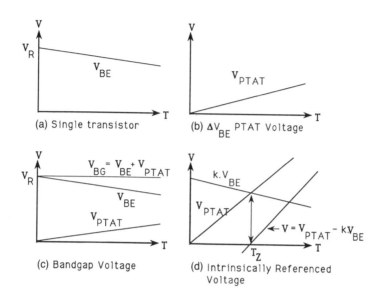

Figure 8.5 Four ways of combining V_{BE} and V_{PTAT}: (a) V_{BE} is used to measure temperature; (b) the difference of two V_{BE} values is used to measure temperature in PTAT sensors; (c) adding a multiple of a linearly increasing PTAT voltage to compensate for the linear decrease in V_{BE}; (d) shifting the zero of a PTAT sensor to any desired temperature T_z by subtracting V_{BE} from a multiples of a PTAT voltage, creating an intrinsically referenced sensor.

In the single transistor sensor (Figure 8.5(a)), V_{BE} is used directly to measure temperature, whereas the difference of two V_{BE} values is used to measure the temperature in the PTAT sensors (Figure 8.5(b)). We can construct bandgap references (Figure 8.5(c)) by adding a multiple of a linearly increasing PTAT voltage to compensate for the linear decrease in V_{BE}. We can also shift the zero of a PTAT sensor (Figure 8.5(d)) to any desired temperature T_z by subtracting V_{BE} from a multiple of a PTAT voltage and thus creating an intrinsically referenced sensor. A similar temperature scale offset can also be generated by using the bandgap voltage together with a PTAT voltage.

Though the idea of the single transistor temperature sensor is appealing from the standpoint of sensor simplicity, it has two drawbacks. First, the A/D conversion is demanding because the sensitivity of V_{BE} to temperature $(\Delta V/V_{BE0})/\Delta T$ at 300 K is only about (2 mV/600 mV)/°C or about 1 in 300. This problem is best tackled by using an intrinsically referenced sensor. Second, generating an I_c with a precisely tailored temperature behavior away from the thermal environment of the sensor, in order to have a linear output, demands complex external support electronics. A better solution is to generate the bias currents on-chip using transistor PTAT sources. For special applications, particularly where power usage must be kept to a minimum or where simplicity is required, the single transistor or the diode may still be the best choice.

The PTAT voltage ΔV_{BE} (8.13) depends only on the absolute temperature, the ratio of the two collector currents and the ratio of the two base-emitter interface areas, and not on fabrication details nor on an individual collector current's temperature dependence. The ratio of two collector currents can easily be made independent of temperature by using equal or ratioed currents supplied by current mirrors. The designer can control the PTAT voltage by adjusting the area ratio by repeating the base-emitter structure, by adjusting the current ratio, or by a combination of the two approaches. A simple realization of a self-biased PTAT sensor [3,6] is illustrated in Figure 8.6. The lower-case letters next to the emitters of the transistors indicate area ratios. The current mirror $Q3$–$Q4$ causes ratioed currents to flow in each branch, if we ignore base currents and the Early effect. The voltage across R_1 is the PTAT voltage

$$\Delta V_{BE} = V_{BE1} - V_{BE2} = V_T \ln(qr) \qquad (8.14)$$

where q and r are the area ratios. The output voltage V_{OUT} is also a PTAT voltage:

$$V_{OUT} = IR_2 = (R_2/R_1)V_T \ln(qr) \qquad (8.15)$$

It is important to note that the PTAT voltage V_{OUT} does not depend on the temperature

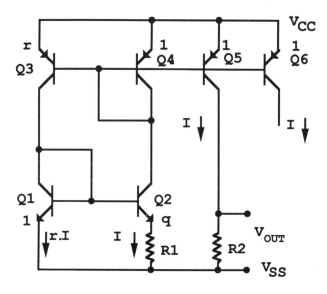

Figure 8.6 Simple PTAT current/voltage generator.

coefficient (TC) of the resistors, so that common on-chip resistors can be used. The branch currents, however, do depend on the TC of the resistor

$$I = V_T \ln(qr)/R_1 \qquad (8.16)$$

The generation of PTAT currents requires low TC resistors, which can be either on-chip thin-film resistors or off-chip resistors.

The simple PTAT circuit (8.6) needs to be improved in several directions to take into account the nonideal properties of real transistors, such as their finite current gain, high-level injection, and the Early effect, before the ideal current of (8.16) is approached. The simple current mirrors and the PTAT generator (Figure 8.6) are inadequate because the transistors in them have unbalanced emitter-collector voltages, so that unequal collector currents flow due to the base-width modulation effect, that is, the Early effect. Even worse, a change in the supply voltage will cause unequal collector current changes in, for example, the current mirror transistors $Q3$ and $Q4$, because their emitter-collector voltages change unsymmetrically. Even a small supply voltage sensitivity can be deleterious for a PTAT sensor because its sensitivity to temperature is so low; for example, a 1°C or 1 K temperature change around 27°C or 300 K represents only a 1 in 300 or .3% change in the absolute temperature or the thermal voltage $(k/q)T$. The standard solution, illustrated in Figure 8.7 for improving the current mirror $Q3$–$Q4$, is to add a second pair of transistors, $Q7$–$Q8$, to isolate the collectors of the mirror transistors $Q3$–$Q4$ from variations in the supply voltage, while balancing their collector-emitter voltages [3,7]. The output impedance of the current sources is greatly increased with the cascode current mirrors.

Another problem with the simple circuit in Figure 8.6 and with the standard cascode current mirror $Q3$–$Q4$, $Q7$–$Q8$ in Figure 8.7 is that the branch currents are unbalanced by the nonideal base currents, which are drawn directly via the diodelike connections from the collector arms. This problem is particularly acute if the current gain of the transistors is low. The standard solution [3] is to use an extra transistor, as illustrated with transistor $Q11$ in Figure 8.7, to supply the base currents. Due to the current gain of $Q11$, the current $Q11$ draws from the collector arm is a small fraction of the previous base currents. The base currents for $Q1$–$Q2$ are also supplied by an extra transistor, $Q12$, which, with its diode connection to the bases of $Q9$–$Q10$ and the emitter of $Q11$, draws no current directly from the collector arms [7]. The transistor pair $Q9$–$Q10$ isolate the PTAT generator ($Q1$–$Q2$ and R_1) from power supply variations, and $Q11$ forces the base-collector voltages of $Q1$–$Q2$ to be equal, so that their collector-emitter voltages are equal to within a few degrees of V_T. This bias arrangement should also be used with the cascode mirror $Q3$–$Q4$, $Q7$–$Q8$.

As is normal with self-biased voltage or current references [3], a start-up circuit and stabilization capacitors for avoiding oscillations are needed to complete the design. To assure good thermal and geometrical matching, especially with the current

298

mirrors and the temperature-sensing transistors, great care must be taken with the transistor layout and the component placement. Due to the cascode current mirrors, a supply voltage of at least 4 V_{BE} is required in Figure 8.7. If the circuit is to function at lower voltages, cascode current mirrors must be avoided. An interesting design that avoids cascode mirrors, while still achieving excellent supply voltage rejection, is presented in the highly tutorial article by Timko [8] on his design of Analog Devices' AD590.

Many other circuit possibilities for generating either a PTAT current or voltage can be found in the literature. Two variations [5,9], which use only *npn* transistors, are sketched in Figure 8.8. The circuits are straightforward to analyze. By looping around the PTAT transistor quad in Figure 8.8(a), we find that

$$V_{R1} = (V_{BE1} - V_{BE3}) + (V_{BE4} - V_{BE3}) = V_T \ln(qr) \qquad (8.17)$$

What is remarkable with this circuit is that the current in the resistor branch is in-

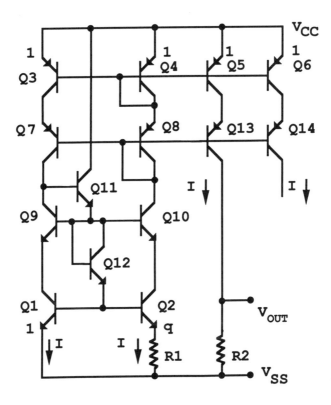

Figure 8.7 Improved PTAT current/voltage source.

dependent of the bias current. Furthermore, it is a PTAT current (I_{PTAT}) if the temperature coefficient of R_1 is null. In Figure 8.8(b), the addition of a second resistor R_2 equal to R_1 adds a V_{R2} term to the equations and forces the sum of the two branch currents to become a PTAT current, again independent of the bias current.

An approach (Figure 8.9) that is also widely used in bandgap literature [10], applies operational amplifier feedback to balance the collector currents instead of a transistor current mirror [11]. The major advantages of this approach are that the sensing transistors do not take their base currents from the balance circuit and that their collector-emitter voltages are equal to within a few degrees of V_T. The op amp compares the two currents that flow through the matched resistors R_1 and R_2 and adjusts the feedback voltage via resistor R_4 on the bases of $Q1$ and $Q2$ until current balance is achieved. When the currents match, the voltage difference between the bases of $Q1$ and $Q2$ is PTAT:

$$V_4 = R_4 \cdot I = V_T \ln(q) \tag{8.18}$$

The sensor's output voltage is then

$$V_{OUT} = V_T \ln(q)(R_3 + R_4 + R_5)/R_4 \tag{8.19}$$

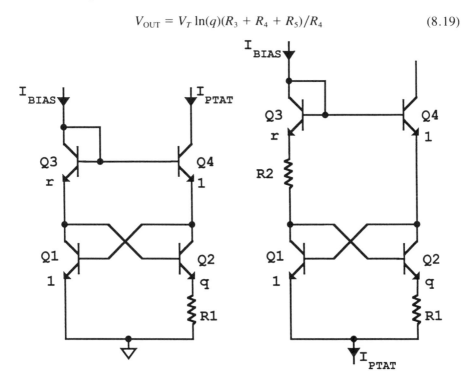

Figure 8.8 Bias-independent PTAT sources: (a) current sink [9]; (b) current source (from [5]).

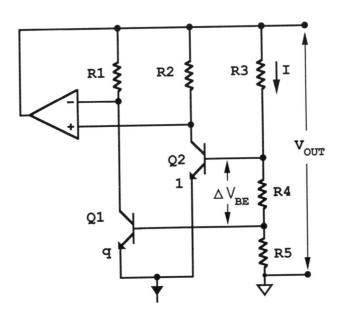

Figure 8.9 VPTAT circuit using op-amp feedback for current balancing [11].

With physically matched resistors, V_{OUT} again depends only on resistance ratios and not on the TC of the resistors. To complete the circuit, the usual start up circuitry and stabilization capacitors must be added.

8.4 INTRINSICALLY REFERENCED TEMPERATURE SENSORS

The temperature sensitivity of PTAT and single transistor temperature sensor, as already indicated, is low in the normal temperature range. A change of 0.1°C at 300 K represents a change of only 1 in 3,000 in a PTAT sensor's output, requiring a 12-bit A/D converter for resolution. On the Celsius scale, which is just the Kelvin scale with a large offset value of 273.15K removed, the same 0.1°C change at what is now 27°C represents a change of only 1 in 270, requiring only an 8-bit A/D converter. It is much less demanding to resolve a 0.4% change rather than a 0.03% change. A shift in the zero of the PTAT voltage, which removes a large constant term and resembles the shift from the Kelvin to the Celsius scale, can be accomplished by subtracting a constant or a linearly decreasing voltage from the PTAT voltage (V_{PTAT}). This is illustrated in Figure 8.5(d), where the zero output temperature T_z represents the zero of an "intrinsically referenced" temperature scale. Two circuits that can accomplish this goal are sketched in Figure 8.10. In Figure 8.10(a), the output voltage V_{OUT} equals $RI_{PTAT} - V_{BE}$, where I_{PTAT} is a PTAT current. In

Figure 10(b), an output current equal to $I_{PTAT} - V_{BE}/R$ is generated by current subtraction. The circuits discussed earlier can be used to generate the I_{PTAT} and the V_{BE}. More generally [12], we can generate intrinsically referenced currents I_1 and bandgap currents I_2 by combining I_{PTAT} currents V_{PTAT}/R with I_{BE} currents generated from V_{BE}, V_{BE}/R, as illustrated in Figure 8.11. The currents are

$$I_1 = \frac{1}{R_1}\left(V_{PTAT} - \frac{R_1}{R_2}V_{BE}\right)$$

$$I_2 = \frac{1}{R_4}\left(V_{PTAT} + \frac{R_4}{R_3}V_{BE}\right)$$

(8.20)

An intrinsically referenced voltage and a bandgap voltage can be generated by passing I_1 and I_2 through resistors. Because these voltages depend only on resistance ratios, no special, low TC resistors are needed. The designer can adjust for the zero voltage output temperature T_z, or the temperature coefficient of the bandgap voltage, by trimming either resistance ratios of ordinary on-chip resistors or the emitter areas in a V_{BE} trim.

When adding or subtracting a V_{BE} voltage from a PTAT voltage, the curvature in the V_{BE} voltage (Figures 8.1 and 8.3) appears. For high-precision, wide-ranging temperature measurement, at least the quadratic nonlinearity of V_{BE} (8.11) must be compensated for, as discussed in Section 8.2.2 in conjunction with Figures 8.2 and 8.3. The circuit techniques developed for curvature correction in bandgap references, such as the addition of extra PTAT or $(PTAT)^2$ currents to the collector current, can be applied here [13,14]. For applications in a narrow temperature range [15], such

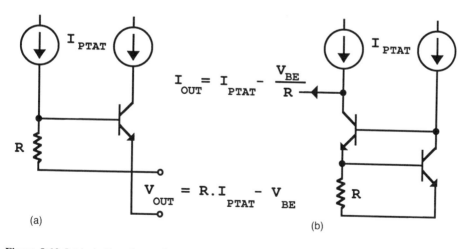

(a)

(b)

Figure 8.10 Intrinsically referenced sources [5]: (a) voltage source; (b) current source.

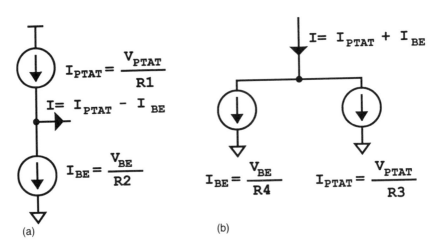

Figure 8.11 Current subtraction and current addition [5]: (a) intrinsically referenced current; (b) bandgap current.

as the biomedical temperature range between 30°C and 45°C, the quadratic term introduces an error of only about 0.02°C, so that a 0.1°C resolution is possible without curvature correction. Curvature correction can be simple in a digital approach.

8.5 TEMPERATURE SENSORS IN CMOS TECHNOLOGY

Digital signals or signals in the frequency or time period domain can be easily used by even the simplest microprocessor. This is one reason for the considerable interest in sensors whose output is directly compatible with a digital system. If a smart sensor with on-chip signal processing and communication is to be realized, CMOS or BiCMOS are the preferred technologies.

8.5.1 Bipolar Transistors in CMOS Technology

Considerable effort has been spent on the design of CMOS bandgap or bandgaplike references [14,16] because in many mixed CMOS analog-digital designs, such as A/D converters, need a stable reference voltage. Techniques developed for bandgap references are directly applicable to temperature sensors. Unfortunately, CMOS bandgap reference and temperature sensor design in a standard CMOS technology is hampered by the absence of thin-film resistors and full function bipolar transistors. Two limited bipolar transistors [17,18] can be realized in any standard CMOS process: a vertical bipolar transistor in which the collector is the substrate, and a more versatile lateral bipolar transistor in which the vertical bipolar transistor unavoidably appears as a parasitic transistor. To avoid repetition, we restrict our discussion in this section to *p*-well, *n*-substrate CMOS processes.

Both bipolar transistors can be illustrated by referring to Figure 8.12, which shows a highly schematic, lengthwise slice through an *n*MOS transistor. For both types of bipolar transistor, the emitter (*E*) is an *N*+ region (the *n*MOS transistor's source [*S*]) in the *p*-well and the base (*B*) is the *p*-well. The collector of the vertical transistor is the substrate (*CS*), whereas the lateral transistor's collector (*C*) is an *N*+ region (the *n*MOS transistor's drain [*D*]) in the *p*-well placed laterally with respect to the emitter, that is, separated from it by the *P*-doped region under the gate. The emitter current of the lateral transistor thus always contains a contribution from the unwanted vertical transistor. The simple vertical bipolar transistor results when the gate (*G*) and the collector (*C*) are eliminated. If the *p*-well is thin and lightly doped, the vertical transistor's current gain can be large, typically 100 or more. The substrate collector *CS* cannot be used by the designer.

In the lateral transistor, the polysilicon gate is used to create two minimally separated *N*+ regions in the *p*-well. When operating the *n*MOS transistor as a lateral transistor, the gate bias must be low enough to prevent inversion under what would normally be the channel of the *n*MOS transistor. The need to bias the gate to prevent inversion represents a drawback of the lateral bipolar transistor, especially for low threshold voltage CMOS processes, in which the polysilicon gate cannot be simply tied to the emitter, but a negative gate voltage must be supplied. The high threshold voltage, parasitic *n*MOS transistor that can be formed from two *N*+ regions and a gate region, in which the gate is just a metal layer over the field oxide layer [18], provides an alternative for low threshold processes. In this case, the metal gate can be tied directly to the emitter. The lateral transistor, with its base width determined by the gate width, competes with the vertical transistor for emitter current. The lateral bipolar transistor action can be optimized by enclosing a minimum-size *N*+ emitter region (source [*S*]) with a minimum-width gate region and the *N*+ collector region (drain [*D*]). The emitter current $I_E - V_{BE}$ characteristics of vertical bipolar transistors closely match the theoretical $I_c - V_{BE}$ form (8.9), because the current gain

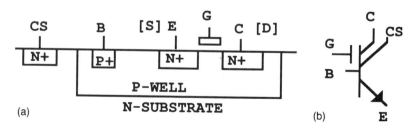

Figure 8.12 CMOS MOS transistor considered as a lateral/vertical transistor: (a) vertical slice; (b) symbol for lateral transistor [17].

with thin, lightly doped p-wells is large enough that the nonideal base current contributions to the emitter current are small. The lateral bipolar transistor's $I_c - V_{BE}$ curves also agree well with the theoretical form. The current splitting ratio $(1:2$ to $1:3)$ between the lateral mode and the vertical mode is nearly constant [19]. Care must be taken in using these transistors to avoid high-level injection.

8.5.2 Integrated Temperature Sensors in CMOS Technology Based on Bipolar Transistors

The bipolar circuits presented earlier can be used in CMOS designs if the constraints on the CMOS bipolar transistors are taken into account. The design constraint with vertical bipolar transistors is that their collectors are tied to the substrate and cannot be used. The simple self-biased design in Figure 8.13(a) uses diode-connected, vertical bipolar transistors with their collectors tied to the substrate, that is, Vdd, to generate the PTAT voltage. The matched nMOS transistors $M3$–$M4$ form a current mirror that forces equal currents I through $Q1$ and $Q2$. $Q2$ is a multiple of $Q1$. The matched pMOS transistors $M1$–$M2$ ideally have equal gate-source voltages so that their sources, nodes A and B, will be at the same voltage. The voltage developed

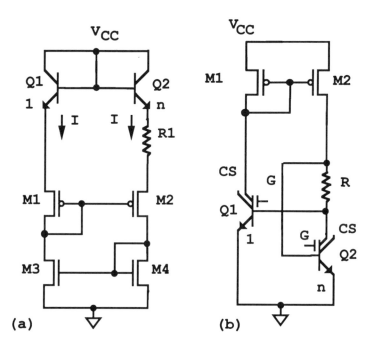

Figure 8.13 Self-biased sources, using (a) vertical bipolar transistors [3] and (b) lateral bipolar transistors.

across the resistor R then equals ΔV_{BE}, or $V_T \ln(n)$, so that the current I will be a PTAT current if R has a zero temperature coefficient. As in the bipolar case (Figure 8.7), the supply voltage rejection and the quality of the circuits $M1$–$M2$ and $M3$–$M4$ can be enhanced by using cascode structures [3] to avoid channel-length modulation effects.

The value of the emitter current in lateral bipolar transistor designs cannot be critical because the emitter current always contains a contribution from the unwanted vertical transistor. The simple self-biased circuit in Figure 8.13(b) takes this constraint into account by tying the emitters to ground. The circuit is a variation of Figure 8.6, in which the emitter resistor was moved to a collector arm and the feedback loop was modified accordingly [17]. The feedback voltage developed across the resistor is again a multiple of the thermal voltage. Current mirrors can be used to bring out the I_{PTAT} currents with both sources.

In Figure 8.14, another self-biased design, similar to the design used by Krummenacher and Oguey [20] in their CMOS temperature sensor, uses an approach reminiscent of the design in Figure 8.13(b), except for the use of more complex current feedback. The temperature-sensing transistors $Q1$ and the feedback $Q3$ form a ratioed current mirror. The current in $Q1$ is then equal to I/r, where I is the current in $Q3$ and the feedback resistor R. With $M1$ and $M2$ matched geometrically, the same

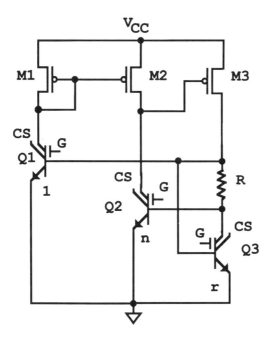

Figure 8.14 PTAT current/voltage source with lateral bipolar transistors [20].

current I/r is forced through $Q1$ and $Q2$, so that the difference in their base-emitter voltages, that is, the voltage across the resistor R, is again a multiple of the thermal voltage. Transistor $M3$ closes the feedback loop. With a large current ratio r, the currents in the temperature sensing transistors $Q1$ and $Q2$ can be kept low to avoid high-level injection even while I is large.

Several methods exist to generate V_{BE}/R currents. In Figure 8.15(a), the circuits of Figures 8.13 or 8.14 can be used to supply $Q1$ with a PTAT collector current I_{PTAT} to generate a V_{BE} voltage. The base-emitter voltage of the vertical bipolar transistor $Q1$ is impressed across the resistor $R1$ with the help of an op amp and the buffer transistor $M1$ to generate the I_{BE} current $V_{BE}/R1$. A simple self-consistant V_{BE}/R circuit [3] cannot be used here as the collector current needs to have a power-law temperature behavior if the discussion of Section 8.2.2 is to apply. Generating an I_{BE} current ($V_{BE2}/R2$) is simpler with the lateral bipolar transistor (Figure 8.15(b)) because the I_{PTAT} current can be fed directly to the lateral collector in the usual bipolar manner; that is, as in Figure 8.10(b). $M2$ shields the base of $Q2$ from output voltage fluctuations. The lateral bipolar transistors may, however, require a negative gate bias voltage. In BiCMOS processes, these difficulties disappear, because at least one type of fully functioning transistor would be available.

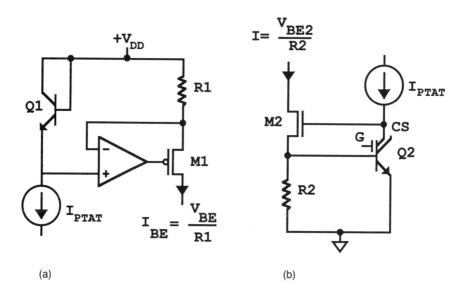

(a)　　　　　　　　　　　　(b)

Figure 8.15 I_{BE} current generators [20]: (a) vertical bipolar current source; (b) lateral bipolar current sink.

8.5.3 Temperature Sensors Based on MOS Transistors and Resistors

The p-well resistors, polysilicon resistors, and MOS transistors are also available for analog design in any standard CMOS process. The p-well resistors generally show a nonlinear temperature dependence following a T^x power law. The exponent x generally falls between 1.8 and 2.5, depending on the doping level of the p-well. In a 3 μm CMOS process that we have used extensively, the exponent has stayed between 1.9 and 2.0 over a period of several years and many fabrication runs. The p-well resistors suffer from nonlinearity, a large production spread in absolute value, and sensitivity to supply voltage variations. In the same CMOS process, the resistance of polysilicon resistors, which are very heavily doped, increases nearly linearly with a temperature coefficient (TC) of only about 800–900 ppm/°C around room temperature. The temperature dependences of both types of resistors are illustrated in Figure 8.16 [21]. In a system in which the temperature-sensitive resistors were incorporated into RC relaxation oscillators and on-chip single-poly EPROMs can be used for correction, the nonlinearity of the p-well resistors proved not to be an obstacle to making a satisfactory sensor [23]. With a quadruple-linear-segment correction, as discussed in Section 8.3.3, a precision of ±0.1°C over the military temperature range has been achieved [21].

The lateral bipolar transistor action discussed in Section 8.5.1 fits into a continuum of nMOS transistor action; as the gate-source voltage is decreased, the transistor passes from the MOS strong-inversion regime through the MOS weak-inversion regime to the lateral bipolar regime. Weak-inversion transistors also exhibit a diode-like exponential current-voltage relationship, which has been used to construct bandgaplike references and temperature sensors [19,21,22]. However, the effects of leakage currents start to become significant in MOS transistors operated in weak inversion at temperatures as low as 60°C, so that they are impractical for wide ranging temperature measurement [21].

The drain currents of MOS transistors in strong inversion have a complex temperature dependence. In saturation, the drain current I_D of an nMOS transistor is given by

$$I_D = D\mu(V_{GS} - V_t)^2 \tag{8.21}$$

Here μ is the surface channel mobility, D is a constant depending on the device geometry and construction, V_{GS} is the gate-source voltage, and V_t is the threshold voltage. The mobility and the threshold voltage have very different effects on the I_D temperature dependences. With $\mu \propto T^{-m}$, $m \approx 1.5$, and

$$V_t = V_{t0} - \beta(T - T_0) \qquad \beta \approx 2 \text{ mV/K} \tag{8.22}$$

as the temperature increases, the mobility μ will tend to decrease the drain current, whereas the decrease in the threshold voltage V_t will tend to increase the drain current. I_D has a positive temperature coefficient (*TC*) at low V_{GS} voltages, a nearly zero temperature coefficient (ZTC) at the *ZTC* voltage, and a negative temperature coefficient at higher V_{GS} voltages. Our ability to effectively select a TC by biasing the gate-source voltage of a MOS transistor makes possible analog temperature correction and the construction of simple, self-biased voltage references [21]. Thermal sensor designs using the temperature dependence of the mobility and the adjustable TC of the drain current for curvature correction have yielded devices with a linearity of better than 0.3°C over the military temperature range. The long-term stability of a MOS transistor's threshold voltage V_t is not as good as the long-term stability of the bipolar transistor's V_{BE} characteristics, because mobile ions in the gate oxide can cause small shifts in V_t and the effective mobility. For nondemanding temperature measurements (±1°C) this should be no problem.

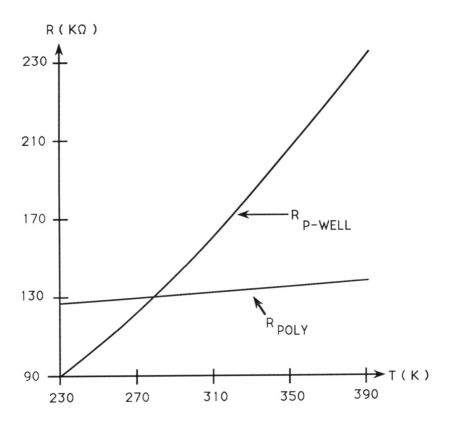

Figure 8.16 Poly and *p*-well resistors in a standard CMOS process.

8.5.4 Temperature Sensors with Digital Output

To make the output of IC temperature sensors compatible with digital systems requires either a pulse-width, a frequency-modulated, or a digital number [12]. Generating a frequency output with a value equal to a temperature reading in Hz, such as 3,000 Hz representing 300.0 K, requires expensive low-temperature-coefficient resistors [24]. Much of analog IC design is based on the observation that component values cannot be precisely predicted, but that component ratios can be accurately matched. We can exploit the matching properties of components to make sensors with a simple to use, digitally compatible output by making the temperature signal proportional to ratios of resistances, currents, or voltages. One way this has been done in a bipolar technology is with a duty cycle oscillator [15]. The concept is illustrated in Figure 8.17. If ΔV is the voltage difference between the Schmitt's thresholds, C is the capacitance, and I_1 and I_2 are two different currents, the ratio of the discharging time P_2 to the charging time P_1 will be equal to the current ratio:

$$P_2/P_1 = (C\,\Delta V/I_2)/(C\,\Delta V/I_1) = I_1/I_2 \tag{8.23}$$

PTAT and V_{BE} currents are generated when V_{PTAT} and V_{BE} voltages are impressed across resistances. If we take for I_1 the "intrinsically referenced" current of (8.20) and Figure 8.11 and if we take for I_2 the "bandgap" current (8.20), the period ratio becomes

$$P_2/P_1 = \frac{V_{PTAT}/R_1 - V_{BE}/R_2}{V_{PTAT}/R_3 + V_{BE}/R_4} = \frac{R_4}{R_1}\left[\frac{V_{PTAT} - (R_1/R_2)V_{BE}}{V_{BE} + (R_4/R_1)V_{PTAT}}\right] \tag{8.24}$$

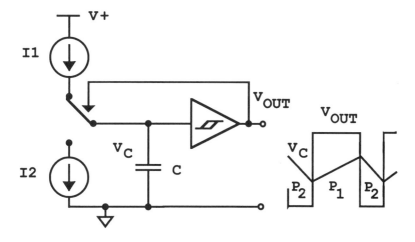

Figure 8.17 Current-to-frequency or period oscillator and its signals [15].

The ratio R_1/R_2 determines the temperature T_z at which the period ratio is 0. The ratio R_4/R_1 can be adjusted to yield the bandgap voltage. Because only ratios of resistances with matching temperature characteristics are involved, the individual resistor's temperature dependencies cancel, so that ordinary on-chip resistors can be used for trimming. An added benefit of this method is that we can shift the resistance ratio R_4/R_1 slightly away from the value needed for a zero TC bandgap voltage to realize a small linear term in the denominator that can compensate for the quadratic nonlinearity of the V_{BE} term in the numerator [25]. To illustrate the idea, assume that I_1 and I_2 have the quadratic and linear forms shown in (8.24). Their ratio can then be approximated by

$$
\begin{aligned}
I_1/I_2 &\cong \frac{I_{10}(1 + B\Delta T + C(\Delta T)^2)}{I_{20}(1 + D\Delta T)} \\
&\cong \frac{I_{10}}{I_{20}} [1 + (B - D)\Delta T + (C - BD + D^2)(\Delta T)^2 + \ldots
\end{aligned}
\tag{8.25}
$$

If the quadratic term in I_1 is low compared to its linear term, we can eliminate the quadratic term on the right by adjusting D by trimming the resistor ratio $R4/R1$.

A major advantage of a ratiometric approach is that the stringent conditions, that I_1 should be linear and I_2 should be constant, can be relaxed, and replaced by the condition that their ratio should be a linear function of the temperature. We can thus use two currents, I_1 and I_2, which taken individually are nonlinear but whose ratio is a linear function of the temperature.

By working in a CMOS technology, we can complete the A/D conversion and carry out numerical corrections on-chip. If I_1 and I_2 now drive separate relaxation oscillators, that is, current-to-frequency (I/F) converters, with periods P_1 and P_2, respectively, then oscillator I_2 can be used to numerize the current ratio I_1/I_2 by counting the number N of pulses of oscillator I_1 in a fixed number N_o of pulses of the oscillator I_2. If the current ratio varies linearly with the temperature, adjusting N_o is equivalent to setting the slope—that is, the pulse-number per unit-temperature-interval ratio, for example—to 10 pulses per °C or 100 pulses per °F or as desired. The conversion to a temperature scale can be completed by including a digital offset M_o in the output counter (Figure 8.18) that adjusts the value in the output counter so that its 0 coincides with the 0 of the temperature scale.

The scale conversion factor N_o and the scale offset factor M_o can be stored, for example, in an on-chip EPROM [20,21]. If the ratio of I_1/I_2 is not linear, the design in Figure 8.18 can be easily generalized to implement multiple-linear-segment corrections. Alternatively, we could use the output count N to generate an address in an EPROM look-up table. The usual digital data processing techniques can be applied to the output signal. This approach has been tested in a CMOS temperature sensor using p-well and polysilicon resistor relaxation oscillators with a quadruple-

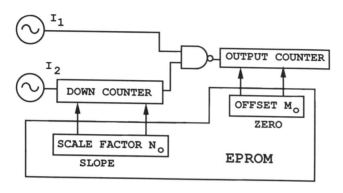

Figure 8.18 Simple dual oscillator numerization circuit [21].

linear-segment correction and also with a MOS-current-oscillator design. A drawback with the dual oscillator approach is that on-chip cross talk can be a problem.

If an outside clocking oscillator is available [7,20], only a single current-to-frequency converter is needed. The clocking oscillator need not be particularly sophisticated as long as it is stable over a measurement period. The principle of the data conversion is the same as with two oscillators, if a two-step approach is adopted. In the first time period, I_2 is switched into the current to frequency (I/F) converter, and the number of clocking oscillator pulses N_C in N_o pulses of I_2 is counted. This time period is then reproduced in the second time period by running the clock counter for N_C cycles. During this period, I_1 is switched to the I/F converter and its pulses are counted in the output counter. A CMOS approach [20] using this technique with a lateral bipolar transistor and three-point calibration has resulted in a sensor with digital output linear to within $\pm 0.5°C$ over the military temperature range. This approach has the advantage that the same I/F converter is used for both I_1 and I_2 and that either a system clock or a very compact, low-power clocking oscillator can be used, thus keeping low the overall chip-current use, and the attendant self-heating.

8.6 POLYSILICON RESISTORS

Polycrystalline silicon films consist of small crystallite grains separated by grain boundaries. The resistivity of polysilicon is a complex function of crystallite grain size, doping concentration, and temperature [25,26]. At low doping concentrations, the properties of the grain boundaries and the associated defect-generated trapping states dominate electrical conduction. Mobile charge carriers are trapped at the boundaries and become immobile, establishing potential barriers. The resistivity at low doping levels can be modeled by

$$\rho = \rho_0 \exp[E_B/kT]$$

$$(8.26)$$

where E_B is an activation energy reflecting the barrier height. As the doping increases, the barrier energy E_B initially increases until all the traps are filled. As more impurity atoms are introduced, the potential barrier and its associated dipole layers are squeezed and reduced. At very large impurity concentrations or large grain sizes, the barriers are less important and the polysilicon resistivity is principally determined by the bulk resistivity of the grains. The resistivity of polysilicon [27], as shown in Figure 8.19, can thus be controlled by changing the doping from a lightly doped regime, in which the resistivity is very large and decreases exponentially with increasing temperature, to a heavily doped regime of bulk conduction, in which the resistivity is low and increases slightly with temperature. With the resistivity ρ and its thermal behavior very sensitive to both the doping concentration and the thermal processing history, precise control of the resistivity, and thus its temperature characteristic, is difficult to achieve, especially for high-resistivity polysilicon.

Polysilicon films grown on dielectric films such as SiO_2 or Si_3N_4 do not depend on diode isolation, so they can be used at much higher temperatures than diodes or implanted resistors. They are also affected much less by substrate bias voltages than implanted resistors and have very low parasitic capacitances. Polysilicon resistors show excellent long-term stability, especially when passivated with Si_3N_4 to inhibit impurity migration and when implanted with boron to avoid carrier segregation [27]. Polysilicon resistors follow Ohm's law closely at low electrical fields.

Lightly doped polysilicon resistor's exponential dependence on inverse temperature can be used to make thermistorlike thermal sensors. Polysilicon resistors can be linearized with networks of polysilicon resistors of different temperature coefficients by employing techniques similar to those used with thermistors. A highly

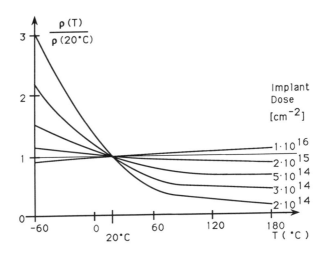

Figure 8.19 Normalized resistivity change of boron-doped polysilicon with temperature (after [27]).

linear temperature sensor with a TCR of 3,500 ppm and a conformity to linearity of better than 1%/FS between −60°C and 180°C has been reported in the literature [27]. Polysilicon resistors' high thermal sensitivity and compatibility with standard IC processing steps make them attractive for applications such as resistors for microbridge silicon hot-wire anemometers [28] or for temperature-corrected pressure sensors [29].

8.7 SUMMARY

Even with our rather complete understanding of the thermal behavior of bipolar and MOS transistors, the design of IC thermal sensors remains an art; requirements such as high temperature sensitivity, long-term stability, excellent supply voltage rejection, simple calibration, complete linearization, and signal processing all have to be met while the chip operates over a large temperature range and uses extremely little power to avoid self-heating.

Bipolar temperature sensors and bandgap voltage references are based on what is probably the most studied active-device characteristic, the base-emitter voltage V_{BE}–collector current I_c relationship. The simplest, and commercially most successful, IC temperature sensors generally use ΔV_{BE} and have a voltage or current output in units such as mV/K or $\mu A/K$. More sophisticated intrinsically referenced sensors combine V_{BE} and ΔV_{BE} on the same chip in a fashion analogous to bandgap circuits to eliminate a large offset similar to that between the Kelvin and the Celsius scale that make difficult A/D conversions of the voltage or current output.

We can make the output of an IC temperature sensor more compatible with digital systems if it is converted to a frequency or time period output by adding current-to-frequency converters, such as RC relaxation oscillators, on the same chip. To complete the conversion to a numerical output, low-power CMOS processing is called for. Because simple EPROMs can be constructed even in a single-poly CMOS process, "soft" linearization and calibration, and data conversion with numerical techniques can also be carried out on the same chip. The problem of working in a CMOS technology is the absence of full-function bipolar transistors. Some sophisticated solutions involving vertical and lateral bipolar transistors, resistors, and MOS transistors have been proposed. BiCMOS offers the promise of a nearly ideal solution to the construction of the "smart" temperature sensor: bipolar sensing and CMOS data processing. For higher temperatures (200–300°C) and in micromachined thermal sensor applications, doping-controlled polysilicon-resistor-based sensors are promising.

REFERENCES

[1] *Omega Complete Temperature Measurement Handbook and Encyclopedia*, Vol. 26, Stamford, CT, Omega Engineering Inc., 1989, pp. U-1–24.

[2] T. C. Verster, "The Silicon Transistor as a Temperature Sensor," in *Temperature, Its Measurement and Control in Science and Industry*, Vol. 4, ed. H. H. Plumb, Pittsburgh, ISA, pp. 1125–1134.

[3] P. R. Gray and R. G. Meyer, *Analysis and Design of Analog Integrated Circuits*, 2d ed., New York, John Wiley & Sons, 1984, pp. 289–292, 730–737.

[4] Y. P. Tsividis, "Accurate Analysis of Temperature Effects in I_c-VBE Characteristics with Application to Bandgap Reference Sources," *IEEE J. Solid-State Circuits*, Vol. SC-15, 1980, pp. 1076–1084.

[5] G. C. M. Meijer, "Thermal Sensors Based on Transistors," *Sensors and Actuators*, Vol. A10, 1986, pp. 103–125.

[6] A. Ohte and M. Yamagata, "A Precision Silicon Transistor Thermometer," *IEEE Trans. Instrum. Meas.*, Vol. 26, 1977, pp. 335–341.

[7] G. C. M. Meijer, R. Van Gelder, V. Norder, J. Van Drecht, and H. Kerkvliet, "A Three-Terminal Integrated Temperature Transducer with Microcomputer Interfacing," *Sensors and Actuators*, Vol. A18, 1989, pp. 195–206.

[8] M. P. Timko, "A Two-Terminal IC Temperature Transducer," *IEEE J. Solid-State Circuits*, Vol. SC-11, 1976, pp. 784–788.

[9] R. C. Dobkin, "Input Supply Independent Circuit," U.S. Patent 3,930,172, 1975.

[10] A. P. Brokaw, "A Simple Three-Terminal IC Bandgap Reference," *IEEE J. Solid-State Circuits*, Vol. SC-9, 1974, pp. 388–393.

[11] *Linear Data Book*, Santa Clara, CA, National Semiconductor Corporation, 1982, pp. 9-25–32.

[12] G. de Haan, G. C. M. Meijer, "An Accurate Small-Range IC Temperature Transducer," *IEEE J. Solid-State Circuits*, Vol. SC-15, 1980, pp. 1089–1091.

[13] R. A. Pease, "A New Fahrenheit Temperature Sensor," *IEEE J. Solid-State Circuits*, Vol. SC-19, 1984, pp. 971–977.

[14] B. S. Song and P. Gray, "A Precision Curvature-Corrected CMOS Bandgap Reference," *IEEE J. Solid-State Circuits*, Vol. SC-18, 1983, pp. 634–643.

[15] G. C. M. Meijer, A. J. M. Boomkanp, and R. J. Duguesnoy, "An Accurate Biomedical Temperature Transducer with On-Chip Microcomputer Interfacing," *IEEE J. Solid-State Circuits*, Vol. SC-23, 1988, pp. 1405–1410.

[16] J. Michejda and S. K. Kim, "A Precision CMOS Bandgap Reference," *IEEE J. Solid-State Circuits*, Vol. SC-19, 1984, pp. 1014–1021.

[17] E. A. Vittoz, "MOS Transistors Operated in the Lateral Bipolar Mode and Their Application in CMOS Technology," *IEEE J. Solid-State Circuits*, Vol. SC-18, 1983, pp. 273–279.

[18] M. G. R. Degrauwe et al., "CMOS Voltage Reference Using Lateral Bipolar Transistors," *IEEE J. Solid-State Circuits*, Vol. SC-20, 1985, pp. 1151–1157.

[19] E. Habbekotte, "Silicon Temperature Sensors," *Bull. ASE/Ucs*, Vol. 76, 1985, pp. 272–276.

[20] P. Krummenacher and H. Oguey, "Smart Temperature Sensor in CMOS Technology," *Sensors and Actuators*, Vols. A21–A23, 1990, pp. 636–638.

[21] J.-Y. Zhu, W. Rasmussen, S. Richard, and D. Cheeke, "Ultrastable Integrated CMOS Oscillator," *Int. J. Electronics*, Vol. 70, 1991, pp. 433–441.

[22] B. J. Hosticka, J. Fichtel, and G. Zimmer, "Integrated Monolithic Temperature Sensors for Acquisition and Regulation," *Sensors and Actuators*, Vol. A6, 1984, pp. 191–200.

[23] W. Rasmussen, J.-Y. Zhu, S. Richard, and D. Cheeke, "CMOS Intelligent Temperature Sensor," *IEEE Proc. 33rd Midwest Symp. Circuits and Systems*, Calgary, Canada, 1990, pp. 849–852.

[24] B. Gilbert, "A Versatile Monolithic Voltage-to-Frequency Converter," *IEEE J. Solid-State Circuits*, Vol. SC-11, Dec. 1976, pp. 852–864.

[25] J. Y. W. Seto, "The Electrical Properties of Polycrystalline Silicon Films," *J. Appl. Phys.*, Vol. 46, 1975, pp. 5247–5254.

[26] N. C.-C. Lu, L. Gerzberg, C.-Y. Lu, and J. D. Meindl, "Modeling and Optimization of Monolithic Polycrystalline Silicon Resistors," *IEEE Trans. Electron. Dev.*, Vol. ED-28, 1981, pp. 818–830.

[27] E. Obermeir, P. Kopystynski, and R. Niessl, "Characteristics of Polysilicon Layers and Their Application to Sensors," IEEE Solid-State Sensor Workshop, Hilton Head Is., SC, 1986.

[28] Y.-C. Tai and R. S. Muller, "Lightly Doped Polysilicon Bridge as an Anemometer," *Transducers '87*, IEE of Japan, 1987, pp. 360–363.

[29] P. Kopystynski and E. Obermeier, "An Interchangeable Silicon Pressure Sensor with On-Chip Compensation Circuitry," *Sensors and Actuators*, Vol. A18, 1989, pp. 239–245.

Chapter 9
Planar Silicon Photosensors

Jon Geist
Sequoyah Technology

9.1 INTRODUCTION

The general topic of photosensors is so large that many books have been devoted to it, and only a most superficial overview of the whole topic can be covered in a single chapter. In fact, it takes at least a chapter just to introduce the major types of photosensors and related literature. References [1–7] are a good start on an overview of the whole subject. This chapter will focus on planar silicon photodiodes, the type most readily integrated with signal processing electronics on a single chip.

Six major aspects of photosensor operation are associated with integration. The first four are the traditional areas of photosensor performance: the region of the radiation spectrum to which the photosensor responds, the accuracy with which the output electrical signal can be related to the input optical signal, the minimum optical signal that can be detected, and the speed with which the optical signal is transduced into an electrical signal for further processing. The last two aspects are specific to integrated microsensors: the level of integration as measured by the number of photosensors on a chip, and the level of integration as measured by the complexity of the on-chip signal processing.

It should be pointed out that the optimum technology for maximizing the performance of a sensor in one or more of the traditional areas of performance is not necessarily the optimum technology for maximizing its performance in the areas of sensor integration. Consequently, three different approaches may be taken to developing integrated photosensors for any particular application. The first approach is to develop the photosensor in an existing sensor technology that is optimum from the point of view of the traditional measures of performance for a particular application and to develop circuitry in this same technology. Attempts to develop infrared focal plane arrays and associated on-chip electronics using HgCdTe and InSb technologies are excellent examples of this approach [7]. This approach is quite difficult if the optimum sensor technology is not well suited for producing either circuits or multiple sensors.

The second approach is to extend the performance of an existing integrated circuit technology to achieve a high, if not optimum, level of sensor performance in a particular application. The application in the X-ray spectral region of silicon photodiodes designed for use in the visible region [8], the application of planar silicon technology to produce photodiodes that are optimized for use in the vacuum ultraviolet and X-UV spectral regions [9], the extension of planar silicon technology to the fabrication of metal silicide CCD arrays for use in the infrared region [7], and the extension of planar silicon technology to the fabrication of blocked impurity band detectors for use in the low-background infrared region [10,11] are examples of this approach.

The third approach is to develop a new technology that is optimum with respect to the sensor and signal processing requirements of the particular application. The development of III–V compound semiconductor technologies for optoelectronics and higher speed electronics is an important example of this approach [6]. In addition to developing a new technology totally from scratch, it is possible to develop a new technology as a hybrid of an existing circuit and existing photosensor technology. HgCdTe focal plane arrays coupled by indium bump technology to silicon CCD arrays for readout are an example of this approach [7].

With a few important exceptions, which have already been mentioned, the second approach is by far the most practical for integrating photosensors because silicon technology currently dominates both integrated circuit and photosensor technology. Since its invention in 1954 [12], the planar silicon photodiode has become the most widely used photosensor in the visible and near visible spectral regions, as well as the most widely used device for solar conversion in both terrestrial and space applications. The major limitation of planar silicon technology, or any silicon technology for that matter, is that it cannot produce photodiodes that respond to photons with photon energies less than about 1.1 eV, the bandgap of silicon, while operating at room temperature. Consequently, HgCdTe and InSb have the reputation of being the material of choice for infrared photosensors. However, for applications requiring focal plane arrays or other types of integrated photosensors, which must be cooled to the temperature of liquid nitrogen or lower to achieve the desired performance, metal silicide [7] and blocked impurity band photodiodes [10,11] fabricated in extensions of planar silicon technology may outperform HgCdTe and InSb, depending upon the details of the application.

Silicon photodiodes are the prototypical integrated sensor. Not only were silicon photodiodes the first sensors to be fabricated in new planar technologies following the development of the latter, they also played an important role in the development of these technologies. For instance, diode rectifiers [13] that could be used as photodiode solar cells [12] were the first devices fabricated to demonstrate the diffusion [14] of boron into *n*-type silicon wafers.

Because the existing germanium photodiode technology was transferred to silicon as soon as good quality silicon became available [1], it is difficult to identify the first application of silicon planar technology to the fabrication of photodiodes. *P-n* junctions passivated by silicon dioxide described in 1963 [15] would seem to qualify even though they were intended to measure high-energy particles. Similarly, silicon photodiode arrays developed for use in vidicons appear to be the first sensors fabricated with planar technology to have more than one sensor on a chip [16]. Finally, imaging was anticipated as an important potential application in the original proposal [17] for the CCD array and demonstrated [18] with one of the first CCD devices, an 8-bit shift register.

Key to understanding the performance of silicon photodiodes are the concepts of photocurrent, reflectance, external quantum efficiency, and internal quantum efficiency. The photocurrent is the current flowing in an external circuit connected to the photodiode as a direct result of photons incident on the active area of the photodiode. Any current flowing in the absence of incident photons is called a *dark current* and must be subtracted from the total current to determine the photocurrent. The external quantum efficiency is the ratio of the photocurrent (electrons per second) to the rate at which photons are incident on the active area of the photodiode (photons per second). The internal quantum efficiency is the ratio of the photocurrent to the rate at which photons enter the active area of the photodiode. The reflectance is the ratio of the rate at which the photons are reflected away from the active area to the rate at which they are incident on the same area. Therefore, the external quantum efficiency is the product of one minus the reflectance and the internal quantum efficiency. The relation of these key photodiode properties to more fundamental material and device properties is presented in detail in Section 9.3.

Silicon photodiodes are particularly outstanding in applications requiring high accuracy. The key concept in such applications is the uncertainty with which the sensor output can be related to the rate at which photons are incident on the active area of the photosensor. For photodiodes, this uncertainty is determined by the uncertainty in the external quantum efficiency.

For a few years following 1980, some high-quality silicon photodiodes were capable of a level of accuracy and precision in the visible spectral region that exceeded that available from calibration against the radiometric standards and techniques in existence at that time [19]. As a result, new calibration techniques based on amplitude-stabilized lasers and silicon photodiode self-calibration experiments were developed, as was a new generation of conventional radiometric instrumentation and techniques [20]. An international comparison of scales of detector spectral response has demonstrated that silicon photodiode self-calibration is capable of uncertainties less than ±0.2% over much of the visible spectral region [21].

In recent years, much attention has been given to ultrahigh-accuracy applications. Radiometers based on one type of planar silicon photodiode have been shown to be capable of interpolating two external quantum efficiency calibrations having uncertainties on the order of ± 0.0003, one at 1.45 eV and another at 2.8 eV, to the spectral region between them with no increase in uncertainty [22]. Also, simulations of experimental results indicate that another type of planar silicon photodiode has an internal quantum efficiency of 0.9997 ± 0.0003 at 2.8 nm [23], and it has been experimentally demonstrated that the reflectance of radiometers based on this type of photodiode can be measured with an uncertainty smaller than 0.0001 [24].

Recently, a new self-calibration technique was developed for the X-ray spectral region. This technique is even simpler than the original self-calibration technique for the visible spectral region. It requires nothing more than recording the photocurrent as the photodiode is rotated in a constant intensity X-ray beam and fitting the results to a three parameter model of the photodiode [25].

In the remainder of this chapter we will introduce a generic planar silicon photodiode structure and review the fundamentals of photosensors in terms of that particular structure. This review will touch on all of the areas of photosensor performance mentioned already except the level of sensor integration from the point of view of the number of photosensors cointegrated on a single chip. This topic, which is best illustrated by CCD array technology, is presented in the next chapter.

9.2 PLANAR SILICON PHOTODIODES

Figure 9.1 shows a common planar structure used for commercial silicon photodiodes. Variations of this structure are readily cointegrated with circuitry in standard planar processes. This structure is most commonly fabricated commercially in a lightly to moderately doped, 200–300 μm thick, n-type silicon wafer (the substrate). The rear of the wafer is heavily doped with an n-type dopant either by high-temperature diffusion or by ion implantation. The front of the wafer is heavily doped with a p-type dopant either by high-temperature diffusion or by ion implantation, in such a way as to produce a shallow junction. The oxide layer at the front of the photodiode

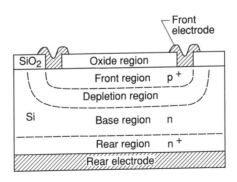

Figure 9.1 A common planar structure used for commercial silicon photodiodes.

is either grown by thermal oxidation of the wafer or by chemical vapor deposition. Metal electrodes are deposited on the p^+ and n^+ regions to make ohmic contacts for electrical connection to the photodiode. The rear electrode is often replaced by a second front electrode attached to an n^+ region formed outside the p^+ region for better compatibility with standard planar processing. A common variation of the structure of Figure 9.1 has an n^+ front region and a p-type base.

The junction between the p^+- and n-type silicon gives rise to a built-in electric field in the depletion region. Consider photons of energy $h\nu$ incident from a direction that makes the angle Θ with the normal to the front surface of the planar photodiode in Figure 9.2. Some of these photons are reflected away from the photodiode at the air-oxide interface, and the rest enter the oxide layer as illustrated in the figure. There, some are multiply reflected between the oxide-silicon and air-oxide interfaces, thus contributing to the reflectance of the photodiode front surface; and the rest penetrate into the silicon, where they are absorbed. The absorption density is given by the absorption coefficient $\alpha(h\nu)$ of silicon. A fraction given by $1 - \exp[-\alpha(h\nu)x_4]$ of the photons that enter the silicon will be absorbed before they reach the rear of the photodiode. Of the remaining fraction $\exp[-\alpha(h\nu x_4)]$, some will be transmitted through the rear surface of the silicon substrate, some will be reflected back toward the front surface, and some more will be absorbed. On the average, the absorbed photons will create $\eta(h\nu)$ electron-hole pairs per absorbed photon, where $\eta(h\nu)$ is the quantum yield of silicon for electron-hole pair production.

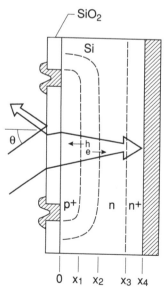

Figure 9.2 Illustration of the physical processes associated with the generation of a photocurrent by the structure of Figure 9.1 (rotated through 90°).

Now consider a photon that is absorbed in the depletion region of the device illustrated in Figure 9.2 to create an electron-hole pair there. The electron and the hole will either drift in opposite directions under the influence of the built-in field until they exit the depletion region, or one or both of them will recombine with a carrier of the opposite type before they exit the depletion region. In the former case, the electron will end up in the n-type silicon and the hole will end up in the p-type silicon, thereby causing a charge imbalance between these regions. If an external short circuit is connected between the electrodes attached to these regions, an electron will flow from the n^+ region to the p^+ region to correct the charge imbalance. If either the electron or the hole recombines before it exits the depletion region, no charge imbalance will be created.

The time it takes carriers to drift across the depletion region is negligible compared to the recombination lifetime. However, this is not necessarily true for transport of carriers in the quasi-neutral regions of p-n junction photodiodes. For instance, if a photon is absorbed deep within the n-type region to create an electron-hole pair, separation of the electron-hole pair is dependent upon the minority-carrier hole diffusing to the edge of the depletion region and being swept across that region before it recombines. What happens to the majority-carrier electron created simultaneously with the hole is unimportant. Diffusion from deep within the n-type region is a very slow process compared to drift across the depletion region, and much more time is available for recombination during diffusion. Thus, this time is not necessarily negligible with respect to the recombination lifetime.

9.3 FUNDAMENTALS OF PHOTOSENSORS

The fundamental concepts of photosensors are the interaction of the material properties with the device structure to determine the spectral variation of the quantum efficiency, noise current, and the response time. With minor variations the model described here is widely used to describe the performance of photodiodes and solar cells made of various materials, yet its validity has been questioned as recently as 1981 [26]. Reference [27] summarizes the physical (quantum mechanical) justification for this model. The model will be presented in terms of the planar silicon photodiode structure described in the preceding section, but the model applies with appropriate structure-related modifications to any type of photodiode or photosensor that behaves like a photodiode, including blocked impurity band detectors.

9.3.1 Material Properties

The ultimate limits of performance that can be obtained with any given photodiode are set by the properties of the material from which the photodiode is fabricated. During processing it is easy to introduce unintentional defects into the photodiode that reduce its performance from the limits set by the properties of defect-free material. In fact, processing the material in such a way as to produce the desired semiconductor device without introducing unintentional defects that compromise its performance is a major challenge of processing technology. However, this is much less of a problem with standard silicon processing than with other semiconductor processing technologies, and it is one of the reasons that silicon continues to dominate markets for which it is less ideal than other semiconductors. This section introduces the most important materials properties for photodiodes.

The quantum yield [28,29] for electron-hole pair production by photons of energy $h\nu$ in a material is defined as

$$\eta(h\nu) = [1 + \beta(h\nu)]\alpha_1(h\nu)/\alpha(h\nu) \tag{9.1}$$

where

$$\alpha(h\nu) = \alpha_0(h\nu) + \alpha_1(h\nu) \tag{9.2}$$

and $\alpha_j(h\nu, x)$ is the jth component of the absorption coefficient spectrum of the material. The partition of the absorption coefficient spectrum into a sum of terms is such that the jth component describes absorption by those processes that create exactly j free-carrier pairs (or j plasmons) per absorbed photon. The quantity $\beta(h\nu)$ describes the creation of extra free-carrier pairs through a cascade of impact ionizations. The impact ionizations involve collisions of the energetic particles created by the absorption process with bound-carrier pairs and are a mechanism through which the energetic particles lose kinetic energy while relaxing to their lowest energy, free-carrier states [30,31]. This process, which is fast compared to the process by which the free carriers recombine into bound-carrier pairs, competes with phonon emission, which is also a fast process. The 1 in the factor $1 + \beta(h\nu)$ corresponds to the creation of one free-carrier pair by the fundamental absorption process described by $\alpha_1(h\nu)$.

Equation (9.2) ignores multiphoton mechanisms and mechanisms creating more than one free-carrier pair as part of the fundamental absorption process, because these mechanisms are negligible at the photon energies and rate densities encountered in most, if not all, practical applications of photosensors.

At room temperature only the excitation of a valence band or core electron into the conduction band to create a free electron-hole pair or to create a plasmon contributes significantly to $\alpha_1(h\nu)$. Figure 9.3 [32–35] plots $\alpha_1(h\nu)$ from $h\nu = 1.05$ eV to $h\nu = 1000$ eV for pure (intrinsic) silicon at room temperature. Many other spectra covering portions of the spectral range of Figure 9.3 have been reported. The agreement between them is surprisingly poor. At the writing of this chapter, $\alpha_1(h\nu)$ is known with an uncertainty of 2% at a photon energy of 1.9587 eV (the 633 HeNe laser wavelength) [36] and less accurately at other photon energies.

At room temperature only the excitation of free electrons or holes to higher kinetic energy contributes significantly to $\alpha_0(h\nu)$ in silicon. Because the number of free carriers can vary over many orders of magnitude in a semiconductor, the important quantity is the absorption coefficient per unit density of free carriers. This quantity, whose units are area, is called the *absorption cross section* for free carrier absorption. Multiplication of the absorption cross section by the density of free carriers gives the absorption coefficient. For silicon, the absorption cross section for free carriers can be approximated by

$$\sigma_0(h\nu) = 7.4 \times 10^{-18} \text{ eV}^2 \text{ cm}^2 (h\nu)^{-2} \tag{9.3}$$

in the vicinity of the silicon band edge.

Figure 9.4 compares the sum of $\alpha_1(h\nu)$ calculated from [33] and $\alpha_0(h\nu)$ calculated from (9.3) for 10^{20} cm^{-3} free carriers with measured values [37] of $\alpha(h\nu)$ for silicon doped with 10^{20} boron atoms cm^{-3}. The data are plotted as a function of the photon wavelength $\lambda = 1.23985$ μm eV$/h\nu$.

Figure 9.5 compares the results of a theoretical calculation [28] and an experimental measurement [38] of $\beta(h\nu)$ for field-free [39] silicon in the spectral region where $\beta(h\nu)$ becomes nonnegligible compared to unity. At higher photon energies $\beta(h\nu)$ approaches 0.273 ± 0.003 pairs per eV of photon energy asymptotically for silicon [25,30].

All of the material properties considered so far are optical properties. Silicon also has electronic properties. However, with a single exception, photodiodes are quite insensitive to these properties. This is not the case with photoconductors, which

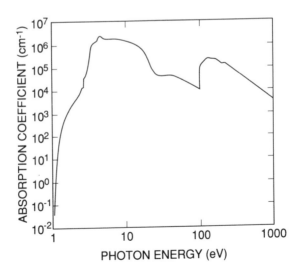

Figure 9.3 Absorption coefficient of intrinsic silicon at room temperature from 1 to 1,000 eV [32–35].

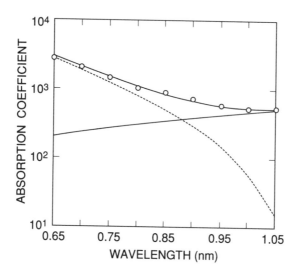

Figure 9.4 The absorption coefficient of 10^{20} free carriers per cm^3 in silicon at room temperature (lower solid line) from (9.3), the absorption coefficient of intrinsic silicon at room temperature (dashed line) from a formula in [33], their sum (upper solid line), and experimental data [37] (open circles) for the absorption coefficient of silicon doped with 10^{20} cm^{-3} boron atoms.

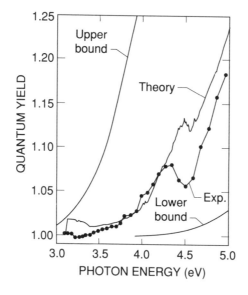

Figure 9.5 Comparison of quantum yield for silicon as calculated (Theory) from a first-principles silicon band structure corrected to give the measured silicon bandgap [28], and as measured (Exp.) on a silicon photodiode [38]. Upper and lower limits for all possible band structures having the same bandgap as silicon are also shown.

are also sensitive to many electronic and nonoptical environmental properties. Indeed, this is the major reason that photoconductivity is an important research tool in physics, as well as the major reason that the photodiode structure provides better photosensor performance in many applications.

Contrary to their relative insensitivity to other electronic properties, photodiodes are quite sensitive to the recombination lifetime of excess carriers in different parts of the photodiode. This quantity, which directly affects the quantum efficiency of the photodiode, is highly dependent upon processing technology. Consequently, it is treated as if it were a device property like a doping distribution, rather than a material property like the absorption coefficient spectrum.

9.3.2 Quantum Efficiency

The external quantum efficiency of a photodiode is the ratio of the photocurrent generated by the photodiode to the photon flux incident on the active area of the photodiode. Subject to a number of simplifying assumptions, the external quantum efficiency of the planar silicon photodiode in Figure 9.2 for photons of energy $h\nu$ and polarization p with respect to the normal to the active area of the photodiode is well approximated by

$$\varepsilon_e(h\nu) = T(h\nu)\eta(h\nu)F(h\nu) \tag{9.4}$$

where $F(h\nu)$ is called the *collection efficiency*, and $T(h\nu)$ is the transmittance of the oxide structure given by

$$T(h\nu) = 1 - \rho(h\nu) - a(h\nu) \tag{9.5}$$

where $\rho(h\nu)$ is the reflectance of the photodiode structure, and $a(h\nu)$ is the absorptance of the oxide structure.

The external quantum efficiency $\varepsilon_e(h\nu)$ is the only quantity in (9.4) that is directly measurable. However, $\rho(h\nu)$ is also directly measurable, so it is common to write (9.4) in the form

$$\varepsilon_e(h\nu) = [1 - \rho(h\nu)]\varepsilon_i(h\nu) \tag{9.6}$$

where $\varepsilon_i(h\nu)$ is called the *internal quantum efficiency* of the photodiode. In spectral regions where $a(h\nu) = 0$, $\varepsilon_i(h\nu) = \eta(h\nu)F(h\nu)$ and $T(h\nu) = 1 - \rho(h\nu)$, but these relations are not valid in general. Usually (9.6) is more useful for the discussion of experimental results, whereas (9.4) is more readily used in theoretical discussions.

9.3.3 Collection Efficiency Models

Subject to the same assumptions needed to write (9.4), $F(h\nu)$ is given by

$$F(h\nu) = \int_0^{x_4} \exp[-\alpha(h\nu)x]\alpha(h\nu)P(x)dx \qquad (9.7)$$

where $P(x)$, which is called the *collection probability function* or the *collection efficiency distribution*, is the probability that an electron-hole pair created at the point x in the silicon will be collected (separated) by the built-in electric field.

The interpretation of (9.4) and (9.7) is straightforward. A fraction $T(h\nu)$ of the photons incident on the front surface of the photodiode is transmitted through the oxide into the silicon. A fraction $\exp[-\alpha(h\nu)x]$ of these penetrate to the point x, where a fraction $\alpha(h\nu)$ is absorbed per unit length at the point x to create $\eta(h\nu)$ electron-hole per absorbed photon. A fraction $P(x)$ of the electron-hole pairs created at the point x is separated by the built-in field to induce a photocurrent in a virtual short circuit across the photodiodes electrodes.

The simplifying assumptions associated with the use of (9.4) and (9.7) to describe the photodiode of Figure 9.2 are as follows:

1. Equation (9.4) will be applied for normal incidence;
2. $P(x)$ is 0 in the oxide;
3. $\alpha(h\nu)$ and $\eta(h\nu)$ are independent of position in the photodiode;
4. The cascade of impact ionizations described by $\beta(h\nu)$ are completed before any significant carrier transport occurs;
5. $\alpha(h\nu) x_4 \gg 1$;
6. No gain is associated with the carrier transport;
7. The photocurrent is linearly related to the incident photon flux.

The assumptions associated with (9.4) and (9.7) introduce little or no error in most applications and are a serious source of error only in special cases.

Assumption 1 can be removed by including the explicit dependence of $T(h\nu)$ on the angle of incidence θ and polarization p with respect to the plane of incidence of the photons on the photodiode and by replacing x by $x/\cos\theta_{Si}$ in the exponential function under the integral in (9.7), where $\theta_{Si}(\theta)$ is angle of propagation of photons in the silicon as a function of the angle of incidence θ of the photons falling on the front surface of the photodiode. Assumption 2 can be removed by adding a term similar to the righthand side of (9.7) to describe collection from the oxide. This term is usually small even with energetic photons [9]. Assumption 3 can be removed by including the explicit dependence of $\alpha(h\nu)$ and $\beta(h\nu)$ on position in the photodiode. This dependence is implicit for $\beta(h\nu)$ through the dependence of this quantity on the local electric field, the dopant concentrations, and the carrier concentrations [39].

Assumption 4 can be removed by replacing $\beta(h\nu)$ by an integral that distributes the carriers resulting from the cascade of impact ionizations with respect to position according to the strength of the electric field at the point x. Assumption 5 can be removed by replacing $\exp[-\alpha(h\nu)x]\alpha(h\nu)$ in the integral by an expression that accounts for multiple reflections between all of the interfaces in the photodiode. Assumption 6 can be removed by multiplying the righthand side of (9.7) by another integral that describes the gain associated with the transport of carriers such as avalanche gain in avalanche photodiodes. Assumption 7 can be removed by letting $P(x)$ and $F(h\nu)$ depend upon the photocurrent and the photocurrent density flowing in the photodiode [40]. Equation (9.7) can also be extended to more complex planar structures by the inclusion of more layers.

The expression obtained when assumptions 1 through 7 are eliminated is very complex and probably has never even been written down in its full generality. In fact, the expression that removes assumption 5 for the case of Figure 9.2 is already very complex in that it depends upon interference effects and the extent to which they are averaged by intentional and unintentional imperfections in the planar structure of the photodiode, such as a lack of parallelism between the front and rear surfaces of the silicon and the roughness of these surfaces. The expression for the case where all of the interference effects are averaged has been presented previously [41], and the expression for the case where none of the interference effects is averaged can be calculated with the same algorithms used to calculate the transmittance of interference filters [42].

Before (9.7) can be used, $P(x)$ must be known. According to its definition, $P(x)$ is the solution to the drift-diffusion (carrier transport) equations [43] for delta-function carrier generation at the point x [44]. $P(x)$ has the properties illustrated in Figure 9.6. In qualitative terms, it increases monotonically from a value $P(0)$ at the oxide-silicon interface to some value (usually 1) at a point x_1 located between the oxide-silicon interface and the junction. $P(x)$ is approximately constant between the point x_1 and a point x_2 that lies in the vicinity of the rear edge of the depletion region. $P(x)$ decreases monotonically from the point x_2 to the rear of the photodiode.

The simplest way to use (9.7) to model $F(h\nu)$ is to use a plausible phenomenological model for $P(x)$. $F(h\nu)$ is more sensitive to $P(0)$ and x_1 than to the shape of $P(x)$ between the points 0 and x_1 [29]. A plausible model for $P(x)$ based on this result postulates that $P(x)$ is linear between 0 and x_1 and that $P(x)$ is described by the drift-diffusion equations for a diffusion length L that is independent of position and carrier concentration between x_2 and x_4. This model for $P(x)$ gives

$$F = P + \frac{(1-P)}{\alpha x_1}[1 - \exp(-\alpha x_1)] - \frac{(x_4 - x_2)}{\alpha L^2}\exp(-\alpha x_2) \qquad (9.8)$$

where $P = P(0)$ [45], and the dependence of F and α on the photon energy $h\nu$ has been suppressed. The most important point about (9.8) is that it depends upon three parameters that must be determined experimentally, P, x_1, and L, and that it has been very useful in modeling the collection efficiency of silicon photodiodes [29, 45, 46].

More complex expressions for $P(x)$ can be obtained from analytic and integral transform solutions to the drift-diffusion equations subject to a number of simplifying assumptions. However, at this level of complexity there is little reason to calculate $P(x)$ as an intermediate step to calculating $F(h\nu)$. It is just as convenient to solve the drift-diffusion equations directly for the photocurrent by using $\exp[-\alpha(h\nu)x]\alpha(h\nu)$ as the generation term. In this case $F(h\nu)$ is the ratio of the photocurrent with the recombination mechanisms turned on to that with all of the recombination mechanisms turned off. Solutions of particular interest include Hovel's [47] extension of Gaertner's [48] analytic solution to the case of a general heterostructure having constant (different) doping concentrations throughout the front region and base, and an integral transform solution [49] that accurately describes the decrease of collection efficiency due to surface recombination. These solutions are no more accurate than (9.8) when applied to real photodiodes that do not satisfy the assumptions made in deriving them, but they are nevertheless useful. The first provides physical insight into the interaction of the material and device parameters in determining the behavior of photodiodes, and the second is useful in verifying the accuracy of numerical models that do not depend upon so many simplifying assumptions.

For the highest accuracy it is necessary to use iterative solutions of finite element (or difference) approximations of the drift-diffusion equations for the device structure in Figure 9.2. To be more accurate than (9.8), it is also necessary to use measured front-region doping distributions and the dependence upon dopant concentration of the various transport properties in the model. PC-1D is an example of

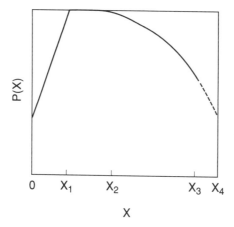

Figure 9.6 Qualitative behavior of the collection probability function $P(x)$ for the photodiode structure of Figures 9.1 and 9.2 and for similar structures.

a convenient one-dimensional, finite-element, semiconductor model, which runs on personal computers, that can be used to determine $F(h\nu)$ very accurately when so desired [50,51].

Solving the drift-diffusion equations numerically allows nonlinearity to be treated at the same time, but this is not as much of an advantage as it might at first appear. The problem is that to accurately model nonlinearity requires a two-dimensional model of the photodiode [52,53], and a more detailed knowledge of the distribution of recombination traps with respect to energy over the silicon bandgap than is normally available [51]. Generally, those applications that do not require linear transduction usually do not require accurate modeling either.

Even when the drift-diffusion equations are solved numerically, $F(h\nu)$ still depends upon three parameters that must be determined experimentally [22,23]. These are the density of charge Q trapped in the oxide, the surface-recombination velocity S at the oxide-silicon interface, and the lifetime τ of the minority carriers in the base. The first two are more closely related to the microphysics of the photodiode materials than $P(0)$ and x_1, but τ and L are equivalent through the relation $\tau = L^2/D$, where D is the diffusion constant for the minority carriers in the base [54].

9.3.4 Photocurrent

The photocurrent from a particular photodiode is given by

$$I = 2\pi \int_{(x,y)\varepsilon S} dx\, dy \int_0^{\pi/2} d\Omega(\theta) \int_0^\infty dh\nu \sum_p \Lambda(h\nu, p, \theta, x, y)\varepsilon_e(h\nu, p, \theta, x, y) \quad (9.9)$$

where $\Lambda(h\nu, p, \theta, x, y)$ is the rate at which photons of energy $h\nu$ and polarization component p relative to the normal to the photodiode surface are incident per unit of projected area per unit of solid angle on the front surface of the photodiode at the point (x, y) from the direction θ; S is the active area of the photodiode surface; and $d\Omega(\theta) = d\theta \sin\theta \cos\theta$ is the differential of projected solid angle. The term $\Lambda(h\nu, p, \theta, x, y)$ is often called the *photon radiance*.

Equation (9.9) is complex in that it involves a four-dimensional integration and a sum over polarization components, but aside from that complexity, its physical interpretation is quite simple.

1. The photodiode has an external quantum efficiency $\varepsilon_e(h\nu, p, \theta, x, y)$ that depends in general upon the position and angle of incidence, photon energy, and polarization of the incident photons.
2. The radiance $\Lambda_p(h\nu, p, \theta, x, y)$ is the density (concentration) of the incident photon flux with respect to all of these quantities.

3. The photocurrent is a linear combination of the product of the radiance and external quantum efficiency over the range of variation of all of the quantities upon which the external quantum efficiency depends.

The applications of (9.9) and other similar integrals representing the response of various physical and biological systems to radiant energy is the subject of radiometry [4,5,27].

A minor problem with (9.9) is that the units of I are electrons per second rather than amperes because the units of $\Lambda(h\nu, p, \theta, x, y)$ are photons per unit of area per unit of solid angle per unit of time. The reason for this is that there is no unit for the same number of photons as there are electrons in a Coulomb or for the same number of photons per second as there are electrons per second in an ampere. The Einstein would be an excellent name for the former, and the Becquerel (after the Becquerel who discovered the photovoltaic effect in electrochemical cells and who was the grandfather of A. H. Becquerel, who won a Nobel Prize for the discovery of radioactivity) for the latter. Unfortunately, photochemists have already defined the Einstein as a mole of photons, and high-energy physicists have already defined the Becquerel as a unit of radioactivity.

9.3.5 Noise Current

Noise limits the sensitivity of photodiodes. Any photocurrent that is less than a few standard deviations of the noise current from the photodiode will be masked by that current and will not be detectable except by averaging over a longer period of time.

The noise current receives contributions from a number of sources. The first and most fundamental source of noise in the output signal from a photodiode is the noise associated with the random arrival of photons on the front surface of the photodiode. This includes the photons associated with any background radiation as well as the photons associated with the signal of interest. The arrival noise in the incident photon flux manifests itself as a shotlike noise in the photocurrent.

The variance of the shot noise in a photocurrent I is given by

$$\langle I^2 \rangle = 2qI\Delta f \tag{9.10}$$

where q is the quantum of charge, and $\Delta f = 1/2T$ is the bandwidth of the current measurement that results from averaging over the time period T [4, p. 132]. Strictly speaking, (9.10) describes particles that obey Poisson statistics, whereas photons obey Bose-Einstein statistics. Thus (9.10) is only an approximation to the arrival noise of photons, even when they are derived from a thermal source characterized by a temperature [4, p. 234]. The errors incurred by approximating photon noise by shot noise are usually small.

Only photons for which $h\nu$ is greater than the bandgap of silicon contribute to the photon noise current. The background radiance of photons with energy greater than about 1 eV from a 300K blackbody radiation environment is negligible. Thus the shot noise associated with the background radiation is also negligible for silicon photodiodes operating in the intrinsic silicon spectral region.

The next source of noise to be considered is the noise current generated by a photodiode in the dark as a result of its temperature. When the bias voltage V applied to the photodiode is 0, then the variance of the thermal noise current i is given by

$$\langle i^2 \rangle = 4kT\Delta f dI(V)/dV \qquad (9.11)$$

where $I(V)$ is the dark current as a function of applied voltage V, $dI(V)/dV$ is the reciprocal of the dynamic resistance, and the derivative is evaluated at $V = 0$. The result in (9.11), which is just Nyquist's result with the resistance replaced by the dynamic resistance, is rigorous [4, p. 192; 55].

It is common to use (9.10) and (9.11) even when $V \neq 0$. In this case, I in (9.10) is interpreted as the sum of the photocurrent and the dark current, $dI(V)/dV$ is evaluated at the value of V that is applied to the photodiode, and the variance of the total noise current is calculated as $\langle I^2 \rangle + \langle i^2 \rangle$. This procedure probably produces a reasonable approximation, but it is not correct. To correctly derive the variance of the current in a biased semiconductor device, it is necessary to consider the microphysics of the generation and recombination occurring in the particular device as illustrated in [4, p. 192; 55]. However, even the result given there is not rigorous when $V \neq 0$ because it is based on an approximate expression for the $I(V)$ curve.

The $I(V)$ curve of a good quality planar silicon photodiode has contributions from the base and depletion regions. The contribution from the base region in a photodiode with a rear diffusion can be derived from the depletion approximation [54, sec. 5.3]. The resulting dynamic resistance at $V = 0$ is given by

$$R_0 = \frac{kTNL \tanh (L/B)}{q^2 N_c N_v \exp(-E_g/kT)} \qquad (9.12)$$

where k is Boltzmann's constant, T is the temperature of the photodiode, N is the dopant concentration in the base, L is the diffusion length for minority carriers in the base, $B = x_4 - x_2$ in Figure 9.2 is the thickness of the base, N_c and N_v are the densities of states in the conduction and valence bands, respectively, and E_g is the bandgap of silicon.

Equation (9.12) shows that for high dynamic resistance, N and L should be as large as possible, while B and T should be as small as possible. Actually N and L tend to be correlated; larger diffusion lengths can be obtained in material with a lower dopant concentration, so the product of $NL \tanh (L/B)$ must be made as large as possible. For any photon energy $h\nu$ of interest, (9.7) establishes a lower limit for

B. If $\exp[-\alpha(h\nu)x_2] \ll 1$, then *B* can be 0; but if this is not the case, $B > 1/\alpha(h\nu)$ is a lower limit. Further decreases in *B* will cause a significant loss of quantum efficiency, and the net effect will be an increase rather than decrease in the minimum detectable photon flux.

When a photodiode is optimized to minimize the contribution to the $I(V)$ curve from the base region, then the noise current will often be determined by the contribution from the depletion region. Unfortunately, this contribution is much more difficult to calculate [54, sec. 5.3], but the conclusions are essentially the same as those based on (9.12).

The electronic circuit used to measure the photocurrent can also introduce a nonnegligible contribution to the noise in the photocurrent under certain conditions. The total variance in the photocurrent is just the sum of all of the individual variances contributing to the noise in the photocurrent. The square root of this quantity is often referred to as the *noise-equivalent photocurrent*.

9.3.6 Response Time

It was mentioned earlier in this section that the diffusion of minority carriers from the rear of the base to the rear edge of the depletion region is a very slow process compared to the drift of electrons and holes across the depletion region. For photons with energies near the silicon bandgap energy, this can be the dominant contribution to the response time of a photodiode.

It is interesting that the response time for this effect has a maximum for intermediate lifetime material. In this case, the photocurrent often displays two time constants as shown in Figure 9.7 [53]. The slow component of the photocurrent is caused by minority carriers being captured and re-emitted from traps in the base region. Because the fraction of the incident radiation that penetrates into the base region increases with increasing photon wavelength, both the magnitude and the time constant of the slow component increase with increased wavelength. The same effect occurs, but with a much shorter time scale, in material that is essentially free of traps due to the different times needed for the minority carriers to diffuse to the edge of the depletion region from different portions of the base.

For large area photodiodes, the response time is often limited by an RC time constant, where R is the resistance of the feedback resistor in a transimpedance circuit, and C is the capacitance of the depletion region, which can be approximated for a step junction as

$$C = \left| \frac{Nq\varepsilon}{2|V_{bi} - V|} \right|^{1/2} \tag{9.13}$$

where N is the dopant concentration in the base, ε is the permittivity of silicon, V_{bi} is the built-in potential of the unbiased junction in the photodiode, and V is the bias

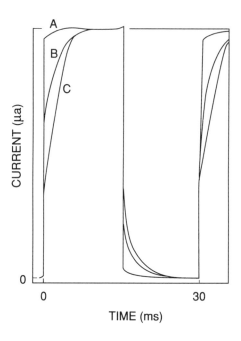

Figure 9.7 Photocurrent from a photodiode having an intermediate density of recombination traps for minority carriers in the base region when irradiated with chopped radiation consisting of 1.77 eV (*A*), 1.38 ev (*B*), and 1.18 ev (*C*) photons (after [53]).

voltage [54,p. 190]. The signs of V and V_{bi} are chosen so that $|V_{bi} - V|$ increases with increasing reverse bias. Equation (9.13) shows that lower dopant concentrations in the base give lower capacitances and that the application of reverse bias can also be used to decrease the capacitance of a photodiode. If enough reverse bias can be applied to extend the depletion region so that $\exp[-\alpha(h\nu)x_2] \ll 1$, then the slow component of photocurrent at low photon energies can also be eliminated.

In any given circuit, once the area of the photodiode is reduced sufficiently, the response time will no longer be limited by the capacitance of the depletion region. Instead it will be limited by the mobility of the majority carriers in silicon along paths that conduct the photocurrent in the photodiode. The carrier mobilities in silicon are small compared to the carrier mobilities in most other semiconductor materials used for photodiodes. This is one reason why silicon is not an important material for ultrahigh-speed electronics or electrooptics, and it is the driving force behind the development of III–V compound-semiconductor technology.

9.4 LEVEL OF INTEGRATION

There are three important aspects to modern sensor integration: the integration of a number of sensors on a single chip, the integration of both sensing and signal processing functions on a single chip, and the integration of separate on-chip functions into a single structure, rather than implementing the separate functions in isolated devices communicating through metal interconnections. The silicon photodiode CCD imaging array was the first integrated sensor to embody all of these aspects on a single chip, and it is such an important example of an integrated microsensor that the next chapter is devoted to it.

CCD imaging arrays represent a high level of integration of the sensor and signal processing electronics in that they multiplex the signals stored on a number of photodiode-capacitors for serial read-out over a few lines. Because the next chapter is devoted to CCD arrays, a different example of a high level of integration will be presented to conclude this chapter. Compared to CCD arrays, this example illustrates an even higher level of integration of the processing electronics, though at a much lower level of integration of the photosensors.

Figure 9.8 is a block diagram for a high-resolution shaft encoder (angle measuring sensor) for an automobile [56]. It is based on a linear array of 10 photodiodes. The array receives the light from an LED source through a slotted disk mounted on the shaft whose rotation is to be measured. In addition to 10 photodiode sensors, the shaft encoder incorporates 10 analog current-detecting circuits, each consisting of 14 linear circuit elements, a digital logic circuit consisting of 94 I^2L gates to process the analog signals, and about 92 other linear circuit elements in LED-driver, supply, and buffer circuits. As a result of the on-chip processing, only three output lines are needed. The direction line uses a bit to encode the direction of motion, the

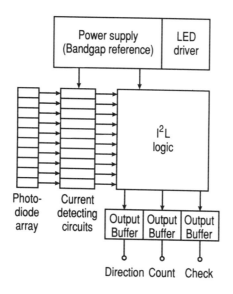

Figure 9.8 Block diagram for an integrated rotary shaft encoder (after [56]).

count line presents a pulse for every third of a degree of shaft rotation, and the check line presents a fail-safe signal to verify that the chip is working properly.

Because the chip is used in an automobile, special care was exercised in designing the circuitry to incorporate temperature compensation. For example, the current-to-voltage converters run on a 2.5V, bandgap-stabilized power supply. The chip was fabricated by adding two implantation steps to a standard bipolar process to make the JFETs for the current-detecting circuits. A schematic diagram for these circuits is shown in Figure 9.9. The JFET implementation is shown in Figure 9.10.

The addition of ion-implantation steps to a standard bipolar process is not a very significant modification of an in-house process, but it would not be possible from a standard bipolar foundry process. This modification of a standard planar process illustrates an important point about the integration of microsensors. Such an extension is justified whenever a new application is large enough to support dedicated

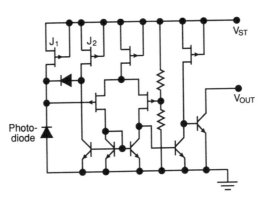

Figure 9.9 Circuit diagram for the current-detecting circuit in the integrated rotary shaft encoder of Figure 9.8: J_1 and J_2 are JFETs (after [56]).

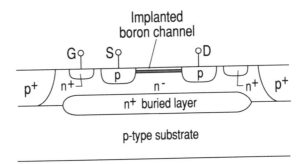

Figure 9.10 Implementation of the JFETs in the current-detecting circuit of Figure 9.9 through the addition of two ion-implantation steps to a standard bipolar process (after [56]).

fabrication lines or additions to an existing in-house line. The advantages of planar silicon technology that can be retained during such an extension often justify the extension as opposed to the adoption or development of a completely new technology based on some other material. However, applications having small markets cannot afford such extensions. Such applications require devising clever ways to realize a microsensor having the desired characteristics without any modifications to some standard foundry process that is capable of providing the desired circuitry.

9.5 SUMMARY

This chapter reviewed planar silicon photosensors primarily from the point of view of the conventional aspects of photosensor performance, but also from the point of view of photosensor integration. Key ideas presented include the following.

Silicon photodiode technology was developed with and contributed to the development of planar silicon IC technology. Consequently, planar silicon photodiode technology is more highly developed and more easily integrated with on-chip circuitry than photosensor technologies based on other materials.

Material properties set the limits to the performance that can be obtained with a photodiode fabricated from a particular material, but unintentional defects introduced during processing usually reduce the performance below these limits. The material properties governing the performance of silicon photodiodes operating in the intrinsic silicon spectral region are the absorption coefficient of silicon, the quantum yield of silicon for the production of free electron-hole pairs, and the lifetime against recombination of the free electrons and holes. Even though the lifetime is a material property, it is highly dependent upon processing, so it is treated as if it were a device property like the front-region doping distribution.

Absorption by free electrons and holes competes with absorption by valence-band electrons to determine the quantum yield of silicon at low photon energies. Impact ionization of valence-band electrons competes with phonon emission as an energy-loss mechanism to determine the quantum yield of silicon at high photon energies. Impact ionization creates extra electron-hole pairs; phonon emission does not. The quantum yield of silicon is very close to unity throughout the visible spectral region.

The performance of photodiodes is characterized by the external quantum efficiency, noise-equivalent photocurrent, and response time. The external quantum efficiency is the product of the transmittance of the photon-receiving (front) surface of the photodiode and the internal quantum efficiency, the latter is the product of the quantum yield for the production of electron-hole pairs and the collection efficiency. Collection (separation) of the photogenerated carriers by the diode junction competes with the recombination of the photogenerated carriers. The collection efficiency is determined by the interaction of the structure of the photodiode with the absorption coefficient of silicon and with the lifetime of the photogenerated carriers.

The transmittance and the collection efficiency can be modeled at various levels of sophistication, depending upon the accuracy required. Currently, three parameters (for instance, the density of charge trapped in the oxide at the front surface of the photodiode, the minority-carrier recombination velocity at the front surface of the photodiode, and the minority-carrier lifetime in the base region) must be determined empirically from measurements on the photodiode to model the collection efficiency at the highest level of accuracy. Similarly, the thickness of the front-surface oxide must be determined empirically to model the transmittance of the front surface at the highest level of accuracy.

Both the noise-equivalent photocurrent and the response time of the photodiode depend upon an interaction of the photodiode structure and the lifetime for minority carriers in different regions of the photodiode. The response time also depends upon the photon energy of the radiation incident on the photodiode through the absorption coefficient spectrum. Like the quantum efficiency, the noise-equivalent photocurrent and response times can be modeled at various levels of sophistication, but at least one parameter, the lifetime, must be determined empirically for each photodiode.

Integration of photosensors with signal processing electronics may be better achieved with silicon than with the material of choice for building the same photosensor as a monolithic device. Integration of silicon photosensors with signal processing electronics may require extension of a standard process or the development of a novel photosensor design that can be fabricated in a standard foundry process capable of providing the needed circuits while still meeting the performance goals for the photosensor.

REFERENCES

[1] G. Bertolini and A. Cochi, *Semiconductor Detectors*, New York, John Wiley & Sons, 1968.

[2] R. H. Kingston, *Detection of Optical and Infrared Radiation*, New York, Springer, 1978.

[3] R. J. Keyes, ed., *Optical and Infrared Detectors*, Topics in Applied Physics, Vol. 19, New York, Springer-Verlag, 1980.

[4] R. J. Boyd, *Radiometry and the Detection of Optical Radiation*, New York, John Wiley & Sons, 1983.

[5] F. Hengstberger, ed., *Absolute Radiometry: Electrically Calibrated Thermal Detectors of Radiation*, San Diego, CA, Academic Press, 1989.

[6] N. V. Joshi, "Modern Photodetectors," Chapter 6 in *Photoconductivity: Art, Science, and Technology*, Optical Engineering, Vol. 25, New York, Marcel Decker, 1990.

[7] D. A. Scribner, M. R. Kruer, and J. M. Killiany, "Infrared Focal Plane Technology," *Proc. IEEE*, Vol. 79, 1991, pp. 66–84.

[8] J. P. Kirkland, T. Jach, R. A. Neiser, and C. E. Bouldin, "*Pin* Diode Detectors for Synchrotron X-Rays," *Nucl. Instr. and Meth.*, Vol. A266, 1988, pp. 602–607.

[9] L. R. Canfield, J. Kerner, and R. Korde, "Stability and Quantum Efficiency of Silicon Photodiode Detectors for the Far Ultraviolet," *Appl. Opt.*, Vol. 28, 1989, pp. 3940–3943.

[10] M. D. Petroff, M. G. Stapelbroek, and W. A. Klienhans, "Detection of Individual 0.4 μm Wavelength Photons via Impurity-Impact Ionization in a Solid State Photomultiplier," *Appl. Phys. Lett.*, Vol. 51, 1987, pp. 406–408.

[11] F. Szmulowicz and F. L. Madarsz, "Blocked Impurity Band Detectors—An Analytic Model: Figure of Merit," *J. Appl. Phys.*, Vol. 62, 1987, pp. 2533–2540.

[12] D. M. Chapin, C. S. Fuller, and G. L. Pearson, "A New Silicon *p-n* Junction Photocell for Converting Solar Radiation into Electrical Power," *J. Appl. Phys.*, Vol. 25, 1954, pp. 676–677.

[13] G. L. Pearson and C. S. Fuller, "Silicon *p-n* Junction Power Rectifiers and Lightning Protectors," *Proc. Inst. Radio Engrs.*, Vol. 42, 1954, p. 760.

[14] C. S. Fuller and J. A. Ditzenberger, "Diffusion of Boron and Phosphorus into Silicon," *J. Appl. Phys.*, Vol. 25, 1954, pp. 1439–1440.

[15] T. C. Madden and W. M. Gibson, "Silicon Dioxide Passivation of *p-n* Junction Particle Detectors," *Rev. Sci. Inst.*, Vol. 34, 1963, pp. 50–55.

[16] M. H. Crowell, T. M. Buck, E. F. Labuda, J. V. Dalton, and E. J. Walsh, "A Camera Tube with a Silicon Diode Array Target," *Bell Syst. Tech. J.*, Vol. 46, 1967, pp. 491–495.

[17] W. S. Boyle and G. E. Smith, "Charge Coupled Semiconductor Devices," *Bell Syst. Tech. J.*, Vol. 49, 1970, pp. 587–593.

[18] M. F. Tompsett, G. F. Amelio, and G. E. Smith, "Charge Coupled 8-Bit Shift Register," *Appl. Phys. Letts.*, Vol. 17, 1970, pp. 111–115.

[19] E. F. Zalewski and J. Geist, "Silicon Photodiode Absolute Spectral Response Self-Calibration," *Appl. Opt.*, Vol. 19, 1980, pp. 1214–1216.

[20] J. Geist, "Current Status of and Future Directions in Silicon Photodiode Self-Calibration," *Proc. SPIE*, Vol. 1109, 1989, pp. 246–256.

[21] E. F. Zalewski, "Comparison of the National Standards of Absolute Spectral Responsivity," *BIPM Comm. Cons. Phot. and Radimometrie*, Vol. 11, 1986, pp. 165–179.

[22] J. Geist, A. M. Robinson, and C. R. James, "Numerical Modeling of Silicon Photodiodes for High-Accuracy Applications: Part III. Interpolating and Extrapolating Internal Quantum-Efficiency Calibrations," *J. Res. NIST*, Vol. 96, 1991, pp. 463–469.

[23] J. Geist, R. Köhler, R. Goebel, A. M. Robinson, and C. R. James, "Numerical Modeling of Silicon Photodiodes for High-Accuracy Applications: Part II. Interpreting Oxide-Bias Experiments," *J. Res. NIST*, Vol. 96, 1991, pp. 471–479.

[24] C. C. Hoyt, P. J. Miller, P. V. Foukal, and E. F. Zalewski, "Comparison Between a Side-Viewing Cryogenic Radiometer and Self-Calibrated Silicon Photodiodes," *Proc. SPIE*, Vol. 1109, 1989, pp. 236–245.

[25] M. Krumrey, "Halbleiter-Photodioden als radiometrische Empfaengernormale im bereich weicher Roentgenstrahlung," thesis, Technische Universitaet Berlin, 1990.

[26] J. Durnin, C. Reece, and L. Mandel, "Does a Photodetector Always Measure the Rate of Arrival of Photons?" *J. Opt. Soc. Am.*, Vol. 71, 1981, pp. 115–117.

[27] J. Geist, W. K. Gladden, and E. F. Zalewski, "Physics of Photon Flux Measurements with Silicon Photodiodes," *J. Opt. Soc. Am.*, Vol. 72, 1982, pp. 1068–1075.

[28] J. Geist and C. S. Wang, "New Calculations of the Quantum Yield of Silicon in the Near Ultraviolet," *Phys. Rev.*, Vol. B27, 1983, pp. 4841–4847.

[29] J. Geist and E. F. Zalewski, "The Quantum Yield of Silicon in the Visible," *Appl. Phys. Lett.*, Vol. 35, 1979, pp. 503–506.

[30] R. C. Alig, S. Bloom, and C. W. Struck, "Scattering by Ionization and Phonon Emission in Semiconductors," *Phys. Rev.*, Vol. B22, 1980, pp. 5565–5582.

[31] J. Geist and W. K. Gladden, "Transition Rate for Impact Ionization in the Approximation of a Parabolic Band Structure," *Phys. Rev.*, Vol. B27, 1983, pp. 4833–4840.

[32] H. A. Weakliem and D. Redfield, "Temperature Dependence of the Optical Properties of Silicon," *J. Appl. Phys.*, Vol. 50, 1979, pp. 1491–1493.

[33] J. Geist, A. Migdall, and H. P. Baltes, "Analytic Representation of the Silicon Absorption Coefficient in the Indirect Transition Region," *Appl. Opt.*, Vol. 27, 1988, pp. 3777–3779.

[34] H. R. Philipp, "Influence of Oxide Layers on the Determination of the Optical Properties of Silicon," *J. Appl. Phys.*, Vol. 43, 1972, pp. 2835–2939.

[35] D. F. Edwards, "Silicon (Si)," in *Handbook of Optical Constants*, ed. E. D. Palik, Orlando, FL, Academic Press, 1985, p. 547.

[36] J. Geist, A.R. Schaefer, J.-F. Song, Y. H. Wang, and E. F. Zalewski, "An Accurate Value for the Absorption Coefficient of Silicon at 633 nm," *J. Res. NIST*, Vol. 95, 1990, pp. 549–558.

[37] W. R. Runyan, "A Study of the Absorption Coefficient of Silicon in the Wavelength Region between 0.5 and 1.1 Microns," Southern Methodist University Report No. 83-13.

[38] F. J. Wilkinson, A. J. D. Farmer, and J. Geist, "The Near Ultraviolet Quantum Yield of Silicon," *J. Appl. Phys.*, Vol. 54, 1983, pp. 1172–1174.

[39] J. Geist, J. L. Gardner, and F. J. Wilkinson, "Surface-Field-Induced Feature in the Quantum Yield of Silicon Near 3.5 eV," *Phys. Rev.*, Vol. B42, 1990, pp. 1262–1267.

[40] J. Geist, "Photodiode Operating Mode Nomenclature," *Appl. Opt.*, Vol. 25, 1986, pp. 2033–2034.

[41] J. Geist, "On the Possibility of an Absolute Radiometric Standard Based on the Quantum Efficiency of a Silicon Photodiode," *Proc. SPIE*, Vol. 196, 1979, pp. 75–83.

[42] O. S. Heavens, *Optical Properties of Thin Films*, New York, Academic Press, 1955.

[43] S. Sze, *Physics of Semiconductor Devices*, 2d ed., New York, John Wiley & Sons, 1981, p. 50.

[44] J. Geist, "Silicon Photodiode Front Region Collection Efficiency Models," *J. Appl. Phys.*, Vol. 51, 1980, pp. 3993–3995.

[45] J. Geist, E. F. Zalewski, and A. R. Schaefer, "Spectral Response Self-Calibration and Interpolation of Silicon Photodiodes," *Appl. Opt.*, Vol. 19, 1980, pp. 3795–3799.

[46] E. F. Zalewski, "The NBS Photodetector Spectral Response Calibration Transfer Program," NBS Special Publication 250-17, Washington, DC, U.S. Government Printing Office, 1988.

[47] H. J. Hovel, "Solar Cells," *Semiconductors and Semimetals*, Vol. 11, 1975, pp. 15–47; or page 800 of [43].

[48] W. W. Gaertner, *Phys. Rev.*, Vol. 116, 1959, pp. 84–87.

[49] J. Geist and H. Baltes, "High Accuracy Modeling of Photodiode Quantum Efficiency," *Appl. Opt.*, Vol. 28, 1989, pp. 3929–3939.

[50] P. A. Basore, "PC-1D Version 2: Enhanced Numerical Solar Cell Modeling," 29th IEEE Photovoltaic Specialists Conference, pub. # 0160-8371/88/0000-389, New York, IEEE, 1988, pp. 389–396.

[51] J. Geist, D. Chandler-Horowitz, A. M. Robinson, and C. R. James, "Numerical Modeling of Silicon Photodiodes for High-Accuracy Applications: Part I. Simulation Programs," *J. Res. NIST*, Vol. 96, 1991, pp. 481–492.

[52] A. R. Schaefer, E. F. Zalewski, and J. Geist, "Silicon Detector Nonlinearity and Related Effects," *Appl. Opt.*, Vol. 22, 1983, pp. 1232–1236.

[53] J. L. Gardner and F. J. Wilkinson, "Response Time and Linearity of Inversion Layer Silicon Photodiodes," *Appl. Opt.*, Vol. 24, 1985, pp. 1531–1534.

[54] R. S. Muller and T. I. Kamins, *Device Electronics for Integrated Circuits*, New York, John Wiley & Sons, 1986, p. 236.

[55] R. H. Hamstra, Jr. and P. Wendland, "Noise and Frequency Response of Silicon Photodiode Operational Amplifier Combination," *Appl. Opt.*, Vol. 11, 1972, pp. 1539–1547.

[56] H. Muro, "Optical Angle Detecting Sensor IC with High Resolution," *Proc. 6th Sensor Symposium*, 1986, pp. 295–297.

Chapter 10
Charge Coupled Devices

A. Russell Schaefer
Science Applications International Corporation

J. A. Janesick
Jet Propulsion Laboratory

10.1 INTRODUCTION

In this chapter, we will discuss an example of a class of sensors that make heavy use of and constantly challenge IC technology. These are the optical imaging sensors known as *charge-coupled devices* (CCD). The CCD has evolved from a relatively simple device into a highly complex integrated device suitable for use in an ever-increasing variety of scientific imaging applications. It is beyond the scope of this chapter to present a complete and rigorous detailed description of the CCD; indeed, that could occupy an entire book itself. Rather, our purpose here is to present an introductory overview to the device, indicating the most important factors that influence its performance and some references to more detailed descriptions.

The earliest form of CCD was a simple 8-bit linear shift register built by Bell Telephone Laboratories in 1970 [1]. The usefulness of these devices was quickly apparent to workers in many different areas of electronics, such as memories, logic circuits, and signal processing.

The greatest contribution of the CCD, of course, has been in the field of imaging sensors. In this arena, CCDs have developed to the point of successfully replacing vidicon tubes in commercial and home video cameras. Additionally, they are used in document scanners, camera focusing sensors, and increasingly in robotics, where they find convenient use as an optical input transducer providing spatially quantized data to a computer. Finally, with the recent advent of very large arrays, CCDs have become extremely useful in scientific instrumentation such as satellite- and ground-based surveillance systems, large astronomical telescopes, and medical imaging applications.

The CCD now greatly surpasses photographic film in most characteristics, such as sensitivity, broad spectral range, stability, dynamic range, and linearity. Only in resolution is photographic film still ahead; and this situation is dwindling, with the current advent of devices such as a 4096 × 4096 pixel CCD [2], where each pixel is 7.5 μm square. It should be mentioned that most of the CCD devices are made in silicon but recently some other materials have been used, such as platinum silicide and AlGaAs on GaAs. We will focus on silicon for the remainder of this chapter.

10.2 BASIC CONCEPTS

10.2.1 CCD Charge Storage Fundamentals

A CCD is basically an integrated circuit consisting of an array of photosensor sites that convert incident photons into electrical charge proportional to the amount of photons absorbed. Each of these sites—"picture elements," or pixels—stores its charge in a MOS capacitor, that is, a potential well.

This MOS capacitor is actually fabricated using polysilicon as the transparent "metal" electrode, which is deposited on top of a thin insulating oxide layer (sometimes nitride followed by oxide). This material rests upon p-type silicon epitaxially grown substrate. One can then apply a positive voltage bias to the electrode, which repels majority carrier (holes) from the interface area and forms a depletion region with no free carriers. This depletion region acts as a potential well for minority carriers (electrons). Any minority carriers, whether generated thermally or from photon conversion, will be collected in this well in a thin inversion zone. The lateral extent of the well is defined by a p^+ channel stop, which restricts the well by pinning the potential near 0. The potential in the orthogonal direction is determined by the bias put upon adjoining polysilicon electrodes (gates). Figure 10.1 schematically illustrates the potential wells and charge storage of a CCD. To avoid interaction of signal charge with the oxide-silicon surface and interface states, which have detrimental noise and transfer efficiency effects, most CCDs now employ a "buried channel." This buried channel is fashioned by using ion implantation to form a layer of opposite polarity (n-type) silicon underneath the surface, which shifts the maximum potential of the wells away from the interface and down into the bulk. This layer is electrically contacted with the output (reset drains) of the CCD.

After an image has accumulated for the desired time, the image charge packets from these pixels are moved out. This is accomplished by a clocking procedure in which a series of appropriately timed voltage biases are placed on polysilicon gates in a manner to shift the charge packets down the array of capacitors, a row at a time, into a serial shift register. This shift register is then read out serially by depositing each charge packet at the output gate of a floating diffusion amplifier, where they are converted to proportional voltages that make up a video signal.

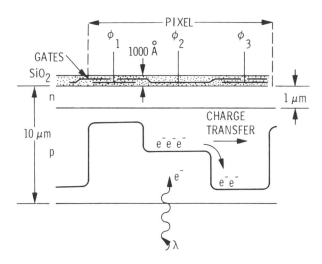

Figure 10.1 Schematic illustration of potential wells and charge storage for a CCD; explanation of ϕ_1, ϕ_2, and ϕ_3.

10.2.2 CCD Architectures

A comprehensive discussion of CCDs is given in [3]. There it is pointed out that four common architectures for CCD imagers have been developed.

1. A simple linear shift register used by slowly scanning the scene vertically past the imager while rapidly reading out the shift register in the horizontal direction;

2. Use of an area (two-dimensional) array as a full-frame imager, in which the scene is imaged onto the entire array and integrated, and then with a shutter closed the array of pixels is read out by shifting one row of pixels in parallel into a serial register where they are read out serially;

3. Use of part of the array to integrate the image, then rapidly transfer it in parallel under an optically masked area, where it can be read out later, while the next image is being integrated—a mode known as *frame transfer operation*;

4. Interline transfer, consisting of a parallel array of line sensors separated by a transfer electrode from masked CCD readout registers that lead in parallel into a single output register. In this mode, the entire image from the line array is shifted simultaneously into read-out registers, and a new image is integrated while the previous image is read out.

All of these modes have their advantages and disadvantages, and practical applications now center around the last three. Modes 2 and 3 lend themselves to scientific applications in which 100% pixel fill factors are important, whereas mode 4 is often used in conventional television video applications, where fast read-out is more important. For the remainder of this chapter we will concentrate only on mode 2 operation, because it adequately illustrates the example of on-chip integration being discussed.

10.2.3 Three-Phase CCD Example

Although several CCD devices will be considered in this chapter, we will focus on the CCD example built by SAIC and fabricated by FORD (now LORAL) Aerospace to address needs for large arrays with good quantum efficiency, low read noise and dark current, good charge transfer efficiency, and low incidence of cosmetic defects (bad columns or pixels). This device is discussed in detail by Janesick et al. [4], who performed some of the original characterization.

This CCD is a 1024 × 1024 area array intended for a number of different applications. The device is an NMOS CCD, with an n-type buried channel ($N_A \approx 3 \times 10^{14}/cm^3$) in p-type 30–50 Ω-cm {100} oriented, epitaxial ($N_D \approx 1.6 \times 10^{12}/cm^3$) silicon. Each pixel is composed of three gates, each connected to a separate charge-moving clocking phase. During image integration, two adjacent phases are biased positive and the third is negative to form the potential well (depletion) in which signal charge collects under two gates. By sequentially bringing the adjacent gate positive and the formerly positive gates negative in the proper phase, it is possible to move the charge in the desired direction with virtually no loss of charge.

The imaging part of the CCD consists of a parallel shifted active area of 18 μm square pixels divided into four 256-row × 1024-column regions, as shown in Figure 10.2. In each of the regions, image charge that has collected in various pixel wells can be clocked, in rows of 1024, up or down toward a serial register at either end of the array. One has individual control over the parallel clocking of each of these four areas, so that each can be individually clocked, allowing optionally for images to be split in the center and clocked into two bordering areas for frame storage and later read-out in the frame transfer mode of operation. At both the top and bottom of the chip are serial horizontal registers into which the parallel rows from the array are dumped, a row at a time. Each of these serial registers is split in the center and independently clocked, allowing the possibility of reading out the serial registers simultaneously through each of the two MOSFET output amplifiers at each end, or through only one end or the other. Therefore, it is possible to read out the image through all four corner amplifiers simultaneously for maximum read-out speed.

Each horizontal register has 20 "dark" pixels at each end, providing baseline-stabilization (zero-level video signal). Figure 10.3 shows a microscopic view of the lower lefthand corner of the CCD, indicating the various output connections and showing the parallel array and serial register polysilicon gates and their respective buses that form the image pixels. Details of the fabrication are discussed in [4].

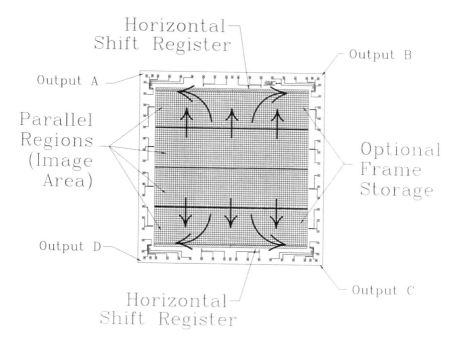

Figure 10.2 Schematic drawing of SAIC/Ford 1024 × 1024 CCD showing optional read-out modes.

Each of the 18 μm square pixels in the parallel array area is composed of three $6 \times 18 \ \mu m^2$ gates, corresponding to the three clock phases. The serial register gates are $6 \times 40 \ \mu m^2$, providing over twice the charge handling capacity of a vertical pixel to assure good horizontal transfer. The last gate at the end of each horizontal register is of twice-normal width ($12 \times 40 \ \mu m^2$) and is independently clocked to provide on-chip pixel summation capability. Thus, two array rows can be summed into the serial register and read out, yielding $36 \times 18 \ \mu m^2$ pixels, or additionally every two serial pixels can be summed into the summing gate to provide the equivalent of 512 × 512—36 μm^2 square pixels.

The output amplifiers are buried-channel MOSFETs connected to a floating diffusion node where signal charge is dumped. The on-chip amplifier is 10×120 μm^2 and was designed based on a previous Ford amplifier that exhibited excellent sensitivity, linearity, and low read noise over the full dynamic range. The floating diffusion is connected to a reset buried-channel MOSFET of $10 \times 16 \ \mu m^2$.

The CCD chip dimensions are 2.2250 cm^2 with the imaging area occupying the central 1.8432 cm^2. The large border is necessary to provide for thinning and backside illuminated operation if desired, as will be discussed later.

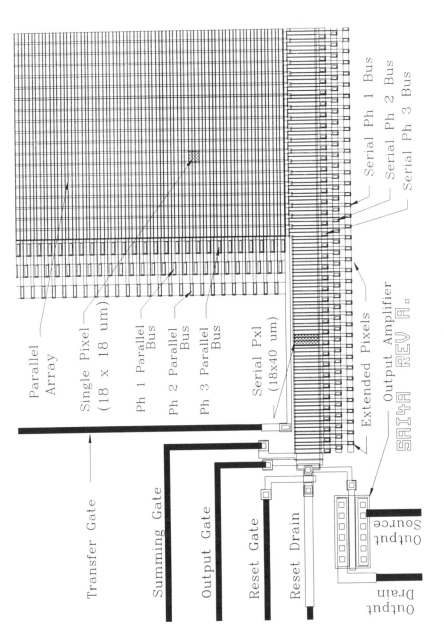

Figure 10.3 Microscopic view of lower left corner of SAIC/Ford CCD.

Many of the superior properties of CCDs come down to capabilities in four basic areas: quantum efficiency (QE), charge collection efficiency (CCE), charge transfer efficiency (CTE), and output read noise. Therefore; we will structure the remainder of the chapter around a discussion of each of these areas in turn. The order of discussion is not associated with relative importance, rather the first two topics lead off from earlier deliberations of similar topics in Chapter 9 of this book (discussions of photosensors), whereas the last two topics are relevant only to CCDs.

10.3 QUANTUM EFFICIENCY

This very important property of CCDs determines the sensitivity of the device to photons incident upon it. Much of the material discussed in the previous chapter about quantum efficiency (QE) of planar silicon photodiodes applies equally well to the CCD. In particular, the external, measurable QE is still described by $QE_{CCD}(h\nu)$ = $T(h\nu)\eta(h\nu)F(h\nu)$, where $T(h\nu)$ is now the transmittance of the polysilicon gate structure and the oxide layers underneath, η is the number of carriers produced per photon absorbed, or quantum yield, and $F(h\nu)$ is the charge collection efficiency. Because charge collection efficiency is an important parameter in itself, with ramifications for not only QE but also modulation transfer function (MTF) and resolution, it will be discussed further in the next section.

For a CCD illuminated from the front, as mentioned in the previous chapter, the transmittance contains both reflectance and significant absorptance components from the gate-oxide structure, which cannot be ignored (most of the scattering and absorption is due to the polysilicon gate itself).

As a result of this absorption and scattering loss term in the polysilicon, front illuminated CCDs have poor QE, especially in the blue and near-ultraviolet spectral regions. The devices themselves take advantage of the excellent material and fabrication properties (such as low surface-recombination velocity and long minority carrier lifetime) currently available in silicon, which results in very high charge collection efficiency, quite similar to that observed in silicon photodiodes. The poor observed QE (typically 5% or less at 400 nm) is due solely to the front surface losses.

To overcome this inherent sensitivity problem, a technique known as *backside illumination* has been developed. In this technique, the substrate of the CCD is etched away or "thinned" in an acid bath until only the epitaxial silicon remains. By using certain procedures, it is possible to absorb incident photons directly into the silicon at the back surface of the CCD unimpeded by the polysilicon gate structure present at the front, resulting in much higher QEs, especially in the blue and near-UV spectral regions.

10.3.1 QE Model

For a rigorous discussion of the details of QE, there are several papers on modeling the various aspects of this question [5–7]. For most applications, a simple approach gives results that are quite adequate. For the case of backside illumination, because the photons need not pass through anything but a nearly transparent oxide layer, one can write, as in the last chapter,

$$QE_{CCD}(h\nu) = [1 - \rho(h\nu)]\eta(h\nu)F(h\nu) \qquad (10.1)$$

Here $\rho(h\nu)$ is the spectral reflectance of the CCD back surface. As described in Chapter 9, the fraction $[1 - \rho(h\nu)]$ of photons incident on the rear surface is transmitted into the silicon. A fraction $\exp(-\alpha x)$ of these penetrate to a depth x, where a fraction α are absorbed per unit distance at the point x to create η electron-hole pairs per absorbed photon. A fraction $P(x)$ of the pairs created at x are separated by the internal electric field and contribute electrons that are stored in the potential well to form an image. This can be expressed mathematically as

$$F(h\nu) = \int_0^{x_{\text{thick}}} P(x)\alpha \exp(-\alpha x)dx \qquad (10.2)$$

where x_{thick} is the total distance of penetration into the silicon. In many cases, we are working in regions where the probability $P(x)$ is equal to 1, simplifying analysis.

10.3.2 Quantum Yield and Photon Transfer Curve

In most cases (throughout the visible and near-infrared spectrum), the quantum yield, η, is numerically equal to 1. Only in the near-ultraviolet region do photons begin to acquire enough energy to create more than one carrier per absorption. For instance, at Lyman-α, 1,216Å, each absorbed photon produces about three electron-hole (e-h) pairs in silicon.

The best way to determine quantum yield is using a method called the *photon transfer technique* [3,8]. In this technique, the signal vs. noise out of a CCD in a test camera is used as a diagnostic to provide absolute calibration data. One can represent the input to the camera in units of incident photons P_i, and the final output signal is S in units of DN (digital number), which is an encoding of the pixel signal value with an analog-to-digital (A/D) converter, typically using 12–16 bits. In addition, we must consider the gain of the camera, A_1, in units of (V/V), and the gain of the A/D converter, A_2, in units of (DN/V). The output sensitivity of the CCD itself is S_v, in units of (e^-/V). We define a constant J that relates photons absorbed to DN as

$$J = 1/(\eta S_v A_1 A_2) \qquad (10.3)$$

Similarly, we define another constant that relates carriers created in the CCD to the output signal, DN, as

$$K = 1/(S_v A_1 A_2) \qquad (10.4)$$

The units of J are absorbed photons/DN, and that of K is e^-/DN. Thus $\eta = K/J$. Obviously, K and J are functions of wavelength; for longer wavelengths, above about 300 nm, where only 1 e-h pair is produced per absorbed photon, $K = J$. In the UV, where $\eta > 1$, then $K > J$.

As discussed in detail in [8], these constants can be determined graphically by plotting a curve (called the *photon transfer curve*), of rms noise σ_N as a function of signal $S(DN)$ for a given array (for example, 20 × 20) of pixels. An example of such a plot is shown in Figure 10.4, for a case in the visible spectrum, where the quantum yield is 1. The abscissa, $S(DN)$, is proportional to the average number of interacting photons per pixel with the array uniformly illuminated at some level. The ordinate, σ_N, is the standard deviation of those 400 pixels from the mean at that exposure level after pixel-to-pixel nonuniformity has been removed, which is easily accomplished by differencing two frames taken at the same light level. The read noise floor, σ_R, indicated in Figure 10.4, represents the intrinsic noise associated with the read-out circuitry; that is, the on-chip source-follower amplifier and any other noise sources independent of the signal level. As the signal is increased, the

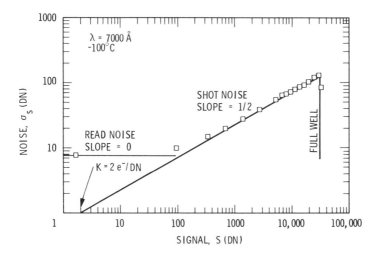

Figure 10.4 Photon transfer curve using 700 nm illumination ($\eta = 1$).

noise eventually becomes dominated by the shot noise of the signal (that is, σ_R becomes negligible) and is characterized by a line of slope $1/2$. The intersection of the slope $1/2$ line on the signal axis (that is, when $\sigma_N = 1$) represents the desired conversion constant $K = 2.0$ e^-/DN (the constant K is evaluated in this example because the wavelength of illumination used to generate Figure 10.4 was 700 nm; the constant J is found in the same manner when using light of wavelengths shorter than 300 nm).

Because S_V, A_1, and A_2 are constant for a given CCD in a given camera system, a decrease in J as the wavelength decreases can be directly attributed to an increase in the quantum yield. Thus, a direct measurement of η is obtained using $\eta = K/J$. A more accurate relationship between J or K and the output signal and variance is shown to be [8]

$$J(\text{or } K_{\lambda>3000}) = \frac{S(\text{DN})}{\sigma_N^2(\text{DN}) - \sigma_R^2(\text{DN})} \tag{10.5}$$

Equation (10.5) provides a powerful and convenient method for calibrating a CCD camera system for absolute gain and for determining quantum yield in the ultraviolet.

10.3.3 Backside Illumination and Thinning

To achieve high CCD quantum efficiency, especially in the blue and near-UV spectral regions, backsided operation is required [3], as discussed earlier. Backside illumination necessitates thinning the CCDs in a manner such that one produces a QE-pinned (essentially 100% internal QE), backside-illuminated device. One of the most successful techniques used is called the *flashgate accumulation approach* [9–12].

This approach dictates three main process steps to be performed upon selected candidate CCDs. First, it is necessary to remove the excess p^+ substrate material down to the epitaxial layer (~ 10–15 μm) of the silicon. This is required because the silicon substrate is optically "dead" material. Only when light is absorbed in the single crystalline region not too far removed from the internal guiding electric field within the CCD is there a chance of carrier production and collection.

Second, growth of a passivating oxide layer on the backside is essential. The instability of QE and the loss of signal charge by recombination at the back surface and in a backside "dead layer" must be avoided. It has been pointed out in the past [10] that the surface of a backside-illuminated CCD represents an incomplete metal-oxide-semiconductor (MOS) device; that is, it has a semiconductor layer and a thin native oxide but no metal. Such an "incomplete" structure is inherently unstable because of trapping and emission of signal carriers in the surface states at the Si/SiO$_2$ interface. This causes the band structure near the surface to vary considerably depending upon its light exposure history, leading to the instability known as QE hysteresis or QEH. Also, as was the case near the front surface of planar photodiodes, these interface states are mostly positively charged, creating a backside potential well about 400 nm deep into which signal electrons diffuse and recombine, resulting in low QE. The low-QE situation is worst for UV photons, which are absorbed almost immediately near the back surface (for example, 30 Å at 250 nm). The solution to this problem lies in the growth of an oxide on the silicon back surface that can be negatively charged to negate the effect of the positively charged interface states.

The third key step for providing good backside response in a CCD is to implement a method of negatively charging the oxide. Several techniques have been developed and are in use today. One is simply to charge the oxide with UV photons. This has the effect of bending the energy bands upward into accumulation, driving signal electrons away from the $Si–SiO_2$ interface, eliminating QEH and providing 100% internal QE, a condition known as the *QE-pinned state*. By flooding the CCD back surface with intense UV light ($\lambda \sim 250$ nm) electrons will be excited to the conduction band of the SiO_2, resulting in a negative charge on the oxide. Although very effective, this technique has the disadvantage of discharging over a period of time unless the CCD is held to temperatures below $-70°C$, which is not always convenient.

Other techniques have been developed to maintain a permanent negative charge. An early method was simply to leave a thin remnant of the diffused boron p^+ substrate material on the epitaxial surface, even prior to growth of an oxide. This will induce the appropriate electric field to propel the signal charge away from the back surface into the front of the CCD, where the collecting well resides. However, the difficulty in controlling thinning well enough to accurately attain just the right thickness of p^+, coupled with the facts that the field strength is not strong enough and the positively charged oxide-induced back surface well still exists, make this method not very satisfactory.

Ion implantation has also been developed as a method of accumulating this back surface. In this technique boron ions are implanted as shallowly as possible into the back surface to produce the desired electric field to accumulate the back surface, just as in the previous technique. This technique suffers in that the doping profiles usually attained again do not provide strong enough electric fields to produce optimum QE pinning, particularly in the UV region. Also, the ion implantation damages the silicon lattice structure, creating bulk traps, which again reduce QE. This damage could be thermally annealed, but it is difficult to attain high enough temperatures after CCD fabrication to anneal the damage and yet not destroy the aluminum gates, diffusions, and so forth. Therefore, this method also has its limitations.

The technique found to be most satisfactory is called *flashgate* [9,10]. In this approach, a thin film (approximately a monolayer) of high work function metal is applied to the backside of the CCD over the oxide. Because the work function of the metal is higher than that of the CCD, it permanently accumulates the back surface with a negative charge and will then repel minority carriers (electrons) away from the back surface toward the desired potential wells under the clock gates at the front of the device. Platinum has proven to be a good material for the flashgate, because it has a high work function and appears to be quite stable. We have found that thermal evaporation is the best method of application to produce a stable thin film of platinum. In the required thicknesses of only a few angstroms, there is negligible optical reflection or absorption, ensuring a high quantum efficiency.

This flashgate process can optionally be followed by a fourth step of applying an antireflection coating to further enhance the QE of the device. Antireflection (AR) coatings are not practical on front-illuminated devices because of the optical effects of the polysilicon gates. One must take care in the application of the AR coating so as not to disturb the negative charge maintained by the flashgate.

10.3.4 Details of Flashgate Theory

In Figure 10.5(a), we show the CCD and metal separated at a long distance, d, with the metal having a larger work function than the CCD. Under these circumstances, the Fermi levels do not coincide, and the system is not in equilibrium. When the metal and CCD are moved closer together as shown in Figure 10.5(b), electrons will tunnel from the CCD to the metal when the physical separation, d, is less than about 3 nm, or about 6 interatomic distances. The flow of electrons creates a negative charge in the metal and an accumulation of holes at the surface of the CCD as indicated in Figure 10.5(c). This generates an electric field within the CCD that raises the potential energy of the electrons with respect to those on the metal until the two Fermi levels coincide, at which point the tunneling current stops. The contact potential that develops after this current flow is simply given by the work function difference between the CCD and metal gate:

$$V_s = (\chi + V_n) - \phi_m \tag{10.6}$$

where V_s is the contact potential (or surface potential), ϕ_m is the work function of the metal, and $(\chi + V_n)$ is the work function of the CCD in which χ is the potential difference between the conduction band and the vacuum level $(2 \cdot \chi = 4.15$ eV for silicon), and V_n is the potential difference between the Fermi level and the conduction band. In actual fact, the full contact potential does not drop entirely within the CCD, but in part develops across the oxide layer on top of the silicon. Details are discussed in [10]. The pertinent fact is that, if the oxide is not of sufficiently high quality, many interface states can exist at the back surface, causing the contact potential to be dropped across the oxide rather than the CCD surface, with the result that the back surface region would remain in depletion with no improvement in QE. Therefore, it is necessary to grow a controlled high-quality oxide to provide accumulation at the back surface.

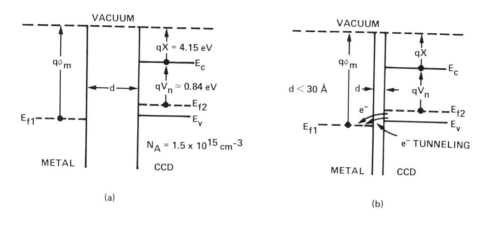

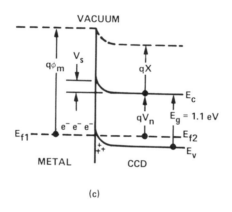

Figure 10.5 Successive stages in the establishment of equilibrium between the CCD and metal for the ideal flashgate. A contact voltage, V_s, develops as electrons tunnel from the CCD onto the metal gate, which has a higher work function than the CCD. (a) The CCD and metal are separated at distance d. (b) Electrons tunnel from the CCD to the metal. (c) The electron flow creates a negative charge and accumulated holes.

Figure 10.6 shows a cross section of a backside-illuminated thinned CCD accumulated by a flashgate. The energy band structure illustrates how photoelectrons generated near the backside are swept to the frontside well by an electric field promoted by an accumulation layer of holes. Photoelectrons generated beyond the critical distance, X_c, will definitely reach the frontside potential collecting well. Electrons generated within X_c may diffuse to the backside and recombine, being lost,

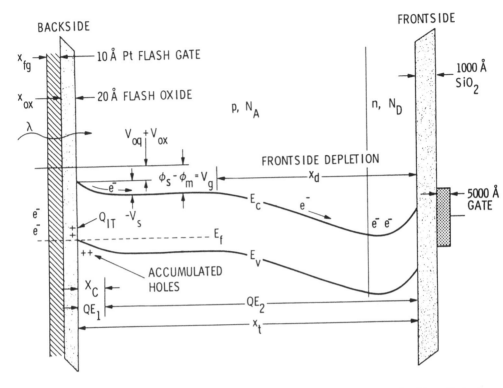

Figure 10.6 Energy band of thinned CCD illuminated at the backside and accumulated by a flashgate.

unless the surface potential, V_s, is of sufficient magnitude to generate an adequate electric field. The relationship between the critical distance, X_c, and electric field generated, E_s, is derived in [13] and shown to be $X_c = D/\mu|E_s|$, where D is the diffusion constant (34 cm^2/s) and μ is the electron mobility (1,500 cm^2/V-s) of silicon.

Assuming a known quantum yield (1 for simplicity), using (10.2) for the charge collection efficiency, we can write the QE in the accumulation region as

$$QE_1 = (1 - \rho) \int_0^{x_c} P_1(x)\alpha \exp(-\alpha x)dx \tag{10.7}$$

The value for probability of carrier collection P_1 can be approximated [5] by assuming, at the surface, $x = 0$, P_1 is 0; at $x = X_c$ the probability of collection P_1 is 1;

for large backside electric fields P_1 increases exponentially with x; and as the field approaches 0 in the CCD P_1 varies linearly with depth as x/X_c. Satisfying these conditions leads to the following collection probability:

$$P_1 = \frac{x}{X_c[\log(|E_s| + 10)]} \qquad (10.8)$$

In the region beyond the critical distance, P_1 is 100%, and so (9.7) reduces to merely the total photon flux absorbed in this region:

$$QE_2 = (1 - \rho)[\exp(-\alpha X_c) - \exp(-\alpha x_t)] \qquad (10.9)$$

where x_t is the total device thickness, and $\rho(\lambda)$ and $\alpha(\lambda)$ are the reflection and absorption coefficients, respectively, of silicon. Both are functions of wavelength, λ. Thus we can combine (10.7) and (10.9) to obtain the overall QE for the CCD: $QE_t = QE_1 + QE_2$.

Most of the cases of interest involve thinned CCDs that are QE pinned; that is, the electric field is strong enough to assure collection of all of the photogenerated carriers. In this case, (10.7) and (10.9) simply reduce to

$$QE_b = (1 - \rho)[1 - \exp(-\alpha x_{\text{thin}})] \qquad (10.10)$$

where x_{thin} is the total thickness of the thinned CCD.

Similarly, for a frontside thick illuminated CCD, we can approximate the QE by writing

$$QE_f = (1 - \rho)[\exp(-\alpha x_{\text{poly}})][1 - \exp(-\alpha x_{\text{epi}})] \qquad (10.11)$$

Here x_{poly} is the thickness of the polysilicon gates (often about 500 nm) and x_{epi} is the thickness of the epitaxial layer. We are assuming that the absorption and reflection coefficients for silicon and polysilicon are approximately the same. The frontside model represented in (10.11) is very approximate. It is difficult to know the exact values of factors such as thickness and optical absorption and reflection characteristics of the doped polysilicon and oxide gate layers and the charge collection efficiency below the depletion region, especially near the epitaxial boundary. Also, the electric field gradient resulting from the doping profile of the depletion region is difficult to predict in advance. Therefore, (10.11) is useful only to estimate frontside QE. Figure 10.7 shows a theoretical QE predicted from (10.10) and (10.11), assuming polysilicon gates of 500 nm thickness and epitaxial thickness of 10 through 50 μm, which are fairly typical values.

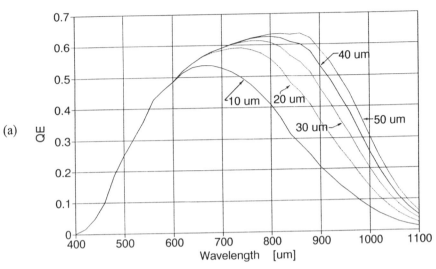

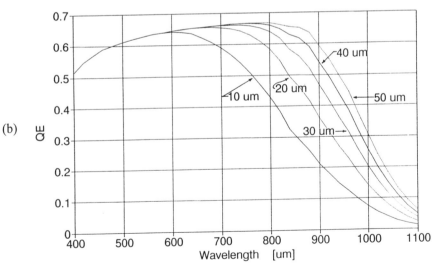

Figure 10.7 Theoretical QE of typical front (a) and back (b) illuminated CCD using (10.10) and (10.11).

10.3.5 Thinning Technology and Backside Performance

Many examples of thinning could be discussed. We will outline the approaches used and results obtained during the last couple years in research programs at SAIC.

To accomplish the actual thinning of CCDs, we employ a series of acid etch solutions composed of various mixtures of acetic, nitric, and hydrofluoric acids. Prior to thinning, a glass substrate is sometimes mounted to the front of the CCD (over the parallel gate array) to hold it flat within a few μm. This avoids problems with unsupported thinned CCDs, which can "wrinkle" to depths of .5–1 mm.

After thinning, the passivation oxide is grown on the back surface, which is essential to stable, high-QE performance in the blue and UV spectral regions. For this purpose, an oxidation furnace that gives a controlled flow of oxygen and water vapor over devices at constant temperature to provide for uniform oxide growth is needed. The temperature must be high enough to allow reasonable oxide growth rates, but not so high as to damage the CCD. A typical temperature range is from 250–280°C.

Following oxidation, the backside is permanently accumulated by using a thermal platinum flashgate deposition performed in a high-vacuum evaporation system.

Finally, one can antireflection coat a thinned device with a two layer MgF_2–ZnS coating, taking care not to destroy the efficacy of the flashgate accumulation. Figure 10.8 shows the measured QE of a bare silicon flashgated device and that of

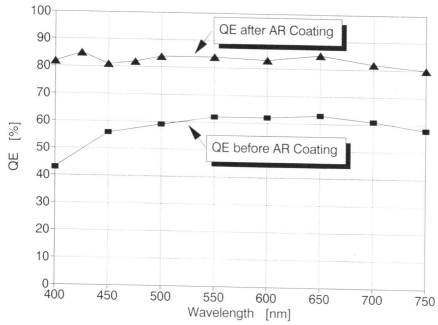

Figure 10.8 Measured QE for a backside-illuminated CCD before and after AR coating.

an AR-coated device. Figure 10.9 shows the same information for a bare flashgated CCD down to 250 nm. Finally, Figure 10.10 gives the QE of a frontsided device; compare it with results from (10.11). Note the comparison between backside-illuminated QE with that of a frontsided CCD, in which the input light is attenuated by the polysilicon gates. It is also interesting to note the level of agreement between the simple theoretical models for QE and the actual measured experimental results. The structure in the graph for the measured QE of the frontsided chip as a function of wavelength results from optical interference upon the passage of the incident light through the polysilicon gate structure on the front of the CCD.

10.3.6 Alternative Approaches for High QE

Although thinning and backside illumination can produce outstanding quantum efficiency performance, as has been demonstrated, the technique is somewhat difficult

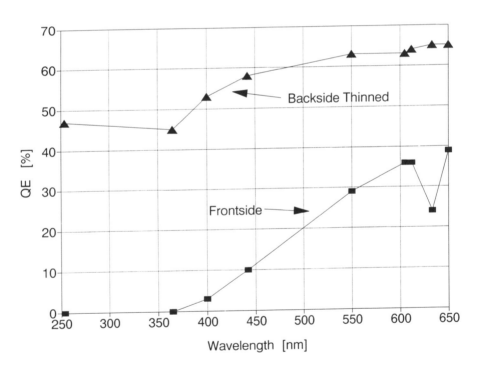

Figure 10.9 Measured QE for a backside-illuminated CCD in the near-ultraviolet region, compared with measured frontsided QE.

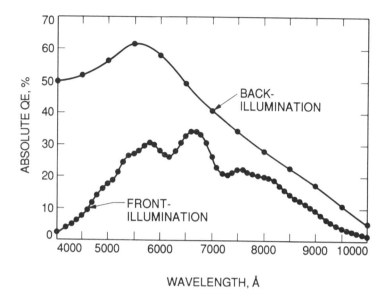

Figure 10.10 Measured QE for a frontside illuminated CCD.

to master and the devices are relatively expensive to produce. Hence, a couple of alternative approaches have been explored, which will be mentioned here only briefly. The first approach is to extend QE of a frontside-illuminated device into the blue and UV spectral regions by applying a phosphor coating over the front gate structure. This phosphor will fluoresce, effectively shifting the wavelength of incident photons into the visible region, where the QE of the frontside device is higher. A good review on this approach is given in [14], in which a fluorescent material that is a polycyclic aromatic hydrocarbon called *Metachrome-2* is described. Figure 10.11 shows the response of a typical frontside-illuminated (thick) CCD coated with this material. It is applied as a thin film, typically 300 to 400 nm thick. As can be seen, QE in the UV is substantial, but not equal to that obtained with a backside-illuminated CCD. In practice, however, for situations in which this technique produces adequate response, devices are more easily produced this way. One potential drawback is possible slow photo-oxidation of the phosphor when exposed to intense UV illumination, leading to a moderate loss in sensitivity. The other alternative technique, originated and described by Janesick [15], is referred to as *open pinned-phase* (OPP), in which some of the polysilicon gates are omitted, increasing the sensitivity of the device.

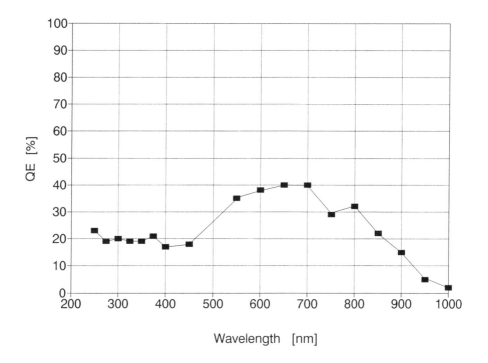

Figure 10.11 QE curve for a frontsided CCD coated with Metachrome-2 [14].

10.4 CHARGE COLLECTION EFFICIENCY

Charge collection efficiency (CCE) measures the ability of a CCD to collect all the charge generated from a single photon event into a single pixel. This, of course, is extremely important if a CCD is to produce an accurate and faithful reproduction of the original scene. Also, of course, as seen in the previous section, good collection efficiency is essential to high overall quantum efficiency.

Basically three regions [16] influence the charge collection process: the depletion region, which includes the charge-carrying buried channel and depleted bulk epitaxial layer beneath it; an undepleted, field-free, neutral bulk below this; and a region of high recombination that could be the surface of an unpassivated thinned CCD or the low-lifetime substrate material of an unthinned, frontside-illuminated device. If a charge is generated within the depletion region associated with a given potential well, there is a very high probability that it will be collected in that pixel. However, if it is generated within the undepleted (field-free) bulk, there is a good probability that the charge may spread into surrounding pixels. And finally, a charge generated within the third region probably will not be collected at all.

Figure 10.12 shows a cross section of a typical thinned CCD, illuminated from the back. Actually, for photons that can penetrate the gate structure, similar results would apply for illumination from the front, because the minority carrier formed from the photon absorption is subject to the same electric fields in any case.

As was the case with planar photodiodes, certain thinned CCDs have been tested [16] and verified to have 100% internal quantum efficiency. A thinned CCD needs three external bias voltages, as shown in Figure 10.12, to establish the proper electric field conditions throughout the photosensitive volume necessary to collect all of the photogenerated carriers: V_{bg} = backside gate voltage, V_{np} = frontside depletion voltage, and V_{fg} = frontside gate voltage. V_{bg} is required to form the accumulation layer on the back of the CCD to propel photoelectrons generated near the back forward into the front collection well, as discussed in the last section. It has been shown that the flashgate-induced potentials must create a field greater than V/cm to prevent backside recombination loss.

V_{np} is used in conjunction with V_{fg} to control the electric field within the CCD nnel. V_{np} can be controlled by applying a bias to the reset drain of the CCD, and

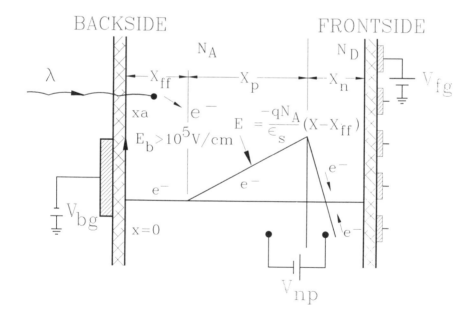

Figure 10.12 Electrical field in CCD, showing a field-free region x_{ff}, the p-depletion region x_p with acceptor concentration N_A, and the n-depletion region x_n, with donor concentration N_D. The indicated electric fields in the depletion regions are produced by voltages V_{np} and V_{fg}, and the backside reflecting field E_b is produced by voltage V_{bg}. Backside illumination is illustrated, but frontside illumination is also possible and advantageous in some instances.

V_{fg} is adjusted by means of the voltage levels of the clocking bias applied to the polysilicon gates on the front of the device. The best CCE is obtained if, during integration of signal charge, V_{np} and V_{fg} are adjusted so that the frontside depletion region extends back to the backside accumulation region.

The depth of the frontside p-depletion region can be calculated [10] by

$$x_p \simeq \left[\frac{2\varepsilon_s N_D}{qN_A(N_A + N_D)} (V_{bi} + V_{np}) \right]^{1/2} \tag{10.12}$$

where ε_s is the permittivity of silicon (1.04×10^{-12} F/cm), N_D is the donor concentration of the n-channel (for example, 1.5×10^{16} cm^{-3}), N_A is the acceptor concentration of the p-region (for example, 1.5×10^{15} cm^{-3}), q is the electron charge (1.6×10^{-19} C), V_{bi} is the built-in potential of the np junction, and V_{np} is the potential of the CCD channel relative to the substrate. Note that the depletion depth is inversely proportional to the square root of the N_A doping.

In [10], equations are developed to calculate the potential distribution as a function of depth into the CCD for different epitaxial dopings. Figure 10.13 shows an example of a plot of the potential distribution as a function of depth for $N_A = 1.5 \times 10^{15}$ cm^{-3} and $N_D = 1.6 \times 10^{16}$ cm^{-3}, assuming an oxide thickness of 100 nm, a substrate voltage of 0 V, and V_{np}, the depletion voltage of 30V. Potentials are shown for 5V intervals of V_{fg}, the frontside gate voltage. Note how the depletion depth increases slowly with increasing V_{fg}.

One test of the CCE is an examination of the modulation transfer function (MTF) of the device. MTF is a measure of the ability of a CCD to respond to sinusoidal spatial modulations of the signal intensity as a function of spatial frequency. Because the CCD is a discrete sampling device, the best MTF obtainable is given by [3] the function sinc($\pi f / 2f_N$), where f is the scene spatial frequency in units of cycles/millimeter, and f_N is the Nyquist frequency of the CCD, given by the expression $f_N = 1/2d$, where d is the pixel spacing in millimeters. For example, in a CCD with 18 μm square pixels, the Nyquist frequency is $1/0.036$ mm $= 27.8$ cycles/mm.

It can be seen that the best MTF obtainable, if $f = f_N$ and MTF $=$ sinc($\pi/2$) is 0.63. Figure 10.14 shows a plot of MTF vs. resolution for an 18 μm square pixel CCD with $N_A = 1.5 \times 10^{15}$ cm^{-3}, a depletion depth $x_p = 3.5$ μm, and a 10-μm thick epitaxial layer for various wavelengths of incident light. Note that the best (Nyquist) MTF degrades with increasing wavelength because the redder light penetrates much deeper into the device under the depletion region. As was discussed in the last chapter, the absorption coefficient for silicon in the red is much less than in the UV, blue, and green parts of the spectrum, resulting in mean-free photon paths before absorption that are much longer in the red and infrared spectral regions.

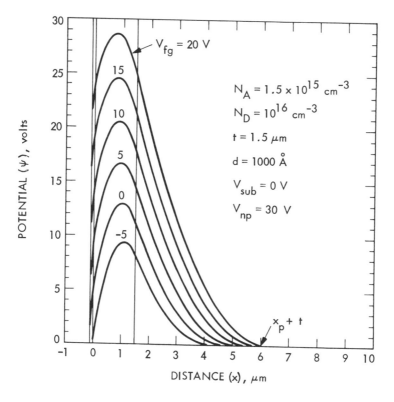

Figure 10.13 Potential distribution through a typical CCD as a function of distance for different V_{fg} leaving V_{sub} and V_{np} fixed, assuming $N_A = 1.5 \times 10^{15}$ cm^{-3} and $N_D = 1.6 \times 10^{16}$ cm^{-3}.

One can deduce from these observations that a thicker epitaxial layer with higher resistivity (lower N_A), so that the depletion depth x_p is increased, is necessary to obtain good red CCE for a frontside-illuminated CCD. Of course, for a backside device, blue photon collection will be enhanced with thinner epitaxial thicknesses.

Another important test for CCE is X-ray response. During the past few years, techniques have been developed [3,17–19] for X-ray characterization and calibration of CCDs. High-energy X-ray photons (0.1–10 nm), when absorbed by silicon, have a quantum yield much higher than 1 and create many electron-hole pairs per photon absorbed. This happens according to the relation $\eta = E_\lambda/3.65$, where η is the ideal quantum yield in electrons, and E_λ is the X-ray photon energy in eV.

One of the more convenient and commonly used X-ray sources for this testing is an Fe^{55} source. It emits photons from the manganese emission line at 5.9 keV, $\lambda = 0.21$ nm. By the preceding relation, each photon will generate 1620 e^-. These 1620 carriers are contained in a small (<0.5 μm) cloud at essentially a point within

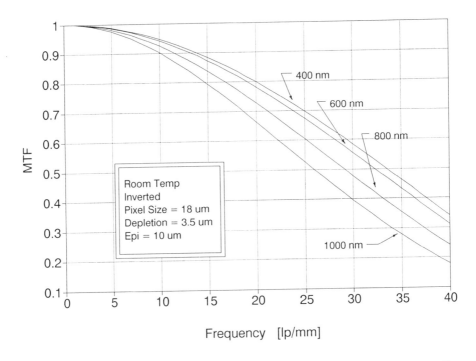

Figure 10.14 Plot of MTF vs. resolution for an 18 μm square pixel CCD with $N_A = 1.5 \times 10^{15}$ cm^{-3}, a depletion depth $x_p = 3.5$ μm, and a 10-μm thick epitaxial layer for various wavelengths of incident light.

the CCD where they were generated. Ideally, the carriers from one X-ray "event" should be confined to the single pixel where the photon was absorbed, and surrounding pixels will contain no charge. However, if the X-ray was absorbed in the field-free region below the depletion region, random diffusion of the charge may allow some of it to spread to adjoining pixels before being swept into the collecting well.

The figure of 1620 e^- per absorbed X-ray photon is subject to a small uncertainty due to a finite amount of energy that is transferred to the silicon lattice by non-e-h pair processes, giving rise to a small statistical difference in the number of e-h pairs actually generated. This uncertainty is characterized by the Fano factor, originally formulated to describe the uncertainty of the number of ion pairs produced in a volume of gas following the absorption of ionizing radiation [20]. As discussed by Janesick in [21], this factor implies an uncertainty about 13 e^- rms in silicon absorption for Fe^{55} X-rays. As was pointed out in that discussion, CCD technology has progressed to the level where their properties can actually be used to measure and verify the Fano factor for silicon.

Figure 10.15 shows a map indicating close up 208 pixels from a CCD exposed to Mn X-rays from an Fe^{55} source. Each tiny dot represents one signal electron that was measured at the CCD output. Completely dark pixels are those in which all 1620 e^- were contained, and they are called *single-pixel events*. The squares that are not completely dark and consist of several pixel clusters are called *split events*. In those cases the 1620 e^- diffused into neighboring pixels that share the charge. The amount of charge splitting depends upon where the photon is absorbed. Those absorbed within the front depletion region will remain contained and produce single-pixel events, whereas photons absorbed in the field-free region beyond the depletion zone will create charge clouds that thermally diffuse outward until reaching the sharply defined potential wells at the lower boundary of the pixel well. At that point, the charge cloud may split between several pixels, leading to the split events. A few split events can be created within the depletion region because of a photon absorbed directly on a boundary between two or more pixels.

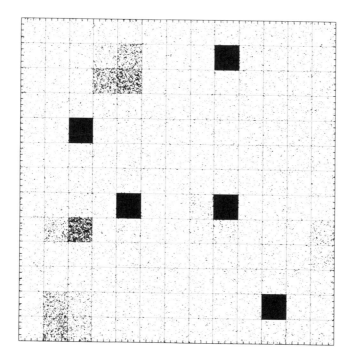

Figure 10.15 A pixel map displaying Fe^{55} Mn X-ray events that generate 1620 e^- for each interaction. Each dark square represents a pixel containing all of these electrons with no loss.

10.5 CHARGE TRANSFER EFFICIENCY

After the signal charge is collected within a pixel, this charge packet must be transferred to the output amplifier of the CCD without loss. The charge transfer efficiency (CTE) of a chip is the ability of the device to transfer charge from one potential well to the next. Silicon processing and design technology have progressed to the point where this is now an incredibly efficient process, making feasible large CCDs with millions of pixels. For well-made buried-channel CCDs, the efficiency will be 0.999999/transfer or even higher [21]. Any charge that is not collected is either lost through recombination or, more likely, dribbled out as a deferred charge tail in later, trailing pixels.

In theory, three mechanisms must be present to obtain efficient charge transfer in a CCD: thermal diffusion, self-induced drift, and the fringing field effect. The size of the charge packet determines the relative importance of each of these. Both thermal diffusion and fringing fields are important for transferring small amounts of signal charge, whereas self-induced drift caused by the mutual electrostatic repulsion of the carriers within a packet dominates charge transfer for larger charge packets. Experience has shown, however, that the most important influence on CTE is a factor called the *spurious potential pocket*. This is the loss of charge during transfer due to improper potential well shape or depth beneath the pixel. Those effects have been found to occur due to channel width variation, polysilicon edgelifting, and boron lateral diffusion [22,23]. Also, the formation of potential pockets, probably in regions of overlapping polysilicon gates, has been studied [24] and experimentally correlated with charge-transfer losses. Even in today's high-quality CCDs, these pockets can ultimately limit performance, if present in any significant quantity.

Several techniques have been developed to measure CTE [21]. One technique is called *X-ray event stacking*. In this method the CCD is flooded with a low intensity of soft X-rays such as manganese 5.9 keV X-rays from an Fe^{55} source as discussed in the previous section. During this time, many lines or columns (several hundred) of raw data are accumulated. The photons of a single X-ray line will generate approximately equal amounts of charge (1620 e^- for the Fe^{55} 5.9 keV). When these rows of data are superimposed on one plot, as in Figure 10.16, the single-pixel events tend to cluster to form a line, if the CCD is functioning so that the majority of X-rays absorbed result in single-pixel events. The split and partial events are spread out in between the single event line and the dark current line. By overclocking, so that extended pixels not in contact with the parallel part of the array are read, one can also determine the read noise, as shown in the figure. If the CTE is less than nearly perfect, this line will have a tilt because pixels shifted more times are losing more charge. Note also that this provides a system calibration, because the single-event line occurs for 1620 e^-, giving a calibration for the signal in digital numbers. In the figure, there is a slight tilt, corresponding to a loss of 50 e^- in 1,044 (1,024 + 20 extended pixels) transfers.

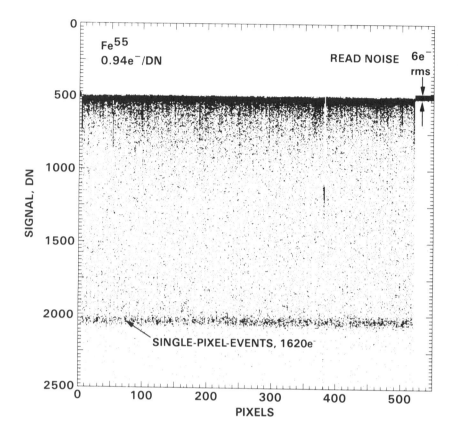

Figure 10.16 An Fe^{55} X-ray line trace response generated by a 1024 CCD. The strong signal at 1620 e^- represents the single-pixel event line. Other signals are due to split and partial events.

CTE can be calculated from the following relation:

$$CTE = 1 - q_d/(q_0 \times N) \qquad (10.13)$$

where q_d is the net deferred charge, q_0 is the initial charge produced by the X-ray photon, and N is the number of transfers. In the example from Figure 10.16, q_d is 50 e^-, q_0 is 1,620 e^-, and N is 1,044 transfers, implying a CTE of 0.99997. This particular device has a horizontal CTE deficiency; in vertical CTE it showed 0.99999. The problem could stem from bulk state traps, which would show up in the higher speed clocking used for horizontal read-out compared to the vertical clocking rate.

Another useful test for CTE is called the *extended pixel edge response* (EPER) [21] *method*. In this technique, the CCD is continuously (no shutter) exposed to a low-level flat-field signal, say, about 16,000 e^-. By overclocking and reading out the extended pixels in the horizontal register, one can observe the last pixel of the array, which is brighter (in this instance it would be about 37,500 e^-) because of charge that diffuses in from the border of the CCD. By averaging many lines (~1,500) and observing the extended pixels immediately following the border pixel, one can see any deferred charge lingering in these first extended pixels. The CTE again is given by (10.13).

The EPER method is a convenient way to measure CTE performance for the CCD. However, it is important to note that this approach represents only a relative measurement of CTE and, therefore, should be augmented with the X-ray method just discussed, which provides absolute CTE measurement.

Figure 10.17 shows a stacked X-ray plot of a CCD with a problem that now fortunately is rare. The CCD has a large single-pixel trap within its horizontal register, which displaces about 1,000 e^- as deferred charge.

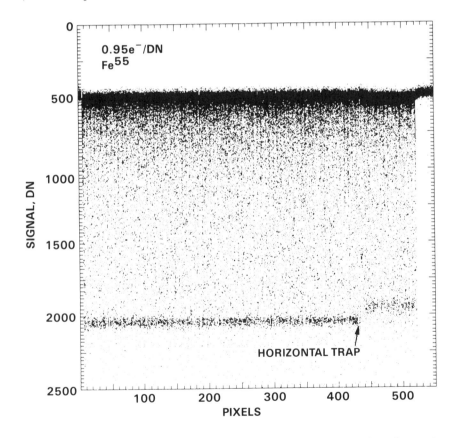

Figure 10.17 A 1024 CCD with a large single-pixel trap within its horizontal register, fortunately now a rare problem.

It is often useful to plot a histogram of the X-ray data, with the number of occurrences on the vertical axis and the signal in DN on the horizontal axis. Figure 10.18 shows such a plot, using a relatively old Fe^{55} source, in which the K_α line is very evident at 1,620 e^-, and the K_β line is weakly present at 1,778 e^-. As discussed in [21], the width of the X-ray line is limited by the Fano factor in this high-quality CCD and not by the read noise. If one examined the stacked X-ray plot, one would observe that the width of the single-event line was greater than that of the over-clocked read noise baseline.

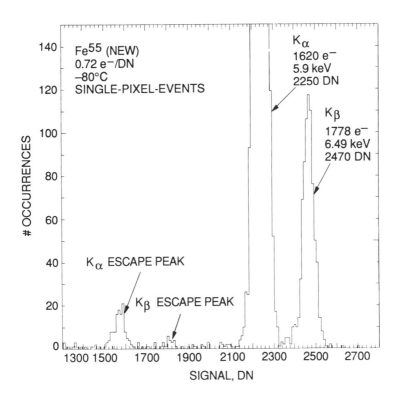

Figure 10.18 An expanded histogram showing K_α and K_β lines in a Fano-noise limited CCD.

10.6 READ NOISE

The read noise of a CCD is a very important parameter. Unfortunately we lack the necessary space to devote to an extensive treatment here—that would require a separate chapter in itself—so we will be content to summarize the key points here [2].

As shown in Figure 10.19, essentially three regimes of noise are to be considered. At high signal levels, the total noise can be dominated by pixel-to-pixel

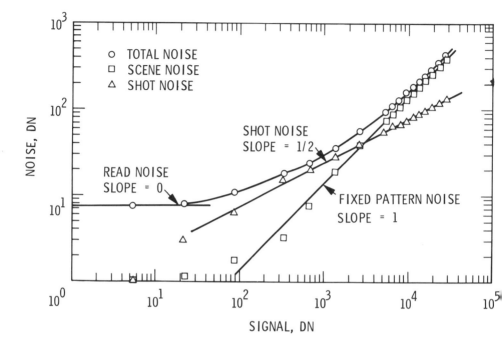

Figure 10.19 Typical plot of noise as a function of input signal for a 20 × 20 pixel subarray, showing the three regimes of read noise, signal shot noise, and pixel-to-pixel fixed pattern noise.

sensitivity variations, also called *fixed pattern noise*. This can be highly dependent upon the exact clocking voltages applied to the array to move the charge packets. Fixed pattern noise can be minimized by choosing the proper operating voltages; and in today's devices it is often not too worrisome. When present and plotted log-log on a photon transfer curve (noise in DN vs. signal in DN), fixed pattern noise will have a slope of 1. When characterizing the performance of a device with a photon transfer curve, the effect of fixed pattern noise can be removed by differencing two frames, thus subtracting it out, because the CCD is a very linear device.

When fixed pattern noise is negligible, the statistics of the signal itself (shot noise) will dominate at intermediate and higher signal levels. As pointed out in the last chapter, the photon noise of a signal can usually be approximated by Poisson statistics, and one writes that the variance of the shot noise in a photocurrent I is given by $\langle I^2 \rangle = 2qI\Delta f$, where q is the quantum of charge, and $\Delta f = 1/2T$ is the bandwidth of the current measurement resulting from averaging over the time period T. This leads to the property on a photon transfer curve that the shot noise slope is $1/2$. A CCD running shot-noise limited (signal-to-noise ratio limited by photon statistics in the signal) is the best one can achieve with any photon detector.

At the lowest illumination levels, the CCD is limited by noise sources intrinsic to the device, forming the so-called read noise floor. As documented in [2], they are trapping-state noise, reset noise, background noise, charge-transfer noise, and output amplifier noise.

Trapping state noise occurs because of interaction between charge packets and surface or bulk states. Buried-channel construction eliminates interaction with the surface states, and modern devices have so few bulk states because of excellent defect control in fabrication that this source is practically negligible.

Reset noise is the uncertainty in the voltage to which the output transistor of the CCD is reset after the charge in a pixel is read. Using the correlated double sampling (CDS) technique [25], one samples the output voltage twice: once after the reset pulse has switched off and once after the signal charge has been read out, effectively eliminating this noise source, which is correlated over the time span of a pixel.

Background noise can originate from any of several sources, such as "fat zero," which in older devices was often added electrically to fill spurious potential pockets; dark current; residual image, which stems from electrons trapped at the front Si–SiO_2 interface and usually occurs only if a CCD has been overexposed or the buried channel becomes undepleted; and luminescence of various CCD structures during operation. Most of these luminescence sources are minor or can be avoided; however, in some CCDs a low impedance between clock gates within a pixel or between the clocked gate and signal channel can result in a blemish luminescence that can saturate the CCD. These are defects that can sometimes be reduced by appropriate clock phasing or adjustment of substrate or gate voltage in integration.

The most important background noise is from dark current, which is a thermally generated charge that obeys Poisson statistics. Therefore dark current noise is given by the square root of the dark current. This dark current can be reduced by cooling (roughly a factor of two for every 8°C reduction) or by a technique called *inversion*, in which the clock voltages are held negative during integration.

Channel inversion [26] is easily accomplished by biasing the array clocks sufficiently negative to pin the surface potential at the Si–SiO_2 interface to substrate potential. At this point, holes supplied by the channel stops are attracted beneath the gate, inverting the n-channel and populating the Si–SiO_2 interface states with holes. This blocks the thermal "hopping" of electrons from valence to interface to conduction levels, reducing dark current by large amounts. To maintain a charge storage well when all three clock phases are inverted, a weak p implant is placed under phase 3 of the CCD. This will maintain a barrier for charge collection even when all clocks are negative. This technique is called *multi-pinned-phase* or *MPP technology*. Figure 10.20 shows the dramatic reduction in dark current by a factor of 30 that this approach allows. This, in effect, makes practical the operation of a CCD at room temperature with reasonably long integration times. In addition to reducing dark current, MPP technology produces other benefits, such as eliminating residual

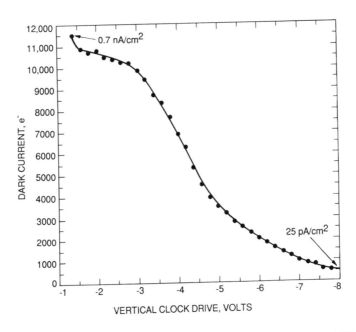

Figure 10.20 Dark current generation as a function of negative vertical bias for a 1024 MPP-CCD biased from the noninverted state to fully inverted state.

image, reducing blooming effects of overexposure, and reducing pixel-to-pixel non-uniformity effects. The only major disadvantage of MPP is reduction of the full well, the amount of charge that can be stored in a pixel. For the CCD with an 18 μm square pixel, for example, the optimum full well running partially inverted (phase 3 at +2.5V, phases 1 and 2 at −8V) is about 350,000 e^-, compared to 110,000 e^- when running fully inverted (all phases at −8V). This is not such a serious problem, because one can tolerate the higher dark currents of partial inversion when working with higher signal levels that necessitate a large full well.

Charge transfer noise comes from two sources: finite transfer inefficiency in the charge transfer process, which is practically negligible in modern CCDs; and spurious charge, the other drawback of MPP operation. This charge is generated when the array is momentarily taken out of inversion during line transfer to the serial register. The holes that had been trapped in interface states are released with enough energy to create e-h pairs in the silicon by impact ionization. These "spurious" electrons are then collected in the nearest potential well. By using a trilevel clocking scheme with an intermediate level, the generation of spurious charge can be significantly reduced.

Ultimately, the read noise of the CCD is limited by the noise of the on-chip output amplifier. The output structure is typically a single MOSFET stage, or sometimes a dual source-follower MOSFET amplifier with an active load. The theory for read noise using correlated-double-sampling (CDS) is described in [21] and is rather complicated. Experimental results for the 18 μm 1024 pixel CCD, Figure 10.21, show that it has a read noise of 6.5 e^-, taken from a photon transfer curve. This device has a white noise floor of 7 nV/$\sqrt{\text{Hz}}$. Measurements were made using a 4 μs clamp-to-sample time. The $1/f$ corner frequency was 20 kHz, and the sensitivity is 0.8 V/e^-. Modern CCDs have recently gone below a read noise floor of 3 e^-, and using special output amplifiers that allow repetitive nondestructive sampling, read noises less than 0.3 e^- have been attained [27].

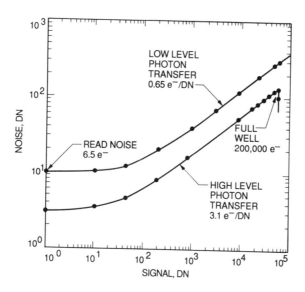

Figure 10.21 Photon transfer curves generated by the 18 μm 1024 CCD showing a read noise of 6.5 e^-.

10.7 SUMMARY

In this chapter we have discussed an integrated circuit, the CCD, which has evolved from basically simple beginnings into a highly complex integrated device suitable for use in an ever-increasing variety of scientific imaging applications. The basic concepts and the fundamentals of how it works are covered. The various CCD architectures have been discussed. A three-phase CCD has been described in detail as an example of a large megapixel scientific-grade CCD array.

A detailed discussion of the four areas of CCD performance includes quantum efficiency, its basic ideas, models, quantum yield, photon transfer curve, backside illumination and thinning of CCDs, flashgating, backside-QE performance, and alternative approaches to high QE; charge collection efficiency, electric fields in devices, depletion depth, MTF, X-ray response, and split events; charge transfer efficiency, X-ray measurement techniques, extended pixel edge response, and their importance to performance; and finally read noise, noise level regimes, pattern noise, shot noise, dark current and dark current noise, read noise floor contributions, MPP and inverted operation, correlated double sampling, and up-to-date noise performance of current scientific CCDs.

ACKNOWLEDGMENTS

So many people have contributed so much to the development of the CCD that it is impossible to fairly acknowledge the work they have done. In particular, we would like to acknowledge the efforts and useful discussions with coworkers John Cover, R. Varian, P. Rose, S. Yang, D. Doliber, R. Smith, and the outstanding technical assistance of R. Larsen, P. Bertling, and J. Bierman at SAIC. At FORD Aerospace (now Loral) we wish to acknowledge the very fine design work and craftmanship of R. Bredthauer, C. Chandler, and J. Pinter. At the Jet Propulsion Laboratory, we want to acknowledge the diligent efforts of T. Elliott and many others.

REFERENCES

[1] W. Boyle and G. Smith, "Charge Coupled Semiconductor Devices," *Bell Syst. Tech. J.*, Vol. 49, 1970, pp. 587–593.
[2] J. R. Janesick, T. Elliott, A. Dingizian, R. Bredthauer, C. Chandler, J. Westphal, and J. Gunn, "New Advancements in Charge-Coupled Device Technology—Sub-electron Noise and 4096 × 4096 Pixel CCD's," *SPIE Proc. on Charge Coupled Devices and Solid State Optical Sensors*, Vol. 1242, 1990, pp. 223–237.
[3] J. R. Janesick, T. Elliott, S. Collins, M. M. Blouke, and J. Freeman, "Scientific Charge-Coupled Devices," *Opt. Engr.*, Vol. 26, 1987, pp. 692–715.
[4] J. R. Janesick, T. Elliott, R. Bredthauer, J. Cover, A. R. Schaefer, and R. Varian, "Recent Developments in Large Area Scientific CCD Image Sensors," *SPIE Proc. on Optical Sensors and Electronic Photography*, Vol. 1071, 1989, pp. 115–133.
[5] J. Janesick and D. Campbell, "Quantum Efficiency Model for the CCD Flash Gate," IEEE Int. Electron Devices Meeting, 86CH2381-2, 1986, pp. 350–352.
[6] M. M. Blouke, G., Womack, and W. A. Delamere, "Simplified Model for AR-Coated Backside-Illuminated CCDs," *SPIE Proc. on Electronic Imaging Science and Technology*, Vol. 1447, 1991.
[7] C. M. Huang, J. R. Theriault, E. T. Hurley, B. W. Johnson, B. E. Burke, J. A. Gregory, and B. B. Kosicki, "Quantum Efficiency Model for p⁺Doped Back-Illuminated CCD Imager," *SPIE Proc. on Electronic Imaging Science and Technology*, Vol. 1447, 1991.
[8] J. R. Janesick, K. P. Klaasen, T. Elliott, "Charge-Coupled-Devices Charge-Collection Efficiency and the Photon-Transfer Technique," *Opt. Engr.*, Vol. 26, 1987, pp. 972–980.
[9] J. R. Janesick, D. Campbell, T. Elliott, and T. Daud, "Flash Technology for Charge-Coupled-Device Imaging in the Ultraviolet," *Opt. Engr.*, Vol. 26, 1987, pp. 852–863.

[10] J. R. Janesick, T. Elliott, T. Daud, and D. Campbell, "The CCD Flash Gate," *Proc. of SPIE Instrumentation in Astronomy VI*, ed. D. L. Crawford, Vol. 627, 1986, pp. 543–582.

[11] A. R. Schaefer and R. H. Varian, "Large Area Megapixel CCD Arrays," *Proc. of 33rd Midwest Symp. on Circuits and Systems*, Calgary, Canada, 1990.

[12] A. R. Schaefer, R. H. Varian, J. Cover, and R. Larsen, "Megapixel CCD Thinning/Backside Progress at SAIC," *SPIE Proc. on Electronic Imaging Science and Technology*, Vol. 1447, 1991.

[13] J. R. Janesick, T. Elliott, T. Daud, and J. McCarthy, "Backside Charging of the CCD," *Proc. of SPIE*, Vol. 570, 1985, pp. 46.

[14] G. R. Sims, F. Griffin, and M. P. Lesser, "Improvements in CCD Quantum Efficiency in the UV and Near-IR," *SPIE Proc. on Optical Sensors and Electronic Photography*, Vol. 1071, 1989, pp. 31–42.

[15] J. R. Janesick, "Open Pinned-Phase CCD Technology," *SPIE Proc. on Optical and Optoelectronic Applied Science and Engr.*, Vol. 1159, 1989.

[16] J. R. Janesick, T. Elliott, S. Collins, T. Daud, and D. Campbell, "Charge-Coupled Device Advances for X-Ray Scientific Applications in 1986," *Opt. Engr.*, Vol. 26, 1987, pp. 156–166.

[17] J. Janesick, T. Elliott, H. Marsh, S. Collins, M. Blouke, and J. McCarthy, "Potential of CCD's for UV and X-Ray Plasma Diagnostics," *Rev. Sci. Instrum.*, Vol. 56, No. 5, 1985, pp. 796–801.

[18] J. Janesick, T. Elliott, J. McCarthy, H. Marsh, S. Collins, and M. Blouke, "Present and Future CCDs for UV and X-Ray Scientific Measurements," *IEEE Trans. Nucl. Sci.*, Vol. NS-32, No. 1, 1985, pp. 409–416.

[19] R. Stern, K. Liewer, and J. Janesick, "Evaluation of a Virtual Phase Charge-Coupled Device as an Imaging X-Ray Spectrometer," *Rev. Sci. Instrum.*, Vol. 54, No. 2, 1985, pp. 198–205.

[20] U. Fano, "Ionization Yield of Radiations: II. The Fluctuations of the Number of Ions," *Phys. Rev.*, Vol. 72, No. 1, 1947, pp. 26–29.

[21] J. R. Janesick, T. Elliott, R. A. Bredthauer, C. E. Chandler, and B. Burke, "Fano Noise Limited CCD's," *SPIE Proc. Symp. on Optical and Optoelectronic Applied Science and Engr, X-Ray Instrumentation in Astronomy*, 1988.

[22] G. Taylor, "An Evaluation of Submicrometer Potential Barriers Using Charge-Transfer Devices," *IEEE J. Solid-State Cir.*, Vol. SC-15, No. 4, 1980, pp. 644–648.

[23] C. Chen, K. Venkaleswaran, J. Seto, and F. Amelio, "The Effect of Meterpoly Structure Variation on Charge Transfer Efficiency of a Buried Channel CCD," IEEE CH1504-0606, 1979, pp. 606–609.

[24] J. Janesick, and T. Elliott, "Spurious Potential Pockets and Pixel Nonuniformity in CCD's," Ninth Santa Cruz Workshop in Astronomy and AstroPhysics: Instrumentation for Ground Based Optical Astronomy, Lick Observatory, 1987.

[25] D. Barbe, "Imaging Devices Using the Charge-Coupled Concept," *Proc. IEEE*, Vol. 63, 1975, pp. 38–67.

[26] J. Janesick, T. Elliott, G. Fraschetti, S. Collins, M. Blouke, and B. Corrie, "Charge-Coupled Device Pinning Technologies," *SPIE Proc. on Optical Sensors and Electronic Photography*, Vol. 1071, 1989, pp. 153–169.

[27] C. E. Chandler, R. A. Bredthauer, J. R. Janesick, J. A. Westphal, and J. E. Gunn, "Sub-Electron Noise Charge-Coupled Devices," *SPIE Proc. on Charge-Coupled Devices and Solid State Optical Sensors*, Vol. 1242, 1990, pp. 238–251.

Chapter 11
Sensors for the Automotive Industry

R. Frank
Motorola

11.1 INTRODUCTION

The automotive industry has been specifically chosen to describe sensor applications for two reasons: it represents an extremely harsh environment for electronic components, and it provides high-volume applications that in turn drive the development of new sensing technologies. New systems being developed for automobiles are the result of government legislation (emissions and safety regulations) and the need to have a competitive advantage (antilock brake systems, air bags, and entertainment systems). Many of these applications require a wide variety of sensing inputs and use sensor technologies that are also found in medical, industrial, and instrumentation applications.

A few of the factors influencing the increased use of electronic systems in automobiles are improved performance, product differentiation, system diagnostics for improved serviceability, systems integration for improved reliability, reduced emissions, improved safety and convenience, communications, and improved manufacturability. A simplified model of an automotive control system is shown in Figure 11.1. Input, either from vehicle sensors (or switches) or driver actuated switches, are converted to a form that can be used by the microprocessor. In certain cases the sensor is the feedback element for the closed loop control system. The interface requirements are minimal or nonexistent for simple switched inputs but can require amplification, calibration, buffering, temperature compensation, or analog-to-digital (A/D) conversion for many sensing inputs. The complexity and number of components can be considerably reduced if custom integrated circuits and integrated sensors are developed to provide the interface from the sensing element to the microprocessor.

Parameters that can and need to be sensed in the car vary from system to system. Certain common measurements such as wheel speed are required for several

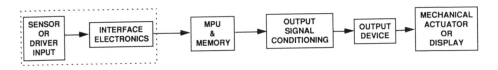

Figure 11.1 MCU-based automotive control system.

systems. Figure 11.2 [1] shows automotive sensor volumes in 1990 for three of the major measurement requirements. The numbers exclude commodity sensors such as temperature and standard vehicle speed sensors. (Temperature sensors account for as much as one-fifth the total volume in 1990.) The total number of motion-position, pressure, and acceleration sensors is projected to double by 1996. This growth is caused by the extension of control systems that are presently only on limited production vehicles to higher volume car lines and in some cases almost the entire fleet. In addition, new systems that are still in the development phase will achieve a level of high-volume production. Acceleration sensors are projected to experience the highest growth.

The automotive environment is recognized to be one of the more difficult applications for electronic systems and microelectronic sensors. The environment includes a wide temperature range, high-humidity conditions, and the need to withstand several chemicals, operate under high electromagnetic interference (EMI) conditions and yet cause relatively low levels of radio frequency interference (RFI), despite extremely wide voltage swings. At the same time, the acceptance of electronics is extremely customer driven and demands low cost and high reliability. Projected sales volumes can decrease (or increase) quickly, depending on customer acceptance of the system, the vehicle in which the system is offered, and external economic factors that affect purchasing decisions for high-cost items such as automobiles.

A major differentiating factor for automotive usage is whether the location of sensors is in the passenger compartment or under the hood. Table 11.1 (after SAE J1211) shows the variation that can occur to several variables depending on the location of the electronic component. High temperature and humidity can significantly reduce the useful life, especially in underhood applications of electronics. The SAE (formerly, Society of Automotive Engineers) has developed "Recommended Environmental Practices for Electronic Equipment Design—SAE J1211" to address the unique problems of automotive electronics.

The system voltage in passenger vehicles is nominally 12V. However, several normal and abnormal voltage variations occur, which must be taken into account at either the component or system level. Normal operating charging systems regulate the output of the alternator to provide sufficient voltage to keep the battery charged under various temperature and load conditions. This voltage can range from 16 V

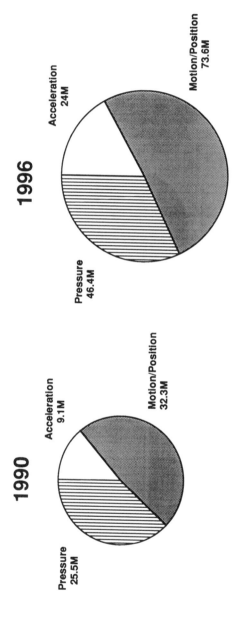

Figure 11.2 Sensor volumes in the North American automotive market. The total sensor market is expected to more than double by 1996.

Table 11.1
The Automotive Environment

Variable	Underhood	Passenger Compartment
Storage temperature	−40°C to +150°C	−40°C to +85°C
Operating temperature	−40°C to +125°C	−40°C to +85°C
Vibration	15g, 10 to 200 Hz	2g, 20 Hz
Humidity	Up to 100%	Up to 100%
Chemicals	Salt spray, water, fuel, oil, coolant and solvent immersion	—
Thermal cycling	>1,000 cycles from −40°C to +125°C	>1,000 cycles from −40°C to +85°C
EMI protection	Up to 200 V/m	Up to 50 V/m

when very cold (−40°C) to about 12V when maximum underhood temperatures are reached. Electronic assemblies on the car must be able to withstand reverse battery conditions (−12V), jump starts from tow trucks (24V), short-duration voltage transients that can easily exceed ±100V, and long-duration (>400 ms) alternator load dump transients, which can be as high as 120V. To prevent damage due to excessive voltage, microelectronic sensors with sufficiently high breakdown voltage (to withstand the variety of automotive voltages) must be used, or the systems they are used in must provide a regulated and protected power bus.

The increasing electronic complexity of modern vehicles has increased the possibility of EMI causing problems between various vehicle systems. In addition, electronics can be susceptible to and radiate RFI causing poor radio reception and malfunctioning vehicle systems with a resulting no-fault condition found during service diagnostic procedures. The potential problems in this area have been addressed by SAE with several information reports and a recommended practice as shown in [2].

Recommended Environmental Practices for Electronic Equipment Design, SAE J1211

Performance Levels and Methods of Measurement of Electromagnetic Radiation from Vehicles and Devices, SAE J551

Performance Levels and Methods of Measurement of EMR from Vehicles and Devices (Narrowband RF), SAE J1816

Electromagnetic Susceptibility Procedures for Vehicle Components (Except Aircraft), SAE J1113

Vehicle Electromagnetic Radiated Susceptibility Testing Using a Large TEM Cell, SAE J1407

Open Field Whole-Vehicle Radiated Susceptibility 10 kHz–18 GHz, Electric Field, SAE J1338

Class B Data Communication Network Interface, SAE J1850

Diagnostic Acronyms, Terms and Definitions for Electrical/Electronic Components, SAE J1930

Failure Mode Severity Classification, SAE J1812

Guide to Manifold Absolute Pressure Transducer Representative Test Method, SAE J1346

Guide to Manifold Absolute Pressure Transducer Representative Specification, SAE J1347

Some possible applications for sensors in existing and future automotive systems are indicated in Table 11.2. This shows a potential of over 100 sensors for 18 different automotive systems. Many of the same parameters are required for different systems, and this will change the way that sensors are used in future vehicle systems. In some cases, such as manifold absolute pressure and mass air flow in engine control systems, one measurement will be preferred over another possible measurement. The list does not include sensors that could be used in measuring current in brushless dc motors or sensors required in entertainment systems and systems with little or no sensor usage, such as electronic mufflers and heated seats.

At least 21 sensing technologies can and are being used to provide these sensors.

- Acoustical
- Capacitive
- Doppler radar
- Electrochemical
- Hall effect
- Inclinometer
- Linear variable differential transformer (LVDT)
- Magnetoresistive element (MRE)
- Optoelectronic
- Piezoelectric
- Potentiometer
- Reed switch
- Silicon piezoresistive
- Silicon capacitive
- Surface acoustic wave (SAW)
- Temperature bimetallic switch
- Temperature—transistor V_{BE}
- Thermistor
- Ultrasonic
- Variable reluctance
- Weigand effect

Many of the sensors are digital (actually a switch) and not analog sensors. Certain technologies have also had much greater acceptance and more widespread use in automotive applications, such as thermistors, Hall-effect sensors, piezoresistive pressure sensors, and variable reluctance sensors.

The requirements for automotive sensors can be divided into three areas: the transducer element; interface electronics; and packaging. The basic transducer element is expected to meet the accuracy requirements of the system and be able to

Table 11.2
Sensor Requirements versus Vehicle Systems

System	Parameter	System	Parameter
Engine control	Manifold absolute pressure	NODS (near obstacle detection) collision avoidance	Relative speed vehicle #2 Distance vehicle #2
	Barometric pressure (altitude)	Intelligent highways	(See nods) (See navigation)
	Exhaust gas oxygen		Traffic flow
	NO_x	HVAC (climate	Humidity in passenger
	Fuel composition	control)	compartment (PC)
	Mass air flow		Temperature (PC)
	Engine knock		Chemical composition of
	EGR valve position		air (PC)
	EGR pressure		Air flow (PC)
	Throttle position		Outside air temperature
	Combustion pressure		Chemical composition of
	Exhaust pressure		outside air
	Fuel composition/flex fuel		Motor phase angle sensors
	Fuel pressure	Cruise control	Vehicle speed
	Vehicle speed		Brake switch
	Air charge temperature		Selector switch
	Engine speed	MUX/diagnostics	Multiple usage of sensors
	Crank angle/position	Idle speed control	AC clutch sensor/switch
	Engine coolant temperature		Power steering pressure Shift lever (PRNDL)
	Humidity		switch
	Torque		Engine speed
	Oil quality	Electronic	Throttle position
	Fluid condition	transmission	Shift lever position
	Engine noise	(continuously	(PRNDL)
	Ambient temperature	variable	Trans oil pressure
Anti-skid brakes	Wheel speed	transmission)	Shaft speed
	Brake switch		Transmission oil
	Brake pressure		temperature
	Deceleration		Vacuum modulation
	Fluid level	Driver information	Vehicle speed
	Fish tailing		Engine speed
Traction control/ abs	Engine speed		Oil pressure
	Steering wheel angle		Fuel level
	Steering wheel rate of change		Oil level Oil quality/condition
Air bags	Deceleration		Coolant pressure
	Bag pressure		Coolant condition
	Vehicle speed		Coolant temperature
Suspension	Vehicle height (wh-body displ)		Ambient air temp Coolant level

Table 11.2
(Continued)

System	Parameter	System	Parameter
	Steering wheel angle		Windshield washer level
	Steering wheel direction		Transmission oil level
	Acceleration		Tire pressure
	Vehicle weight		Tire surface temperature
	Road surface conditions		Battery fluid level
	Side slip angle		Rain sensor
	Speed of yaw angle		Sun sensor
	Lateral displacement	Memory seat	Driver selector switch
	Vehicle speed		Lumbar pressure
	Pneumatic spring pressure		Seat position
	Ride selector switch	Navigation	Gyroscope
Electronic power	Steering wheel torque		Steering wheel turns
steering (also	Hydraulic pressure		Geomagnetic (magnetic
electronic assisted)	Vehicle speed		field) sensors
	Steering wheel position		Distance traveled
4-wheel steering	Vehicle speed		
	Steering wheel angle		
	Front-rear wheel angle		
Security and	Illegal access indication		
keyless entry	Remote access indication		

tolerate the normal overrange occurrences that could be anticipated in certain failure modes of other components. The output should be stable over the operating temperature range and the lifetime of the vehicle. Electronics is frequently required to enhance the output signal, calibrate the zero and sensitivity, compensate for the effects of temperature and other variables, tolerate electromagnetic interference, provide protection against short circuits, reverse battery and overvoltage conditions, and is increasingly being asked to provide diagnostic and self-test features. The ease of interface for digital processing also requires the output of an analog sensor to be converted through an (integral) analog to a digital interface or a serial communication interface. The sensor package required for the automotive environment must be small, easy to handle and mount in the application, extremely robust and capable of withstanding thermal cycling, thermal shock, vibration, various chemicals, and high levels of humidity, and provide the EMI protection for the sensor and interface electronics.

Microelectronic sensors are extremely well suited to satisfy the requirements of increased reliability, easier electronic interface, and self-diagnostics and to reduce the system's cost. Therefore inherent limitations in these devices must be resolved by appropriate sensor design and signal conditioning. Below are a number of undesirable characteristics of semiconductor sensors that have to be addressed [3].

- Nonlinearity
- Drift
- Offset
- Temperature dependence of offset
- Temperature dependence of sensitivity
- Nonrepeatibility
- Cross-sensitivity to temperature and strain
- Hysteresis
- Low resolution
- Low sensitivity
- Unsuitable output impedance
- Self-heating
- Unsuitable frequency response

11.2 SENSING TECHNOLOGY IN VEHICLE SYSTEMS

New technologies can provide a sensing solution for an automotive application only if they can meet performance criteria, form factor (packaging) requirements, program development timetables, and pass the required reliability qualification tests. The decision on the most appropriate technology can then be reduced to a purchasing decision—the lowest cost approach. Some examples regarding current sensing techniques will be discussed in this section.

Electronic engine controls were the first area of on-board computer usage in automobiles. Early systems provided controlled delivery of fuel, based on engine load, engine speed, and engine temperature. Federal and local (for example, California, Denver) standards for emission levels and corporate average fuel economy (CAFE) have necessitated that the precise control provided by electronic systems be incorporated in all passenger vehicles sold in the United States. The most sophisticated systems incorporate several input sensors such as an oxygen (O_2) sensor, coolant temperature sensor, manifold absolute pressure (MAP) sensor, vehicle speed sensor, throttle position sensor, engine speed reference or distributor reference (rpm), mass air flow sensor, air intake temperature sensor, crankshaft sensor, detonation (knock) sensor, throttle body temperature sensor, throttle idle switch, altitude (or barometric) pressure sensor, and a wide open throttle switch. The requirements for these sensors and the number of sensor inputs required for the control system are being increased to meet legislated demands for tighter emission controls and higher fuel economy.

The electronic control unit processes the input signals to provide proper fuel, spark, and engine speed-load control and actuates several output drivers such as the ignition coil, injector drivers, and emission-related solenoids and valves. Figure 11.3 [4] shows a typical engine control system with all of the inputs that are required for

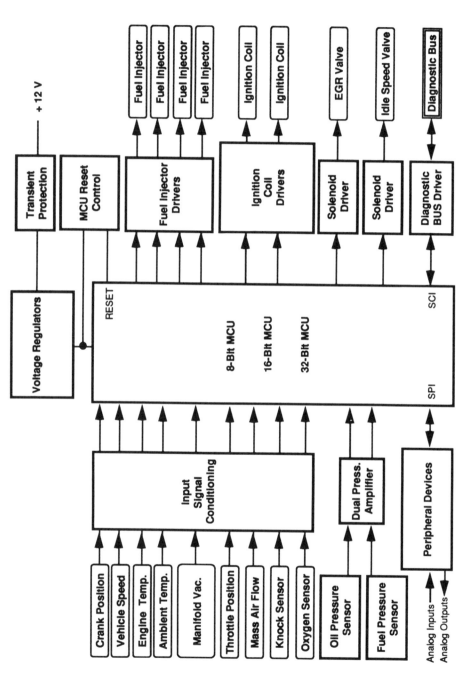

Figure 11.3 Engine control system.

meeting current emissions standards and providing improved fuel economy and drivability. All of the sensors in this system, except the oxygen sensor, are or possibly can be microelectronic sensors. The microelectronic sensors can also integrate the input signal conditioning functions.

11.2.1 Manifold Absolute Pressure

The first and currently largest volume automotive application for micromachined sensors is for sensing manifold absolute pressure and barometric absolute pressure (BAP) required in electronic engine control systems. The measurement of absolute pressure has been accomplished with LVDT (linear variable differential transformer) sensors, aneroid chambers, capacitive ceramic pressure sensors, and piezoresistive and capacitive silicon sensors. The last two will be discussed here as representatives of the new technology.

11.2.1.1 Piezoresistive Pressure Sensor

The most frequently used approach to semiconductor pressure sensors utilizes the piezoresistive effect in silicon. Sensing elements are either a four-element Wheatstone bridge or a single element positioned to maximize the sensitivity to shear stress.

The latest development in semiconductor pressure sensors is shown in Figure 11.4, where the circuitry necessary to adjust the offset and span and to compensate for temperature effects on both offset and span, as well as to provide an amplified signal, has been integrated on the same chip with the micromachined diaphragm. The sensor signal is provided from a single piezoresistive element located at a 45° angle in the center of the edge of the square diaphragm, and this allows for an extremely linear measurement. The offset voltage and full-scale span of the basic sensing element vary with temperature but in a highly predictable manner. This approach requires a minimum number of external components for amplification to provide a usable output signal.

Two silicon wafers are used to produce the piezoresistive silicon pressure sensor (Figure 11.5). The top wafer is etched until a thin, square diaphragm, approximately 25 μm in thickness, is achieved. The square area is extremely reproducible, as is the 54.7° angle of the cavity wall. The top wafer is attached to a support wafer by a glass frit to provide a structure that is isolated from mounting stresses. An absolute reference is obtained by bonding the top wafer to the solid bottom wafer in a vacuum. In addition to the basic sensing element, an interactively laser-trimmed four-stage network is also produced in a single monolithic structure.

The photomicrograph of the fully signal conditioned pressure sensor is shown in Figure 11.6. The additional circuitry is accomplished using the silicon area that is required to provide the mechanical support for the diaphragm. The die size for

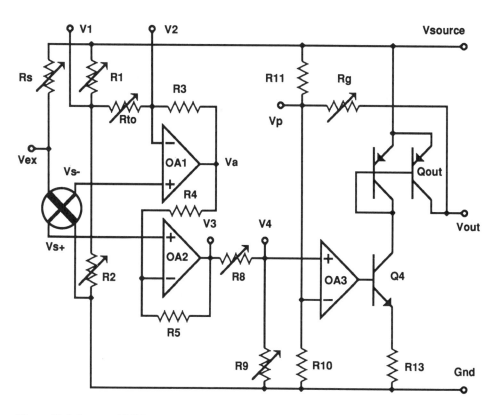

Figure 11.4 Integrated MAP sensor circuit.

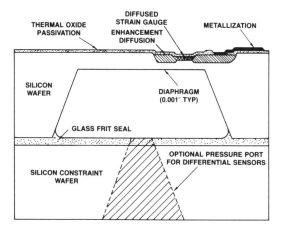

Figure 11.5 Cross section of piezoresistive pressure sensor.

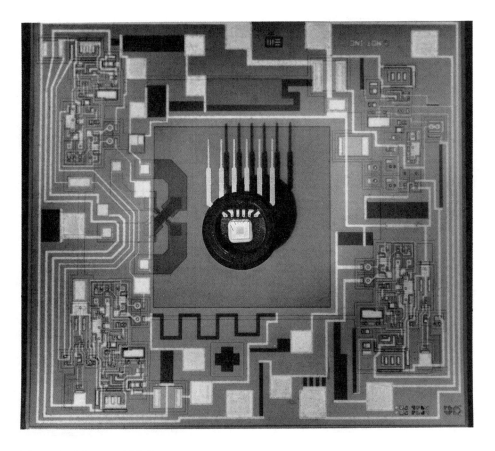

Figure 11.6 Integrated MAP sensor.

the uncompensated sensing element is 3048 μm $\times$ 3048 μm, and the die size for the fully signal conditioned unit is 3685 μm $\times$ 3302 μm.

The cost is minimized when the laser trimming is performed at the earliest stage of assembly. Automatic pattern recognition allows for high throughput in assembly. The assembly techniques, equipment, and statistical process controls are developed for high-volume production.

The laser trimming process can be used to provide a wide variety of transfer functions. Once the trim procedure is completed, it yields a device with a zero-pressure offset voltage of 0.5V nominal and a full-scale output voltage of 4.5V when connected to a 5.0-V supply. Therefore, the output dynamic range due to an input pressure swing of 0–15 psi is 4.0V. The output is ratiometric with the supplied voltage, and a maximum supply voltage of 10V is allowed.

The performance of an automotive version of the sensor for measuring manifold absolute pressure is shown in Figure 11.7. The performance at five temperatures is

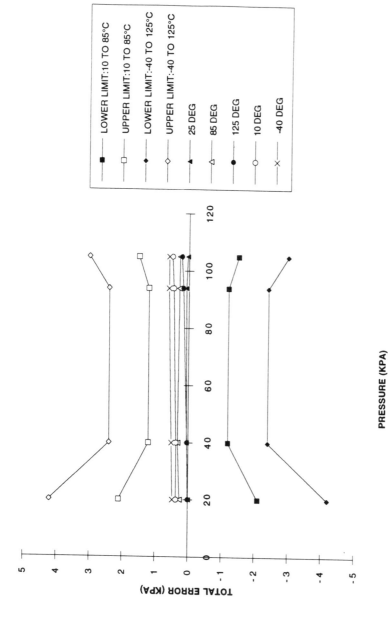

Figure 11.7 Performance of integrated MAP sensor versus temperature.

plotted versus the error band commonly specified for automotive applications from −40°C to 125°C. The performance at −40°C and 125°C is even within the limits required for +10°C to 85°C operation.

11.2.1.2 Capacitive Absolute Pressure Sensor

A silicon capacitive absolute pressure (SCAP) sensor has also been designed to measure barometric or manifold absolute pressure for an automotive engine control system [5]. The plates of the capacitor are formed by a silicon diaphragm and a metallized, glass substrate. The diaphragm in the SCAP sensor is a 5 μm cavity etched into a 127-μm thick silicon wafer. Holes are laser drilled through the glass substrate to obtain the electrical connection and allow the reference vacuum to be sealed inside the cavity. The glass substrate is anodically bonded to the silicon top wafer to achieve a hermetically sealed structure. Solder bumps seal in the reference vacuum and provide the external connections. Additional external circuitry is used to produce a variable frequency output that can have a direct interface to a microcontrol unit (MCU). The sensing element is 45 mm^2 and achieves a full-scale capacitance of approximately 39 pF. The capacitance change over the operating pressure range of 17 to 105 kPa is 7 pF, resulting in an output frequency change from 20 to 180 Hz. The variation of the capacitance with temperature is −30 to 800 ppm/°C and the linearity is typically 1.4%.

11.2.2 Tire Pressure Sensor

One of the traditional measurements necessary for safe vehicle operation is tire pressure. Electronics has recently made it possible to measure tire pressure at each wheel and transmit the data to the passenger compartment for continuous driver awareness of the status of each tire. A system such as the one in Figure 11.8 [6] utilizes semiconductor pressure and temperature sensors mounted inside the tire.

Power for the system is supplied by a combination of rotating and fixed antennas. A high-frequency alternating current applied to the fixed antenna allows an induced voltage on the rotating antenna to provide sufficient power for the wheel's electronic components. Transmission is interrupted every 150 ms to allow the sensors to measure pressure and temperature and transmit digitally coded information back to the fixed antenna. A control unit decodes the sensor signals 6 times/second and compares this measurement to a reference pressure stored in memory to determine if a critical situation exists. The system can distinguish between an increase in tire temperature caused by increased flexing of an underinflated tire and the heat that results from braking or the tire's response to high outside ambient temperatures.

The key sensor characteristic for a piezoresistive pressure sensor such as the one described in Section 11.2.1.1 is an extremely low-current draw (high-impedance device). Also, tire pressure measurements are typically gauge measurements, but

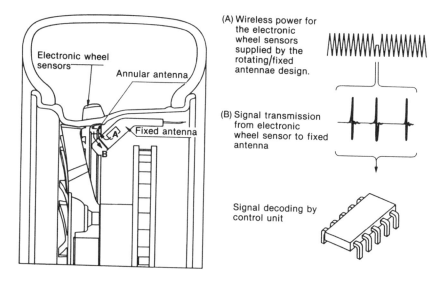

Figure 11.8 Tire pressure sensing.

since the electronic tire pressure is measured inside the tire, an absolute sensor is used.

Other applications for pressure sensing in the automobile that involve both higher and lower pressure ranges and gauge or absolute pressure measurements include the AC compressor, pneumatic lumbar support, air bag pressure, ABS (antilock braking system) brake pressure, engine oil and transmission pressure, fuel injection pressure, coolant system pressure, EGR (exhaust gas recirculation) pressure, pneumatic spring pressure, and combustion chamber pressure. Differential pressure can also be used to measure flow, which will be covered in Section 11.2.4. Pressure sensors can further be used to measure fluid level, as discussed in Section 11.3.2.

11.2.3 Position, Rotation, and Speed Sensing

As indicated in Figure 11.2, motion and position sensing is the highest volume application. The most popular speed sensor in automotive applications is the variable reluctance transducer, which consists of a magnetic pole with a coil of wire wrapped around it. An output voltage proportional to the rate of change of flux in the magnetic circuit produces a pulse when a tooth on a trigger wheel passes in front of the pole. The output voltage can vary from 500 mV to over 80V and requires additional electronic circuitry to eliminate unwanted noise, clamp the output voltage to a safe level, and provide a useful signal to the MCU. Several semiconductor approaches are also being used to detect wheel speed, crankshaft speed, angular rotation, and direction. A comparison of the various techniques is shown in Table 11.3 [7]. In all cases the

Table 11.3

A Comparison of Various Sensors and Their Features

Sensor Type	Zero Speed	Maximum Frequency	Magnets	Nonferrous Wheel	Square Wave Output	S/N	Immunity to EMC
Eddy current	Yes	500 kHz	No	Yes	Yes	High	Poor
Hall effect	Yes/No*	1 MHz	Yes	No	No	Moderate	Poor
Variable reluctance	No	10–100 MHz	Yes	No	No	Low	Moderate
Reed switch	Yes	600 Hz	Yes	No	Yes	High	High
Wiegand effect	Yes	20 kHz	Yes	No	No	High	—
Opto	Yes	10 MHz	No	Any	Yes	Very high	High

*Sensors with capacitor in feedback circuit cannot achieve zero speed.

output of the sensor is a pulse that is counted, or the time between pulses is calculated, to provide the desired speed or angular position. Some of these same techniques are also used to detect linear position. For linear position such as throttle and EGR position, potentiometers are commonly used.

11.2.3.1 Opto Sensing

Properly designed optical sensors have the smallest jitter angle, or variation from actual angular position, for a given diameter sense wheel when compared to other crankshaft position sensors. In addition opto sensors can sense static position and have a constant output with speed. They must be protected from oil and dirt to function properly. In addition, optical sensors have an upper temperature limit that must not be exceeded if the expected life is to be obtained.

Opto sensing techniques using infrared emitters and detectors have been successfully used for several years in rather benign environments to accurately sense rotational and linear motion. Their extension into the automobile has been delayed due to temperature sensitivity and stability problems associated with emitters. Recent developments have resulted in significant improvement, which has allowed their usage in higher (up to 150°C) temperature applications.

Detailed studies of LEDs revealed that under accelerated conditions it was not unusual to experience 15–20% degradation after 1,000 hours of testing. Processing revisions and improvements in process control implemented (by Motorola) to increase LED performance have resulted in extremely stable, long-term reliability. The current transfer ratio (CTR), the ratio of transistor collector current to forward current, was used to measure the LED light output. As shown in Figure 11.9 [8], after 7,500 hours, the average CTR was still almost 98% of the initial value at room temperature.

Improved performance and packaging concepts that take into account the mounting location and surrounding environment have resulted in several custom opto sensors being developed for automotive applications. For example, two pairs of opto emitters and detectors with integrated Schmitt triggers are used to provide steering-wheel rate of turn and direction of rotation in a ride control system (Figure 11.10). The sensor is mounted inside the passenger compartment. Multiple slots in a shutter interrupt the light path between the two pairs of emitters and detectors used to provide the quadrature signal. The opto sensor does not require a magnetic trigger wheel, like most speed sensors. A wheel of stamped steel or any material that will block the light path is acceptable. Schmitt trigger outputs that are integrated on the opto detector provide a signal easily connected to an MCU. The MCU in the ride control system uses an algorithm that determines the straight-ahead wheel position and then calculates the lateral acceleration that will result from the turn angle. If the vehicle acceleration is above a programmed level, the shock absorber damping is switched to firm. The optical steering sensor also provides information that allows the MCU

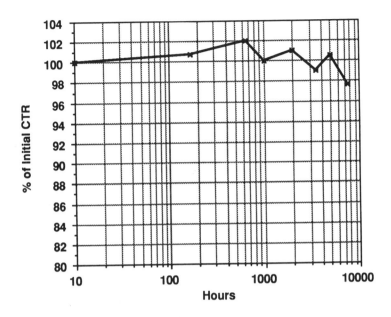

Figure 11.9 Improvement in opto stability.

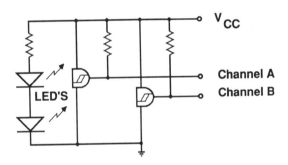

Figure 11.10 Opto sensing for steering-wheel position and direction.

to determine the rate at which the steering wheel is turning. If this rate is too high, then the damping is switched to firm to provide maneuverability for avoiding accidents [9].

Packaging techniques have also allowed opto devices to be used as a shock absorber–mounted vehicle height sensor (Figure 11.11). Reflective sensing from the stainless steel shaft provides information for the MCU in this case. Opto sensing is made possible by utilizing the clean environment that must exist inside the shock

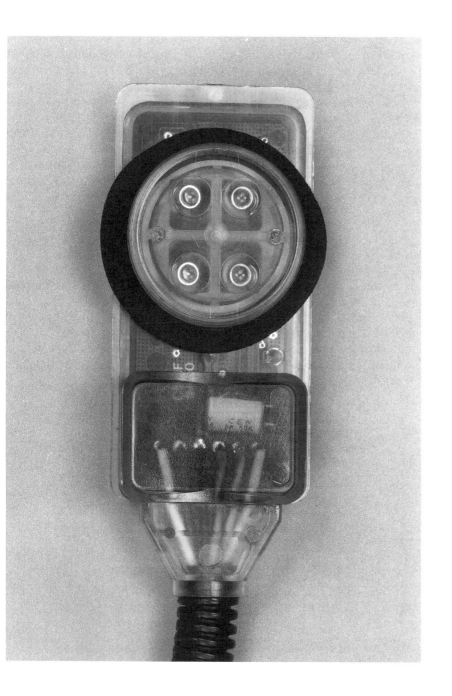

Figure 11.11 Opto vehicle height sensor.

absorber for the stainless steel shaft to move without galling. Dirt and grease are sealed outside the opto sensor by the gasket.

Optical fibers are being considered for noise-free data transmission of multiplex (MUX) signals in critical high-speed data requirements. Low-cost packages have been developed that integrate fiber optic emitters and detectors with the connection to properly trimmed plastic optical fiber. Applications with speeds up to 10 MHz can be addressed.

Fiber optics technology also has several sensing applications, especially where harsh environments are present and isolation is required. Fiber optic sensors have been developed for rotation, vibration, acceleration, acoustic, strain, temperature, and pressure. A possible automotive application is the use of fiber optic gyros to determine the vehicle's position in future intelligent vehicle highway systems (see Section 11.3.7).

11.2.3.2 Hall-Effect Sensing

Another frequently used principle to sense position is the Hall effect. The presence of a magnetic field at a right angle to an epitaxial layer with current flowing through it causes a Hall voltage to be generated [10]. The Hall voltage is sensed by taps in the layer that are at right angles to the current flow, as shown in Figure 11.12 [11].

The output of the Hall-effect device is converted to a linear or digital output. The linear output Hall sensor adds a dc amplifier and voltage regulator to the basic

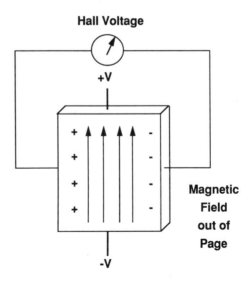

Figure 11.12 Hall-effect principle.

Hall element to provide an output that is linear, proportional to the applied magnetic flux density, and ratiometric (proportional) to the supply voltage. A Schmitt trigger threshold detector with built-in hysteresis added to the linear Hall sensor provides a digital output. The addition of an open collector *npn* transistor allows the Hall switch to be compatible with digital logic. Hall-effect devices have been used for several years as sensors for detecting the speed and position of the distributor trigger wheel in the ignition system.

Addressable Hall-effect sensors have been designed for automotive multiplex wiring systems [12]. The sensor has two pins that connect to the bus and a third pin available for sensing an open or closed switch without requiring a magnetic field. The two-wire bus can power up to 30 sensors that require 2.5 mA each, as shown in Figure 11.13 [10].

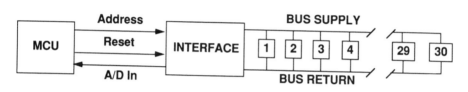

Figure 11.13 MCU addressable Hall-effect sensors.

The presence of the factory programmed address on the bus causes the sensor to transmit a diagnostic signal indicating that it is functioning properly. This signal is followed by a bit indicating the presence or absence of the magnetic field that is being monitored. The chip is addressed by switching the supply voltage between 6V and 9V in a serial pulse train. When the sensor is addressed it increases its quiescent current by 300 μA when the rail is 9V. If a magnetic field is sensed, the current increase is continued when the rail is reduced to 6V.

11.2.3.3 Magnetoresistive Elements

Magnetoresistive sensors are being developed to overcome the limitations of inductive and Hall-effect sensors for linear position, rotational, and speed sensing. Inductive pickups require complex signal conditioning for the signal output and external temperature compensation networks [13]. Hall-effect devices are limited due to temperature drift, temperature range, repeatability, nonlinearity, and offset adjustment.

Magnetoresistive elements (also magnetic resonance elements, or MRE) such as the one in Figure 11.14 [14] utilize four strips of permalloy, a ferromagnetic alloy composed of 20% iron and 80% nickel, arranged in a Wheatstone bridge configuration [15]. The output signal, based on a change in resistance, is proportional to

398

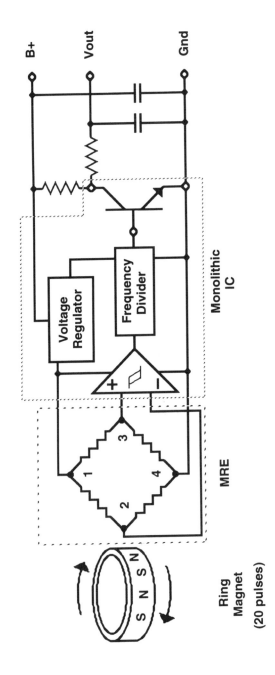

Figure 11.14 Magnetoresistive element (MRE).

the input signal magnitude modulated by the magnitude of the magnetic field. The rotation of a 20-pole magnet ring causes the magnetic field to change. Each rotation of the magnetic ring produces 20 cycles of alternating current output from the MRE. A comparator changes the ac signal to a digital signal, the signal is inverted and then divided into five equal parts by a combination meter. One part is used to drive the vehicle's speedometer and the other four signals are used for inputs to suspension, engine and transmission, cruise control, and power steering.

Linear characteristics up to 175°C have been demonstrated by MRE sensors. An additional application being pursued is replacing the potentiometer, with its associated wearout mechanism, by a noncontact MRE sensor for measuring the throttle position.

11.2.4 Flow

The measurement of flow is performed in automobiles for both liquid and air. One of the more important flow measurements is mass air flow. Karman vortex and hot wire anemometer approaches are used for sensing the mass air flow in engine control systems. A mass air flow measurement in an engine control system provides an improvement in the ability to adjust the fuel flow for improved combustion efficiency over speed-density engine control systems that utilize MAP measurements. However, the mass air flow sensor is more expensive than a MAP sensor. Stricter emission standards are increasing the use of mass air flow sensors. The mass of the air is measured instead of the volume to avoid the effect of temperature and density on the air. Other vehicle flow measurements include fuel flow and air flow in the passenger compartment. Fuel flow is frequently sensed by a turbine in the flow line whose blades are counted by magnetic sensor such as a Hall-effect sensor. Passenger compartment air flow could possibly utilize microsensor technology.

11.2.4.1 Mass Air Flow Sensing

A typical approach for mass air flow involves a hot wire anemometer with a platinum wire as the sense elements as shown in Figure 11.15 [16]. The air temperature probe detects the temperature of the air flowing into the combustion process. The hot wire probe is heated by current supplied from a power transistor (Figure 11.15(a)). The difference in temperature between the probes is detected as a difference in electrical resistance. Electronic circuitry regulates the heating current to the hot wire probe to keep the temperature difference constant at any flow rate. The current required is proportional to the air flow rate. The construction of one of the probes is shown in Figure 11.15(b). One of the main issues in this type of sensor is response time. Improvements that reduce the response time from 50 to 30 ms have been made in recently developed units.

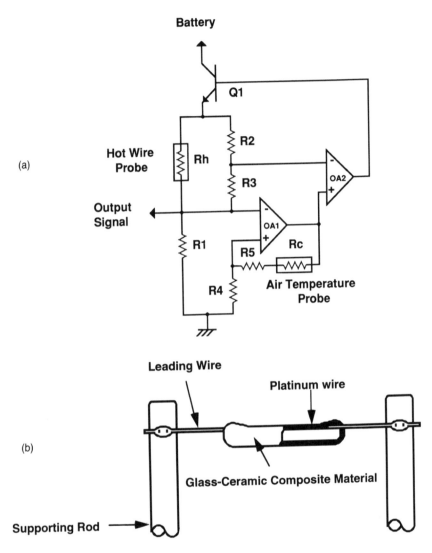

Figure 11.15 (a) Circuit for hot wire anemometer mass air flow sensor and (b) construction of probe.

Microbridge silicon mass air flow sensors have also been developed that utilize temperature resistive films laminated within a thin film (2 to 3 μm) of dielectric material and micromachined cavity as shown in Figure 11.16 [17]. The resistors are suspended over the etched cavity, which also provides thermal isolation for the heater and sensing resistor. Heat is transferred from one resistor to the other as the result

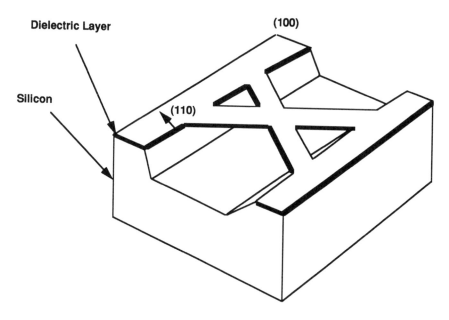

Figure 11.16 Microbridge mass air flow sensor.

of flow. The imbalance in the resistance caused by the heat transfer is directly proportional to the flow. The small size of the resistors allows fast response time
(≤3 ms) and high sensitivity to flow. The device is also useful as a differential or
absolute temperature sensor. Silicon microbridges are being pursued for sensing mass
air flow in future vehicles.

11.2.5 Accelerometers

Sensing acceleration is one of the most intense development areas. Accelerometers
are required for air bag and active suspension systems but can also be used in antilock
brake and ride control systems if the cost is low enough. The most critical and highest
volume application is currently in the area of crash sensors for air bags.

An inflatable restraint (air bag) system is one of the more important new vehicle
systems designed to address the need for improved vehicle safety and satisfy the
legislated requirements of FMVSS208 (Federal Motor Vehicle Safety Standard). Air
bag systems utilize as many as five accelerometers (crash sensors) to provide information to an electronic control unit. The crash sensors are positioned at strategic
locations in the front of the vehicle, and a low arming sensor is located in the passenger compartment. When a crash occurs, the (series) arming sensor is closed and

the rapid deceleration produces a signal from the crash sensor. If this signal is above the calibrated level of a crash event (equivalent to running into a wall at 12 mph) it is allowed to pass the trigger current necessary to ignite the propellant that inflates the bag.

Present systems utilize electromechanical and piezoelectric sensors to provide an air bag deployment signal. These systems have an excellent record. However, the system costs about $400 to $500 for the driver-side-only system and is estimated to cost from $800 to $1,000 when passenger-side units are added. A significant portion of the cost is attributed to the sensors. When light trucks with shorter crash zones, especially vans, are added to the list of vehicles that require air bags, a faster input from the crash sensor will be required. In addition to meeting the system performance levels, reliability requirements, and cost objectives of this application, semiconductor accelerometers are being pursued for the capability to provide a faster trigger, self-test (diagnostic) feature, reduce the number of sensors in the system, and their potential for increased integration [18].

11.2.5.1 Mechanical Crash Sensors

One version of the crash sensor used in current vehicles is shown in Figure 11.17 [19]. The sensor consists of a roller positioned against a calibration backstop by a

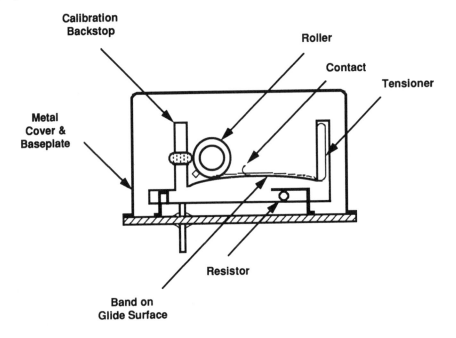

Figure 11.17 Mechanical crash sensor.

flat spring band. Sufficient deceleration causes the roller to move forward and close the contact. Calibration allows the level of deceleration necessary to close the switch to be set at different levels for different vehicles. Other mechanical approaches include a ball in tube that is restrained by either a spring or magnetic force. Deceleration overcomes this force and allows a crash signal to be indicated.

11.2.5.2 Piezoelectric Accelerometers

Ceramic and thin-film piezoelectric sensors solve some of the cost and operational problems associated with mechanical crash sensors. They are analog sensors and not switches like the mechanical units. Figure 11.18 shows a piezoelectric sensor currently used in some vehicles [20]. The sensing element consists of two reversely polarized piezo-oxide bars with two electrodes. The bars are cemented together to form a biomorph element. The cantilever structure is positioned in the module in the direction of vehicle motion. When deceleration occurs, the upper bar is stretched and the lower bar is compressed. Because the bars are reversely polarized a signal doubling occurs. An external stage with an impedance converter, low-pass filter, and amplification are necessary to provide a usable output.

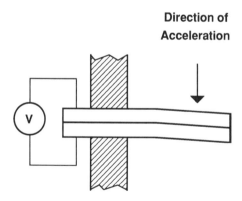

Figure 11.18 Piezoelectric accelerometer (crash sensor).

11.2.5.3 Silicon Accelerometers

Micromachining techniques utilized in the manufacturing of piezoresistive pressure sensors such as the MAP sensor have led to the development of silicon accelerometers. Figure 11.19 [21] shows a piezoresistive silicon accelerometer with micromachined beams and suspended mass. Bulk micromachining allows a very precise

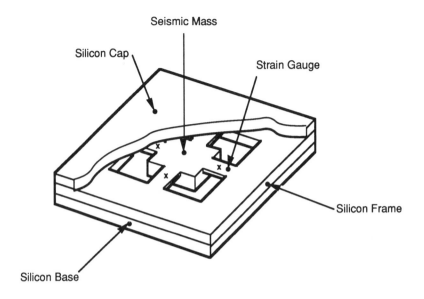

Figure 11.19 Piezoresistive silicon accelerometer.

mass and support structure to be consistently produced in a batch process with several thousand devices in a single-wafer lot. The piezoresistive element(s) of the strain gauge are ion implanted into the suspension arm for maximum sensitivity. The output is in mV/g and the resistive elements are temperature sensitive so additional circuitry is required to provide calibration, temperature compensation, and allow interface with external circuits.

Capacitive approaches for semiconductor accelerometers are also possible, such as the structure shown in Figure 11.20 [22]. In this case, the working structure is obtained by the deposition of sacrificial and structural layers on the top of a silicon substrate. Selective chemical etching is used to etch the sacrificial layer separating the conductive layers to produce the air gap capacitor. Acceleration perpendicular to the substrate surface causes the movable plates of the capacitor to move, producing an output that is directly proportional to acceleration. The output (in fF/g) is converted into a frequency for direct interface to a microcontrol unit.

One of the major advantages of micromachined silicon accelerometers is the ability to integrate self-test features. An electrostatic field applied to an integral capacitor built into a silicon accelerometer can be used to deflect the beam and provide an indication that the accelerometer is working properly and even check the calibration. The ability to check the calibration can be used to test the accelerometer in wafer form, packaged form, and in the final assembly. In the application, the device can be tested for drift and lifetime measurements without having to apply external

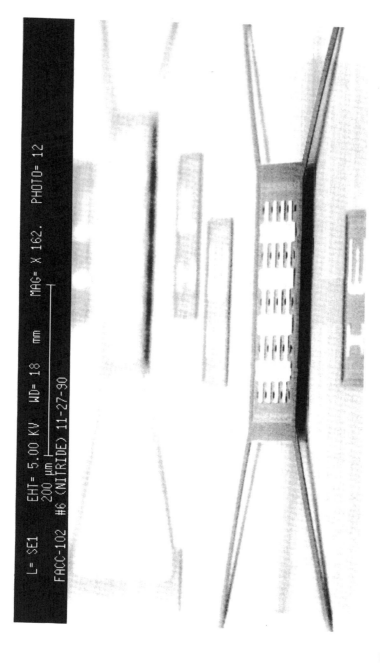

Figure 11.20 Capacitive accelerometer.

forces or the occurrence of a crash. More details on accelerometers can be found in Chapter 12.

11.2.5.4 Knock Sensors

The use of lower octane, lead-free fuels and the improved fuel economy gained by operating just below an engine's knock limit make sensing engine knock important. In addition, turbocharged engines can experience significant damage at high speeds due to detonation if the spark is not controlled. The characteristic frequencies of engine knock are in the range of 3 to 9 kHz [23]. Both resonant and nonresonant (flat bandwidth within 0 to 10 kHz) knock sensors are used in automotive applications. The resonant units detect secondary mode knock that can occur above 10 kHz and provide a higher signal-to-noise ratio. The knock sensor uses a piezoelectric ceramic material, such as barium titanate, to generate a voltage when the sensing mass is subjected to a force due to the acceleration that occurs with engine knock. The amplitude of the generated signal is proportional to the acceleration and therefore to the level of knock.

The sensor has a simple construction (Figure 11.21) and can be manufactured at relatively low cost. The ceramic sensing element is captured between two conductive caps that provide a current path. Engine-generated knock is transmitted to the element from the conductive metal housing through a Belleville washer. The output level is in the range of hundreds of millivolts. The sensitivity to noise requires amplification circuitry external to the sensor. In addition, the need for filtering in the electronic control unit for different engine types and the possibility of requiring more than one knock sensor is spurring development for more direct measurement techniques, such as cylinder pressure, combustion temperature and in-cylinder flame measurements. It is also possible that an integrated sensor could perform the total function even more economically.

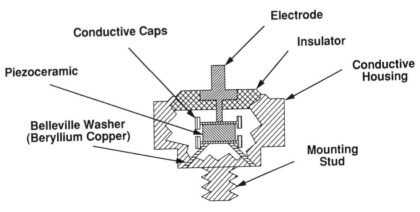

Figure 11.21 Knock sensor.

11.3 FUTURE AUTOMOTIVE SENSING

The increasing complexity of vehicle systems combined with the need for additional sensing inputs with improved accuracy and the automotive environment pose several challenges for sensor designers. Some of the areas that will affect the types of sensors and how sensors will be used in future systems are covered in the remaining sections.

11.3.1 High-Temperature Operation

Higher temperature operation of automobiles, especially for underhood applications ($\geq 125°C$), is pushing the maximum operating temperature of silicon. Silicon sensors are being used more frequently in higher operating temperatures. Alternate semiconductor materials, such as GaAs (gallium arsenide), GaP (gallium phosphorus), GaAs with silicon, and SiC (silicon carbide) are among the possible solutions. These wide-bandgap materials can operate at temperatures that exceed the maximum operating temperature of silicon. Silicon bonding technology is also being pursued as a potential means of obtaining silicon sensors that can survive higher temperature operation. In addition to performance at higher operating temperatures, the stresses placed on the package require sensor manufacturers to address thermal cycling capability with high-temperature extremes.

11.3.2 Liquid Level

Various liquid levels can be measured in a vehicle: engine oil, transmission oil, coolant, windshield-washer fluid, battery, power-steering fluid, brake fluid, and fuel. All of these sensors, except fuel level, could be satisfied by a switch that simply detects that when a predefined minimum level of liquid has been reached so that a driver indication can be provided, either directly illuminating a dash lamp or through a body computer activating an output driver. One way to sense a minimum level of fluid uses a Hall-effect sensor (Section 11.2.3.2) [11]. The magnet is located in a float assembly that keeps the Hall device open as long as the fluid level is sufficiently high. When the minimum level is reached, the Hall switch closes and the integrated Darlington output stage allows current up to a 900 mA peak to flow. Internal current limiting allows the integrated sensor to turn on a dash lamp, such as a Sylvania #168. The circuit also incorporates thermal shutdown, shortcircuit protection, flyback voltage protection, and reverse battery protection. The float and housing material can be chosen to withstand the effects of the fluid being monitored, and the plastic packaged Hall sensor is isolated from the fluid.

Sensing the fuel level in the gas tank is one of the more difficult fluid-level applications. In this case an analog output is required with good accuracy over the entire fuel level. Present systems utilize a potentiometer with a float mechanism. A semiconductor pressure sensor provides an excellent solution for meeting improved performance requirements and avoiding wearout mechanisms; further, they can provide additional features. However, in addition to exposure to gasoline, the sensor package must withstand alcohol and sour gas. Coping with these harsh media has been a problem for sensors that do not incorporate a stainless steel diaphragm; however, stainless steel would not meet the system cost requirements. The use of new materials and techniques continues to be pursued to solve the media compatibility problem.

Other alternatives for fuel tank sensors such as capacitive sensing and ultrasonics have been effective in other harsh liquid level applications. These techniques are also being developed for automotive use.

11.3.3 Chemical Sensing

Chemical sensors can provide an indication of CO or noxious odors in the passenger compartment. The presence of CO could trigger a vehicle alarm, and the presence of noxious odors could be used to automatically change the mode of the heating and air conditioning system from a recirculation of passenger compartment air to allow fresh air into the passenger compartment or the reverse situation depending on the location of the offensive odor. Chemical sensors are being developed for vehicle applications based on thin-film of stannic oxide (SnO_2) whose conductivity changes with the concentration of a specific gas such as CO, H_2S, methane, and other hydrocarbons. The selectivity, operating temperature, and sensitivity of these devices as well as cost are issues that will determine their acceptance in automotive applications.

Chemical sensing already plays an important role in the emissions control system. An oxygen sensor mounted in the exhaust gas provides the closed loop information necessary for the MCU to control vehicle emissions. Sensing NO_x is an area that continues to be investigated as a means of achieving tighter emission controls.

11.3.3.1 Oxygen Sensor

An oxygen sensor is used to monitor the exhaust gas mixture and provide information on the status of the exhaust gas content to the MCU [24]. The control strategy allows the MCU to vary the air-fuel mixture around stoichiometry (14.6 A/F) to increase catalytic converter efficiency, prolong the life of the converter, and provide increased fuel economy. To ensure the fastest response time the oxygen sensor is placed in the exhaust manifold at a point where it can respond quickly to exhaust changes with minimal lag time. The operating temperature inside the manifold can easily exceed 900°C, and in fact, a minimum operating temperature (about 300°C) is necessary to ensure proper performance of existing sensors. External electrical power is frequently applied to an integral heating element in the sensor so that closed-loop operation can be initiated within 30 seconds of starting and during idle.

The sensor is a hollow cone shaped zirconia ceramic (ZrO_2) covered with a porous platinum electrode on both sides (Figure 11.22(a) [25]). The overall effect

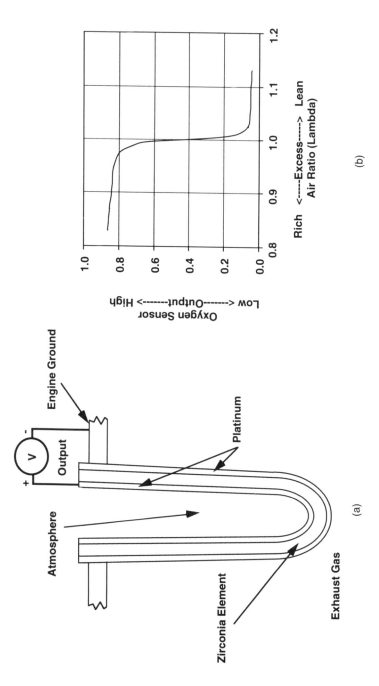

Figure 11.22 Oxygen sensor: (a) construction; (b) output versus air ratio.

of the plating is to create an electrochemical cell that develops a potential difference between the electrodes. The sensor's output voltage (Figure 11.22(b) [25]) is a logarithmic function of the partial pressures of oxygen on either side of the cell wall. One side of the cell is exposed to exhaust gas and the other to atmospheric air. When the engine is operating in a lean mode, the exhaust gas oxygen increases and the partial pressure of oxygen on both sides of the ceramic are almost equal, causing the sensor's output voltage to be near zero volts. As the exhaust gas is enriched, it passes stoichiometry, which is the ideal burn point. Just beyond stoichiometry, oxides (like CO and hydrogen) increase dramatically, which increases the pressure differential across the electrolyte, causing a voltage to be generated (approximately 1V). Recent developments include efforts to increase the response time ($\leq$10 ms) so that individual cylinders and not just the average composition can be detected. One approach uses a thin (1 μm) film semiconducting metal-oxide which is deposited by reactive cathode sputtering on an aluminum oxide substrate [26].

11.3.4 Oil Quality

Oil life sensors, such as the one used on 1991 Buicks [27], are intended to alert drivers to the need to change engine oil and avoid costly engine failures. The Buick sensor was developed by GM Research Labs and uses accumulated engine rpm, time, and engine temperature as inputs to a microcontroller that calculates an effective oil quality and provides an indication. The calculation is really an inferred oil quality and not a direct indication of the actual oil quality (see Section 11.3.5). Ford has also developed a capacitive sensor that measures the change in the dielectric constant of the oil as an indicator to viscosity and therefore oil condition [28]. However, the dielectric constant of different oils varies considerably and as a result the production usage of this approach has not occurred to date. Another approach measures oil viscosity by determining the time it takes for a sample to flow through a small chamber, which is a more direct indication of the actual state of the engine's oil. However, it has proven to be too costly for automobiles but may be applicable for trucks.

11.3.5 Capability to Compute Rather than Sense

The MCU and digital control have been significant driving forces for developing microelectronic sensors. At the same time, an MCU can also limit the number of sensors used in a particular system. The ability of the MCU to calculate from information provided by existing sensors and obtain an inferred measurement can avoid the cost of adding a sensor even though the sensor could provide more accurate data. A paradox occurs when a need exists for sensing redundancy. The cost to benefit must be evaluated for every new sensor that could be added to a system. To demonstrate the importance of this situation in an extremely cost sensitive environment, three specific cases will be examined.

Monitoring engine misfire is one of the requirements of the California Air Resources Board (CARB) regulation for on-board vehicle diagnostics (OBDII), which will be required on 1996 vehicles. California vehicle emissions standards are usually adopted by the federal government's Environmental Protection Agency. As a result, considerable research is being conducted to determine the most cost-effective way to monitor engine misfire [29]. One technique to measure misfire of an individual cylinder utilizes a magnetic sensor in the crankshaft to measure slight deviation in rpm and measurement of torque based on a nonuniformity index. The resulting values are compared to predicted values of power and torque [30]. Another system uses the difference in angular displacement at each end of the crankshaft. The twist of the shaft is determined by measuring the phase differences between the two encoders; this measurement is proportional to torque [31]. In-cylinder pressure measurement is the most straightforward and accurate method of measuring misfire. However, the technique requires relatively expensive sensors, which not only must measure high levels of pressure but withstand combustion temperatures. Also, a pressure sensor must be used in each cylinder [32].

Early requirements to measure manifold absolute pressure and barometric pressure utilized separate sensors. By reading the barometric pressure when the input to the MAP sensor is not affected by manifold pressure (during cranking and at wide-open throttle conditions that other sensors or switches indicate to the MCU) the additional barometric sensor has been eliminated. It should be noted that the manifold absolute pressure measurement is also an indirect measurement compared to measuring mass air flow.

Electronic antilock brake systems demonstrate an alternative to sensing hydraulic pressure. ABS is currently offered on selected models by every auto manufacturer. Use of ABS by some manufacturers is projected for availability on 100% of the vehicles by the mid 1990s. ABS is regarded as the major safety and driving option on present production automobiles. A typical ABS system uses a microprocessor, wheel-speed sensors, semiconductor power control devices, and hydraulic components to automatically pump the brakes rapidly and repeatedly (up to 15 times per second) [33]. This allows the driver to maintain steering ability and retain vehicle stability under difficult road surfaces, such as ice or wet asphalt, and under hard braking conditions.

Most ABS systems use fast-acting solenoid valves activated by power transistors to control the hydraulic pressure applied to the wheels and avoid locking up on slippery surfaces or during panic stops. Sensors measure the speed of the wheels and provide a signal to the microprocessor. The frequency of the signal is proportional to the speed of the wheel. When a wheel lock-up is detected, the MCU drives power FETs to activate solenoid valves that keep each wheel in an optimum slip range.

GM's ABS-VI systems uses three dc electric motors to move pistons inside the hydraulic chambers to bleed off brake fluid when wheel lockup is detected [34]. One of the major advantages of this approach is the capability to obtain information on brake pressure without requiring transducers for measuring hydraulic pressure. In

ABS-VI, the electric current going to the motors is proportional to brakeline pressure. This allows the system to apply optimum brake pressure to stop each wheel without locking. Partial excursions of the piston can be initiated instead of the on-or-off action of solenoids. As a result of this approach, the ABS-VI system has a very low level of brake pedal pulsation. Systems with solenoids can also benefit from information on line pressure because the solenoids could maintain the system pressure at the same level before and after release. However, pressure transducers have to be added to make the measurement because the system does not inherently provide this information.

11.3.6 Vehicle Diagnostics

Vehicle diagnostics are important not only for on-board detection and warning of system faults but for subsequent service for fault conditions. The complexity of electronic systems and the problems of diagnosing system faults have caused the auto industry to develop service diagnostic equipment. Ford's Service Bay Diagnostics System, Chrysler's Mopar Diagnostic System, and GM's Computerized Maintenance System are designed to provide service technicians with the ability to diagnose and correct malfunctions in electronically controlled vehicles on the first try [35]. These systems also allow reduced repair time and increased shop productivity. Continued growth of sophisticated vehicle systems will rely even more on these types of diagnostic units. On-board vehicle sensors provide information through the vehicle's electronic control units. Sensors on the diagnostic equipment provide additional information for verifying calibration and proper function. For example, fault sensing in an inflatable restraint system can provide information on a failed accelerometer that is incapable of providing any signal in the event of a crash and an extremely out of calibration accelerometer that is incapable of providing the correct firing point. An indication to the driver of a system problem and the driver's reaction to the warning by obtaining service to correct the fault are essential.

As indicated in Section 11.3.5, in certain instances sensors provide sufficient information, based on secondary parameters, for MCU algorithms to determine that a fault condition exists and still react to the inputs from other sensors to provide an estimate of how the system should respond rather than strictly going into a default mode. This can provide substantially better performance and safety than simply disabling the system until it is repaired. For example, in an ABS system with four wheel sensors, an open or shorted sensor providing no input can possibly be ignored for certain driving situations especially if other vehicle sensors, such as an indication of braking actuation or deceleration sensor, are providing system inputs that allow alternative calculations to be performed.

The SAE has developed a standard for vehicle diagnostic interfaces as indicated earlier. The International Standards Organization also has a standard for diagnostic interfacing.

11.3.7 RF Sensor Applications

The use of radio frequency (RF) sensing techniques is necessary in some of the most sophisticated systems being developed for future vehicles. Three of these systems are collision avoidance, near obstacle detection (NOD), and intelligent vehicle highway systems (IVHS).

Artificial intelligence simulating human visual, sensory, and judgment functions is the basis of a collision avoidance system [36]. Eight ultrasonic sensors (four in the front and four at the rear of the vehicle) provide information for the range finder in the fuzzy logic control. In addition, two infrared and two touch sensors are mounted on the front and rear bumpers and a color video camera provides long range image detection. When an object is detected that could cause a collision the vehicle's speed is reduced. Blind spot detection and avoiding problems at night and in fog or due to driver fatigue are just a few of the frequently encountered causes for accidents that could be avoided with the use of such a system.

Near obstacle detection is a simplification of a full collision avoidance system. In low-speed traffic maneuvers, such as backing out of a driveway or parking, the presence of an object in the vehicle's path could trigger a warning alarm or automatically apply the brakes. Only short-range ultrasonic, infrared, or radar sensors would be required for this system [37]. A head-up display (HUD) may be part of blind spot detection provided to the driver.

Intelligent vehicle highway systems or smart highways are being evaluated worldwide in an effort to reduce traffic congestion and improve traffic flow in the most populated cities. Improved traffic flow results from the use of traffic sensors and communication links that provide information to the driver regarding the best route to select to minimize the distance traveled. The sensors are 6-ft loops of heavy-gauge wire embedded beneath the pavement that are used to count the traffic flow or receivers that detect infrared signals transmitted from beacons at key intersections. Another aspect of IVHS is driver information that involves bidirectional RF communication with the vehicle. Several potential areas are shown below [38].

- Dead reckoning map matching navigation system with GPS (global positioning satellites);
- Digital traffic information and network link times receiver;
- Dynamic route selection for minimal time/distance travel;
- Color video display for maps, traffic info, guidance and road signs;
- Synthesized voice for traffic info and guidance;
- Map database including turn restrictions, traffic signs and nominal link times;
- Business directory and travel info database integrated with map database;
- Electronic vehicle identification for toll debiting;
- Digital cellular telephone; and
- Semiautomatic MAYDAY using cellular telephone.

Several test systems have been implemented to demonstrate various aspects of IVHS, such as the ADVANCE (advanced driver and vehicle advisory navigation concept) in the Chicago, Illinois, area. Figure 11.23 shows the interaction of the vehicle with global positioning satellites, embedded road sensors and two-way RF communication.

Dead reckoning will require precision magnetic field sensors to provide the actual vehicle location for many of these feature to be implemented. Collision avoidance or near obstacle detection could ultimately be part of IVHS.

In brief, the RF sensing techniques for future vehicles include: SAW (surface acoustic wave) sensors, Doppler radar, sonar, ultrasonics and microwave sensors. Some of these RF sensors are already used in anti-theft and remote vehicle entry systems.

11.3.8 Future Sensor Requirements

The number of systems that are being developed for vehicle usage and the number of sensors (and actuators) required for the proper operation of these systems (Table 11.2) will require changes in the way that future sensors are applied to vehicles. Furthermore, the effect of new computing approaches and the resulting complexity of the vehicle under fault conditions will also have an impact on sensing technology. Figure 11.24 [39] shows the interaction of the various systems with some of the required measurements (sensors).

The combination of signal conditioning with sensing elements that allows a direct interface to and bidirectional communication with the microprocessors is called *smart sensors* [5]. Smart sensors will have built-in intelligence with the capability of performing logic functions. The smart sensors and even smart systems are typically regarded as monolithic microsensors that offer fewer components, improved reliability, and system-level cost reduction but may be sensors and additional interface ICs or a custom MCU combined in a single package. Smart sensors can relieve a heavily used MCU of some of its activity and provide more information than a standard sensor. In addition to their ease of interface, smart input devices can also provide protection, fault detection (diagnostics), and status reporting. Based on the increasing complexity of vehicle systems, diagnostics for end-of-line assembly, driver, and service are being viewed as mandatory for manufacturability, safety, and first-time success on service repairs (see Section 11.3.6). Smart sensors will be an integral part both of more complex control strategies and new system architectures, such as fuzzy logic and neural networks.

Combinations of sensors and sensors with actuators will be necessary on future vehicles. Pressure and temperature and flow and temperature are already possible in monolithic microsensors. The combination of a sensor and an actuator has been classified, in general, as the area of mechatronics, and devices have also been called *sensactors* [5]. Use of such devices will outboard some of the control to decentralized control points in the vehicle.

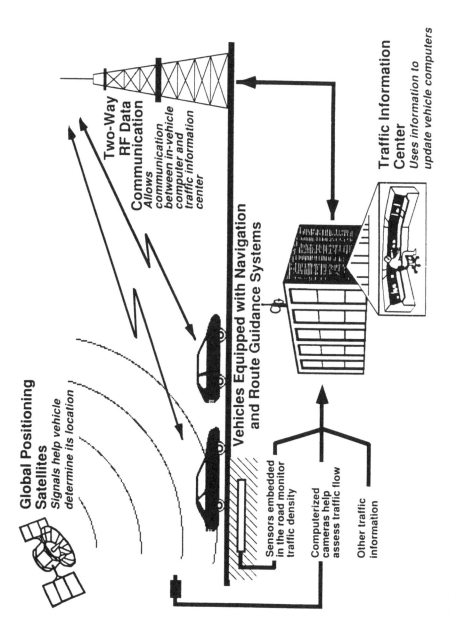

Figure 11.23 Advanced driver and vehicle advisory navigation concept.

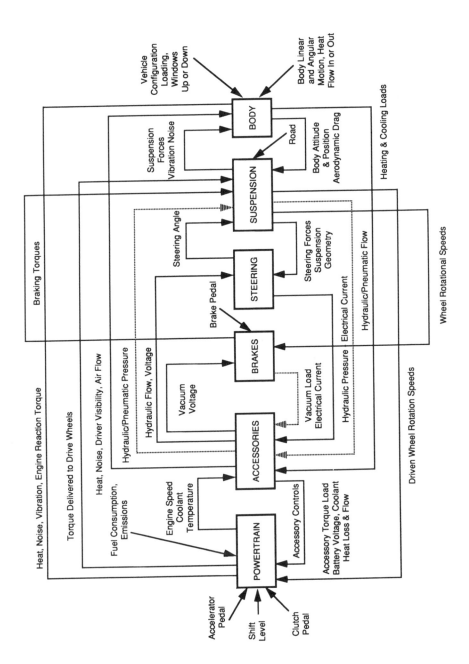

Figure 11.24 Sensor interaction in future vehicle systems.

Multiplexing and multiplex wiring have been implemented in low-volume production vehicles and are considered an important aspect of future vehicles. The SAE has established SAE J1850 as a recommended network interface for MUX systems. A MUX system represents the ultimate integration of vehicle control systems. By utilizing a microprocessor to transfer pulsewidth-modulated input commands over a single data wire at rates of over 5 kB/s to various distributed output devices, the number of wires, weight, and complexity of wiring harnesses can be reduced significantly. This results in easier assembly, higher reliability due to the reduced number of interconnections, and improved diagnostic capability. A fully multiplexed vehicle can easily save over 25 lbs of copper wiring and weight—a saving that translates into improved fuel economy. In addition, system-level simplification will result in a reduced number of components by the sharing of the output from certain sensors, as well as by reduced procurement and vehicle assembly costs. Improved long-term reliability and owner cost will result from the reduction of potential connector failures to half of previous levels. The use of smart sensors will be essential when MUX concepts are applied to a greater portion of the fleet.

MCUs with faster computation capability, higher resolution, fuzzy logic, and neural networks will place additional requirements on the accuracy expectations of vehicle sensors. Systems with 32-bit MCUs are already being pursued by some manufacturers. Fuzzy logic is being applied to collision avoidance systems, cruise control, antilock brake systems, and automatic transmissions [40]. The use of fuzzy sets requires that sensor inputs be assigned values such as high, very high, medium, low, or very low. Each value is given a weight, which allows a decision to be made under circumstances that are imprecise. Neural networks using failure detection filters are being investigated as a real time means of identifying vehicle faults. The computer is trained to recognize the difference in the system between normal operation and under fault conditions, with the capability of observing the whole vehicle system and the dynamic interactions that occur. The full impact of these advanced computational capabilities on sensors is not known at present but the way that sensors are specified in the future is certain to change.

11.4 SUMMARY

The automobile's hostile environment, the high-reliability requirement of automotive systems, and the need for large-volume, low-cost components present both opportunities and significant challenges to sensor manufacturers and system designers. The capabilities of existing sensors are improving at a rapid rate and new technologies are being added or adapted to provide additional approaches to vehicle sensing requirements. The transition of high-precision, low-volume–high-cost sensors to moderate-precision, high-volume–low-cost sensors will provide new sensing approaches for nonautomotive applications that rely on the volume of automotive sales to develop cost-effective technologies that are also rugged and reliable.

This chapter has described a number of vehicle applications for several types of microelectronic sensors in existing and future vehicles. The problems that have been solved and those that must be solved for these devices to be used have been addressed.

REFERENCES

[1] "New and Emerging Markets for Automotive Sensors in North America," Market Intelligence Research Corporation (MIRC) Rep., 1990.
[2] *SAE Handbook*, Vol. 2, *Parts and Components*, Warrendale, PA, SAE, latest edition.
[3] S. Middelhoek, P. J. French, J. H. Huijsing, and W. J. Lian, "Sensors with Digital or Frequency Output," *Sensors and Actuators*, Vol. A15, 1988, pp. 119–133.
[4] "Motorola Semiconductors for Automotive Electronic Systems," Motorola brochure SGBR932S/D, 1990.
[5] J. J. Paulsen, "Powertrain Sensors and Actuators: Driving Toward Optimized Vehicle Performance," Int. Cong. on Transportation Electronics Proc., 1988, Dearborn, MI, pp. 43–63.
[6] W. D. Siuru, Jr., "Sensing Tire Pressure on the Move," *Sensors*, 1990, pp. 16–19.
[7] R. G. Wells, "Non-Contacting Sensors: An Update," *Automotive Engineering*, 1988, pp. 39–45.
[8] Discrete and Special Technologies Group Reliability Audit Report Second Quarter 1990, Motorola Brochure BR923/D Rev 1, p. 35.
[9] R. K. Jurgen, "Detroit '88: Driver-Friendly Innovations," *IEEE Spectrum*, 1987, pp. 53–57.
[10] *Integrated and Discrete Semiconductors*, Allegro MicroSystems Inc., Data Book AMS-500, 1991, pp. 4-64–4-65.
[11] R. Vig, "Protected Hall Effect Contactless Switching of Incandescent and Inductive Loads," *Sensors Expo Proc.*, Chicago, 1990, pp. 205A-1–205A-12.
[12] F. Goodenough, "Multiple Hall-Sensor ICs Talk to Host via Two-Wire, Two-Way Power Signal Bus," *Electronic Design*, 1989, p. 121.
[13] D. J. Strycharz and J. H. Houldsworth, "Contactless Position Sensor for Automotive Applications," 1990 IEEE Workshop on Electronic Applications in Transportation, pp. 32–34.
[14] S. Mizutani and T. Ohtake, "Recent Sensor Technology in Japan," SAE 860410, pp. 35–44.
[15] R. K. Jurgen, "Global '90 Cars: Electronics-Aided," *IEEE Spectrum*, 1989, p. 45.
[16] K. Takahashi, S. Tsuruoka, Y. Nishimura, N. Arai, and H. Tokuda, "Hot Wire Air Flow Meter for Engine Control Systems," SAE SP-805, *Sensors and Actuators*, pp. 1–5.
[17] K. W. Lee and B. E. Walker, "Silicon Micromachining Technology for Automotive Applications," SAE Int. Cong. and Exposition, *Sensors and Actuators*, Vol. SP-655, 1986, pp. 45–53.
[18] M. Baker, L. C. Puhl, E. A. Dabbish, M. Danielson, and R. Frank, "Sensing and Systems Aspects of Fault Tolerant Electronics Applied to Vehicle Systems," SAE P-233 *Proc. Int. Cong. on Transportation Electronics*, Dearborn, MI. 1990.
[19] P. Alling, "Solid-State Sensors to Emerge in Mid-90's," *Automotive Electronics J.*, 1990, p. 14.
[20] D. E. Bergfried, B. Mattes, and M. Rutz, "Electronic Crash Sensors for Restraint Systems," SAE P-233 *Proc. Int. Cong. on Transportation Electronics*, Dearborn, MI, 1990, pp. 169–177.
[21] W. C. Dunn, "Accelerometer Design Considerations," MICRO SYSTEM Technologies 90, Berlin, 1990.
[22] W. Dunn, "Automotive Sensor Applications," 1990 IEEE Workshop on Electronic Applications in Transportation, IEEE CAT 90TH0310-3, pp. 25–31.
[23] W. G. Wolber and P. J. Ebaugh, "Engine Control Sensors: A Status Report," *Automotive Engineering*, Vol. 93 No. 7, 1985, pp. 29–37.

[24] D. L. Hittler, M. A. Boguslawski, L. P. Carrion, and R. K. Frank, "A Microcomputer Emissions Control System," Proc. Convergence 80, SAE SP90, B-5 2.

[25] H. Shiga and S. Mizutani, *Car Electronics*, Japan, ACLA Inc., 1988, pp. 70–71.

[26] John Gosch, "Automotive Gas Sensor Works in 10 mS at a Temperature up to 1000°C," *Electronic Design*, 1991, p. 34.

[27] *The Hansen Report on Automotive Electronics*, Vol. 4, No. 1, 1991, p. 7.

[28] A. Greenberg, "Cost Puts Direct Oil Condition Sensing in Back Seat," *Automotive Electronics J.*, 1990, pp. 9 and 12.

[29] T. Inoue, K. Aoki, and T. Suzuki, "Future Engine Control," SAE P-233, *Proc. Int. Cong. on Transportation Electronics*, Dearborn, MI, 1990, pp. 285–298.

[30] K. G. Tatterson, "U of M Aims to Silence Skeptics with Misfire Measurement Unit," 1990, p. 8.

[31] M. G. Sheldrick, "Interest in Misfire Detection Technology Grows," *Automotive Electronics J.*, 1989, p. 12.

[32] A. Greenberg and M. Boretz, "In-Cylinder Combustion Sensors: Not for Near Term," *Automotive Electronic News* (April 24, 1989), p. 9.

[33] B. Nadel, "Antilock Brakes for $400," *Popular Science*, 1990, pp. 78–81.

[34] P. Frame, "GM's Electric-Motor Powered ABS-VI Bucks Conventional Technology," *Automotive News*, 1990, 14i–15i.

[35] F. J. Gawrowski, "Ford, Chrysler Offer Dealer Diagnostic Systems," *Automotive News*, 1991, p. 16.

[36] B. Harnell, "Fuzzy Logic Guidance Developed by Mazda," *Automotive Electronic News*, 1989, p. 25.

[37] D. H. Walsh, "Electronics for Vehicle Safety in the Near and Intermediate Future," SAE P-233, *Proc. Int. Cong. on Transportation Electronics*, Dearborn, MI, 1990, pp. 165–168.

[38] J. H. Rillings and R. J. Betsold, "An Advanced Driver Information System for North America," SAE P-233, *Proc. Int. Cong. on Transportation Electronics*, Dearborn, MI, 1990, pp. 85–102.

[39] N. A. Shilke, S. M. Rohde, R. D. Fruechte, and J. H. Rillings, "An Automotive Systems Approach," *Automotive Engineering*, 1989, pp. 165–170.

[40] A. G. King, "Nissan Patents 'Fuzzy Logic' ABS, Gearbox," *Automotive Electronic News*, 1989, p. 17.

Chapter 12
Signal Processing for Micromachined Sensors

William Dunn
Motorola

12.1 INTRODUCTION

Silicon, because of its excellent electrical characteristics, is the material of choice in virtually all of the computational and control electronics in use. In addition to that, because of its outstanding mechanical qualities such as high modulus of elasticity, good elastic properties (no hysteresis), machinability in a batch mode, and low cost, silicon is also the prime element of choice for micromachined sensors.

Combining the best of the IC world and the micromachined world has led us to integrated systems—integration of sensors and electronics. The integration process of sensors and IC circuits deals with a broad spectrum of problems, and because of the limits of this chapter, we will focus on some of the problems related to signal processing micromachined sensors. The largest application for micromachined sensors is in the area of sensing mechanical forces such as pressure or acceleration. To sense these measurands, micromechanical structures must be devised for efficient conversion of the measurand into an electrical signal for amplification and conditioning. A number of silicon micromachined [1,2] structures have been devised to that end, using piezoresistive [3], piezoelectric [4], or capacitance [5,6] techniques. Also the corresponding signal conditioning circuits have been designed around these micromachined structures. This chapter will be dealing with those circuits.

12.2 SENSING METHODS

As an introduction to the problem, it is useful to compare devices based on the three basic principles used for the sensing of mechanical forces. These are shown in Table 12.1.

Table 12.1
Methods of Sensing Mechanical Forces

	Piezoelectric	Piezoresistive	Capacitive
Size	Medium	Medium	Small
Sensitivity	Medium	Medium	High
dc response	No	Yes	Yes
Loading effects	High	Low	High
Temperature range	Wide	Medium	Wide
Shift due to shock	Yes	No	No
Self-test features	Possible	Capacitive	Yes
Damping available	No	Yes	Yes

12.2.1 Piezoelectric Sensing

Piezoelectric devices can be cost effective in some accelerometer and pressure sensor applications, but can suffer from offset and temperature-dependent operation. Self-test features are difficult to implement for these devices. The devices usually have a high Q and low damping, which can give them a long settling time. Offsets and temperature effects prevent their use in the frequency range <5 Hz. The piezoelectric device also has a very high output impedance, which makes the device sensitive to load and electromagnetic radiation.

The basic principle of piezoelectric devices is that a piezoelectric material induces a charge or develops a voltage across itself when it is deformed by stress [4]. An accelerometer as shown in Figure 12.1 can be made by placing a layer of piezoelectric material between a mass and a mounting base, with top and bottom contacts. In a similar fashion, a pressure sensor is feasible. The main advantages of piezoelectric sensing are the wide operating temperature range (up to 300°C) and high operating frequency range (up to 100 kHz).

Figure 12.2 shows a typical circuit [7] and the trimming network as used with piezoelectric sensors. The output from the sensor is amplified in a charge amplifier,

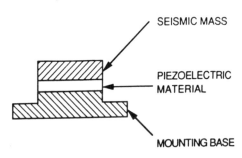

Figure 12.1 Piezoelectric accelerometer.

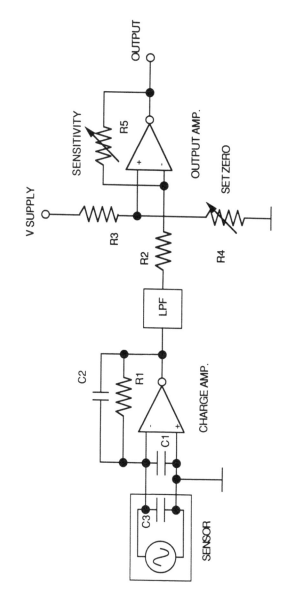

Figure 12.2 Piezoelectric charge amplifier.

which converts the charge generated by the transducer sensor into a voltage that is proportional to the charge. The circuit is a modified voltage amplifier, with feedback via capacitor C2 to maintain the input at 0 or ground potential. This eliminates the effect of stray or ground capacitance, C1. The output voltage from the charge amplifier is fed via a low-pass filter (LPF) to an output amplifier, where it is trimmed for offset by R4 and trimmed for sensitivity by R5. If the input of the amplifier is overloaded due to excessive charges generated by high transient forces on the sensor, the amplifier goes into saturation. The recovery time is long, being set by C2 × R1, which can have a time constant of 1 or 2 seconds. The sensitivity of the sensor is set by the piezoelectric material and the size of the mass (1–200 pC/g). The deviation in sensitivity with temperature is higher with higher sensitivity materials. A typical curve of the sensitivity as a function of temperature for a medium sensitivity material is shown in Figure 12.3. The sensitivity varies ±10% from −40°C to +150°C [8]. Another phenomenon encountered in piezoelectric materials is the pyroelectric effect; that is, changes in the output of the sensor caused by temperature variations. These changes can also be large enough to overload the amplifier at low frequencies (<5 Hz).

12.2.2 Piezoresistive Sensing

The property that silicon changes its resistivity when exposed to stress is called the *piezoresistive effect*. This effect has been in use in pressure sensors for a number of years. The sensing resistors can be either p- or n-type doped regions. The resistance of piezoresistors is very sensitive to strain, and thus to pressure, when correctly placed on the diaphragm of a pressure sensor. Using (100)-oriented silicon material the piezoresistive effect can be described as [3]

$$E = r_0 I \pi_x S \qquad (12.1)$$

where E = electric field; r_0 = unstressed bulk silicon resistivity; I = excitation current density; π_x = piezoresistive coefficient; and S = stress tensor due to force.

It is important to note that π_x depends on crystal orientation. For example, the piezoresistive coefficient in the (111) plane ($\pi_x = \pi_{11}$) can be opposite in sign to π_{12}, where π_{12} is the piezoresistive coefficient in the (112) plane. This anisotropy is used to build a strain gauge in the form of a resistor bridge with four correctly oriented resistors. Because the resistor elements are not positioned at the same location, π_x and S will be slightly different for each resistor, giving rise to nonlinearities in sensitivity to both strain and temperature.

An alternative to the resistor bridge is the transverse voltage strain gauge (X-ducer, which is a trademark of Motorola), shown in Figure 12.4. It is a single resistive element on a diaphragm, with voltage taps $S+$ and $S-$ centrally located on

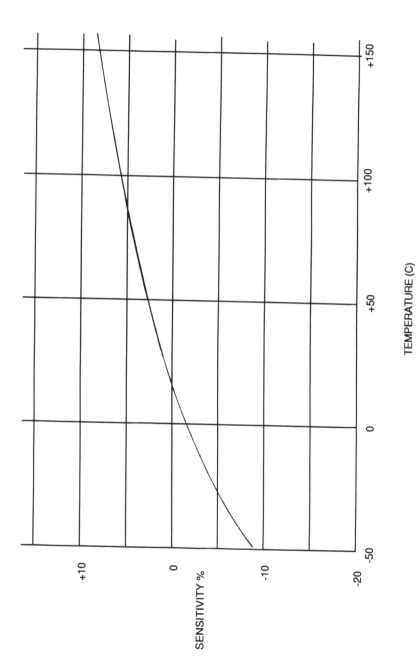

Figure 12.3 Sensitivity versus temperature in a piezoelectric device.

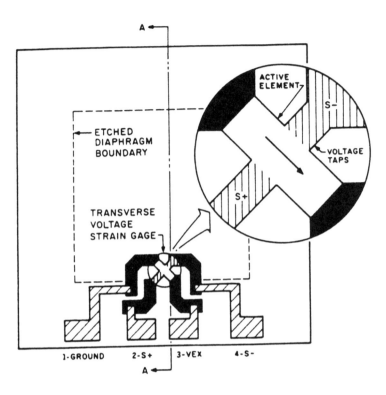

Figure 12.4 Transverse piezoresistive strain gauge.

either side of the resistor. When a current is passed through the resistor, the voltages at the taps are equal when the element is not under strain, but when the element is under strain, a differential voltage output appears. The element is oriented to rely on only one piezoresistive coefficient (π_{44}, which is the shear stress at 45° to the 100-crystal plane). This gives a linear output with strain, as well as improved temperature characteristics compared to the bridge structure. Trimming of the signal conditioning circuits is necessary to set the zero operating point, to set the gain of the system, and to offset temperature effects. The maximum operating temperature is limited to about 100°C to maintain good linearity and accuracy with temperature. This is because of junction leakage currents in the sensing elements and nonlinearity with temperature of the piezoresistive coefficient. Higher operating temperatures have been obtained using oxide-isolated strain gauges [4].

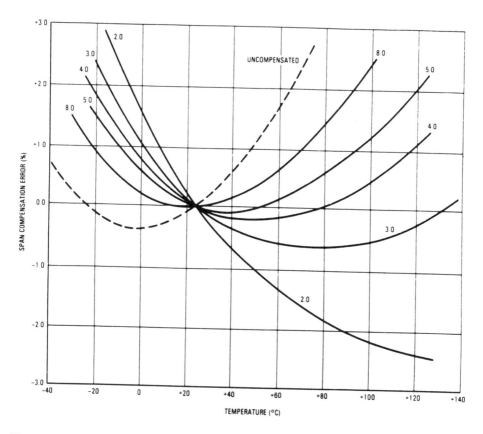

Figure 12.5 Sensitivity vs. temperature in a piezoresistive strain gauge.

Figure 12.5 shows the variation in sensitivity of an X-ducer piezoresistive strain gauge vs. temperature for an uncompensated device and for devices with different values of series resistors added for compensation [9]. The best compensation ratio of resistor value to X-ducer value is between 3 and 4, which reduces the error in the uncompensated strain gauge from about 3%, over the temperature range −20°C to +80°C, to about 1.5%, over the temperature range −20°C to +110°C. This error can be further reduced to about ±0.5%, using a resistor and thermistor combination. A balanced series span compensation is shown in Figure 12.6. The balance resistor network is split so that the output from the X-ducer is at half the supply voltage, enabling direct connection to an amplifier, and the resistor thermistor values are chosen so that their temperature coefficient compensates that of the X-ducer.

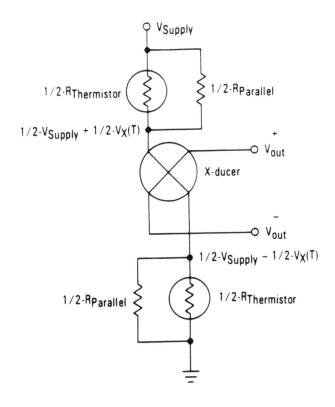

Figure 12.6 Piezoresistive sensor compensation.

12.2.3 Capacitive Sensing

Capacitive sensing has the following advantages over piezoresistive sensing: small size of elements, wider operating-temperature range, ease of trimming, good linearity, and compatibility to CMOS signal conditioning. Capacitive sensing [10] can be done in two different modes of operation, open-loop and closed-loop. Sensing in the open-loop configuration can be accomplished by single capacitance sensing, whereas the differential capacitor measurements can be used in both open- and closed-loop systems. When the differential capacitance sensing method is used to meet the sensitivity requirements, a charge redistribution technique [11] is normally used. Table 12.2 gives a comparison of piezoresistive and capacitive sensing.

Capacitance and capacitive changes can be measured either in a bridge circuit or using switched-capacitor techniques. It is possible to measure capacitance changes of 0.05 fF and lower, under ideal conditions, using switched capacitor techniques

Table 12.2
Comparison of Capacitive and Piezoresistive Sensing
Characteristics and Requirements

	Capacitive	*Piezoresistive*
Trimming	Zero and span	Complex
Temperature compensation	Not required	Required
Temperature range	Wide	Limited
Signal condition	Easy	More complex
Linearity	Good	Poor with temperature
Sensitivity	High	Low
Self-test	Yes	With capacitor

[11]. However, in an industrial environment the lower limit becomes 0.5 fF. A system such as an accelerometer or pressure sensor with a resolution of 1% will, therefore, have a full-scale capacitive change of 50 fF. If the measurement range of the structure is limited to 10% of its allowable movement (for linearity reasons), the actual capacitance could be 500 fF, or 250 fF + 250 fF using differential sensing. Small air gap spacings between two plates on the order of 2 μm or less can be obtained by using surface micromachining, from which a capacitor or plate size of 336 μm × 336 μm (or 237 μm × 237 μm with differential sensing) can be obtained with air as the dielectric.

Any of the capacitive sensing techniques used in a micromachined structure require an ac voltage across the capacitor being measured. This in turn produces an electrostatic field, which generates an attractive force between the capacitor plates, as shown in the single-sided capacitive cantilever structure in Figure 12.7(a). Other forces can be generated by the sensing-current flow, which will produce an electromagnetic field, or by system noise. In the case of system noise, the average noise voltage generated is 0; however, the electrostatic force is proportional to the square of the noise voltage, which is not 0. Using single-capacitor sensing, the sensor capacitance can be compared to a fixed capacitor. The electrostatic field produced by the sensing waveform can be substantial when small capacitor spacings are used. In the case of an accelerometer, this may cause both a deflection of the seismic mass and a zero offset. If the voltage amplitude of the driving waveform is too large, the electrostatic forces can pull and lock the mass down to the substrate, rendering the device inoperable.

Differential sensing, using a movable plate between two fixed plates, as shown in Figure 12.7(b), can be operated in either an open-loop or a closed-loop signal conditioning configuration. In both cases, the electrostatic forces from the two nonmovable top and bottom plates on the movable middle plate will cancel, when the middle plate is in a central position. This greatly reduces any tendency for latch up,

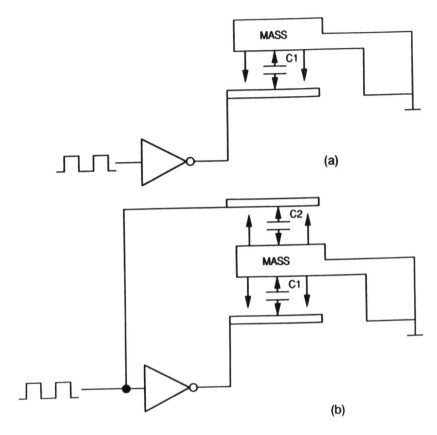

Figure 12.7 Capacitive sensing with electrostatic forces: (a) single-sided capacitive cantilever structure; (b) a movable plate between two fixed plates.

which exists in the open-loop configuration. Also, any electrostatic forces created by noise will tend to cancel. Differential sensing also greatly improves the system linearity and minimizes temperature effects.

Figure 12.8 shows the signal conditioning circuit [12] and trimming scheme used with single-ended capacitive sensing. The first stage of the CMOS signal conditioning circuit uses switched-capacitor filter techniques for converting the capacitance change of the transducer $C1$ into a voltage. Initially, switch $S1$ is at ground potential and switches $S2$, $S3$, $S4$ are closed. Each integrator has unity gain so that each of the negative inputs will be at ground potential. Switches $S2$, $S3$, $S4$ are then opened, and $S1$ is connected to VREF. The output of the integrators is proportional

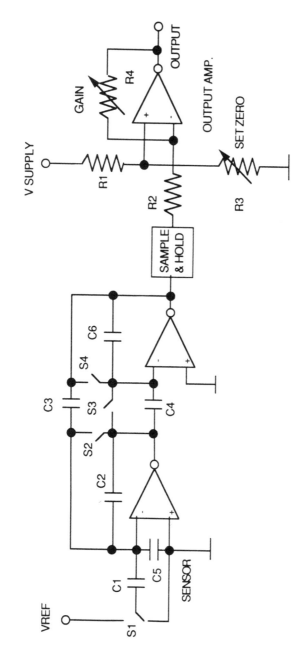

Figure 12.8 Single capacitor sensing block diagram.

to the reciprocal of the difference of the feedback capacitors C2 and C3, and the output voltage is given by

$$V_{\text{out}} = V_{\text{ref}} \frac{C1}{(C2 - C3)} \qquad (12.2)$$

In this converter the input voltage is also maintained at zero potential by feedback via capacitor C2, so that the effect of stray or ground capacitance C5 is eliminated. In this case, because the input or sensing capacitor is switched, saturation of the amplifier or recovery time is no problem. The output voltage of the converter V_{out}, is fed to the output stage via a sample-and-hold circuit, where the signal is amplified and trimmed for zero offset by R_3, and the gain is set by R_4.

In the closed-loop system, the seismic plate is maintained in its central position by electrostatic forces, so that only minimal movement occurs during acceleration. This is achieved by monitoring the capacitance and applying an appropriate voltage to the top or bottom capacitor plate to produce sufficient electrostatic force on the central plate (seismic mass) to balance any acceleration forces on the seismic mass. The servo system feedback is shown in the block diagram in Figure 12.9, which contains the mechanical, electrical, and electrostatic blocks. When the gain of the electrical block is high, the transfer function can be modified, so that the output (V_{out}) is both proportional to the acceleration and independent of the mechanical parameters, except for the mass

$$V_{\text{out}} = \frac{m \times \text{Acc} \times g \times d}{C \times V^2} \qquad (12.3)$$

where m = the mass; g = gravitational constant; C = plate capacitance; Acc = acceleration; d = plate separation; and V = plate voltage.

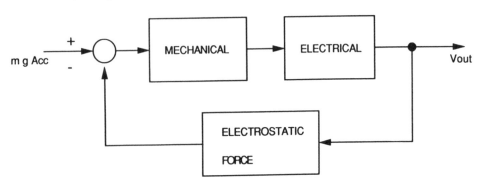

Figure 12.9 Servo loop block diagram.

The closed-loop technique has the following advantages over open-loop sensing: minimized fatigue in the suspension arms, a wide dynamic operating range, a wide temperature range, improved linearity, a sensitivity that depends only on the mass, a reduced bandwidth dependence on film damping. With proper system design, all of the mechanical parameters and variables, except the mass, are reduced to second-order terms, thus both greatly improving the performance of the system and making the sensitivity, bandwidth, and any temperature effects dependent upon the signal-conditioning electronics.

12.3 OPEN- AND CLOSED-LOOP SYSTEMS

In the suspended-mass type of accelerometer, the sensitivity and the bandwidth in the open-loop system depend upon a number of process variables: the width of the suspension arms, the thickness of the suspension arms, Young's modulus, the weight of the seismic mass (density, thickness, dimensions), the spacing between plates, and the damping (slots, gas, pressure). As can be seen, these parameters can give wide spreads in zero offset and sensitivity and in temperature drift. However, in the closed-loop system the process variables become second-order terms, so that in this system, sensitivity, bandwidth (damping), and drift with temperature are minimized.

12.3.1 Different Micromachined Structures—Mechanical Issues

To describe open-loop and closed-loop systems we will use an accelerometer as an example. A number of shapes and configurations for micromachined accelerometers is possible. However, in considering the alternatives, the following requirements must be taken into consideration: sensitivity, stability, material fatigue, resistance to shock, damping, cross-axis sensitivity, linearity, operating temperature range, operating frequency range, and reproducibility.

Three basic methods of suspension are shown in Figure 12.10. These structures have the ability to be designed to meet the preceding requirements. Shown are the cantilever, Figure 12.10(a); the torsion-bar suspension, Figure 12.10(b); and a seismic mass suspended by four cantilevers from a central pillar, Figure 12.10(c). The simple cantilever is the structure with the highest sensitivity for a given size of suspension arm, and thus it can be made smaller for a given sensitivity than any of the other configurations. This is shown in (12.4) and (12.5), which are the acceleration-sensitivity equations for the two-arm cantilever and the four-arm suspension configurations shown in Figure 12.10(a) and (c) [13], respectively. The deflections Δ_{ta} and Δ_{fa} of the mass in each case is given by

$$\Delta_{ta} = \frac{4 \times m \times \text{Acc} \times g \times L^3}{Y \times 2 \times a \times b^3} \qquad (12.4)$$

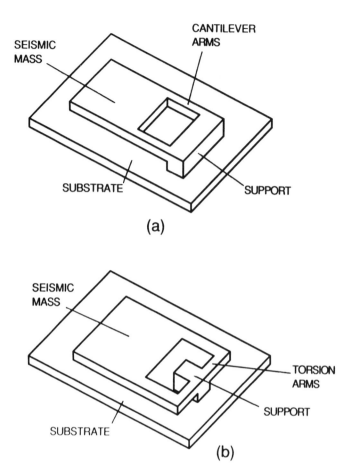

Figure 12.10 Various accelerometer configurations: (a) the cantilever; (b) the torsion-bar suspension; and (c) a seismic mass suspended by four cantilevers from a central pillar.

$$\Delta_{fa} = \frac{m \times \text{Acc} \times g \times L^3}{Y \times 4 \times a \times b^3} \tag{12.5}$$

where Δ_{ta} = two-arm deflection of the mass; Δ_{fa} = four-arm deflection of the mass; L = length of arms; Y = Young's modulus; a = width of arms; and b = thickness of arms. For equal mass and arm size, the cantilever structure is eight times more sensitive than the four-arm structure. In torsion-bar suspension, Figure 12.10(b) [13], the displacement Δ_{tb} of the mass is given by

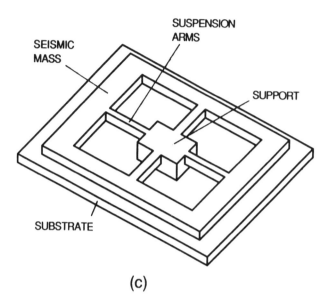

SUSPENSION
ARMS

SEISMIC
MASS

SUPPORT

SUBSTRATE

(c)

Figure 12.10 continued.

$$\Delta_{tb} = \frac{m \times \text{Acc} \times g \times 2 \times L^2}{G \times ab(a^2 + b^2)} \tag{12.6}$$

where Δ_{tb} = torsion-bar displacement of the mass and G = shear modulus. It is difficult to get a comparison of the sensitivity of the torsion-bar suspension versus the cantilever suspension, but in practical devices the size of the torsion-bar devices is usually found to be larger than the cantilever devices.

Two suspension arms are sometimes used, as shown in Figure 12.10(a), to improve the cross-axis sensitivity of the device. The cross-axis sensitivity is defined as the ratio of the output from transverse acceleration to the normal acceleration, and it is typically required to be less than 3%. It can be seen from the preceding equation that, to increase the sensitivity in a structure, the width and thickness of the suspension arms must be minimized, the length increased, the capacitor spacing minimized, and the mass increased. In a flat structure, the long suspension arms required for good sensitivity can be bent, as shown in Figure 12.11, to reduce the overall size of the structure, but the suspension-arm equation must be modified by a torsional term. Under adverse conditions the deflection can become excessive, and the multiple-arm structure can be driven close to one of the fixed plates, in which case the electrostatic force due to the drive voltage becomes very large, as is shown in Figure 12.12. If open-loop sensing is used with this type of structure, latch-up

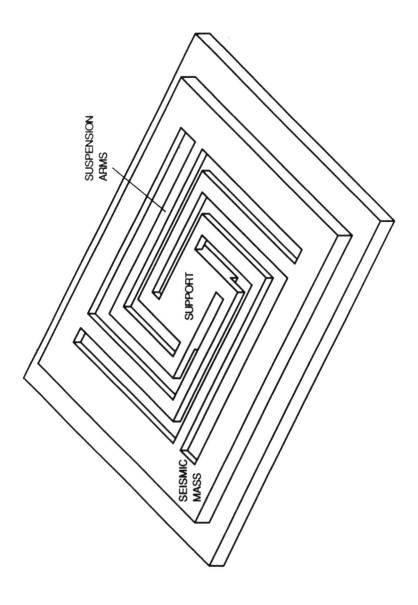

Figure 12.11 Flat accelerometer with curved arms.

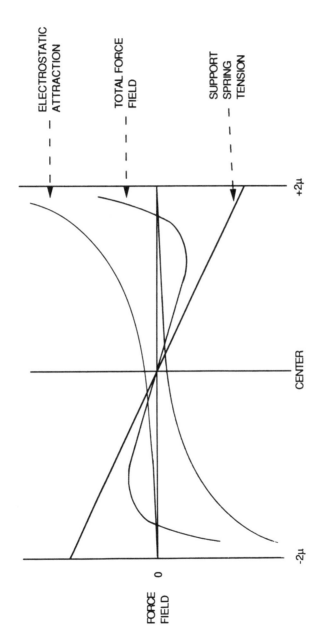

Figure 12.12 Electrostatic and mechanical forces.

can occur unless the deflection is limited by the use of end stops. Figure 12.12 shows the plate displacement vs. spring tension and the electrostatic fields encountered with capacitor spacings of 2 μm. As can be seen, for good linearity the center plate must move less than 10% of the possible deflection for full-scale output. In both the simple-cantilever and the torsion-bar suspension devices, the deflection should also be limited to 10% for good linearity. The moving plate is angled down, and end stops can be used to prevent further movement of the seismic mass, thus limiting the minimum plate separation and hence the electrostatic forces. In this way, latch-up can be prevented with proper design and drive voltage.

12.3.2 Open-Loop Capacitive Sensing

The CMOS signal-conditioning circuit for open-loop differential sensing and the cross section of a capacitive torsion-bar sensor are shown in Figure 12.13 and 12.14, respectively. CMOS has a number of advantages over bipolar, when sensing small capacitance changes using switched-capacitor techniques: very high input impedance, good switch characteristics, low power requirements, and small size.

The output voltage from the converter section in Figure 12.13 is given by

$$V_{\text{out}} = \frac{V_{\text{ref}}(C1 - C2)}{C3} \tag{12.7}$$

which gives an output voltage proportional to the differential capacitance of $C1$ and $C2$. Here, again, the input is referenced to ground by feedback capacitor $C3$, so that

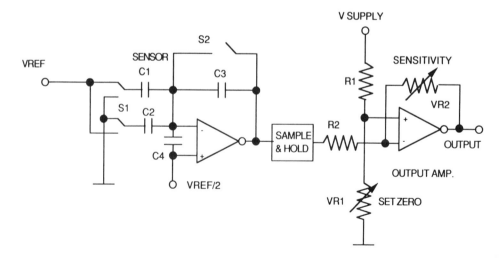

Figure 12.13 Open-loop differential sensor block diagram.

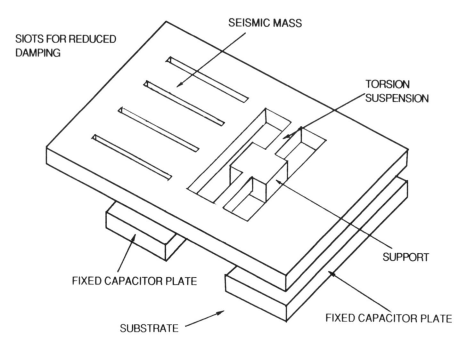

Figure 12.14 Differential capacitive accelerometer.

he effects of stray capacitance are eliminated. The first-stage output is fed via a
ample-and-hold circuit to the output stage, where the signal can be amplified to
give a voltage or a current output, depending upon the system requirements. The
output can also be ratiometric (proportional to the supply voltage), which is some-
imes required when the analog output is digitized by an external A/D converter. If
ingle-capacitor sensing is required, the circuit shown in Figure 12.8 can be used.
To offset processing tolerances of the sensing capacitors, as well as variations in
ystem sensitivity, both zero and gain trimming are required; and this is normally
performed in the output stage. However, an alternative is to individually adjust the
amplitude of the V_{ref}, which is the driving voltage to $C1$ and $C2$. If the circuit is
equired to operate over a very wide temperature range and with an accuracy of no
etter than 2%, no temperature adjustment is required to correct for changes in the
physical properties of the material. Pressure changes can cause changes in the bandwidth of the device, unless it is hermetically sealed.

The driving waveforms for the sensor and the signal conditioning circuits are
hown in Figure 12.15. The clock, which typically has a frequency of 250 kHz, is
divided into four phases, which generate three nonoverlapping phases. These per-
orm the following functions.

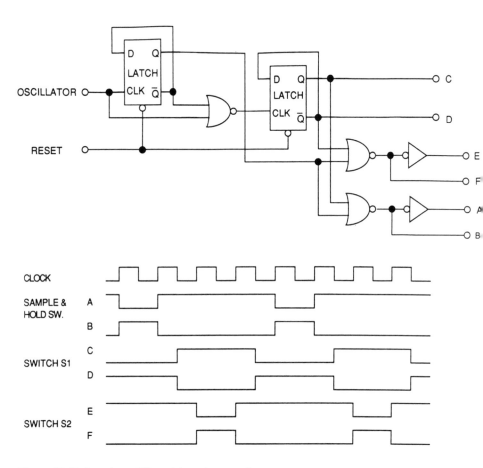

Figure 12.15 Open-loop differential sensing waveforms.

- Reset integrator: sets the input voltage of the integrator to $V_{ref}/2$ by shorting the input to the output.
- Sensor drive: drives the fixed top and bottom capacitor plates in the sensor. The integrator circuit converts the difference in capacitance into a voltage.
- Sample and hold: samples the output of the integrator and stores the voltage on a capacitor to supply dc voltage to the output stage.

12.3.3 Closed-Loop Capacitive Sensing

In a closed-loop system a force due to acceleration can be sensed either with piezoresistors or with capacitive techniques. However, the capacitive sensing has advantages and we will focus on this technique. Figure 12.16 shows the block diagram of a

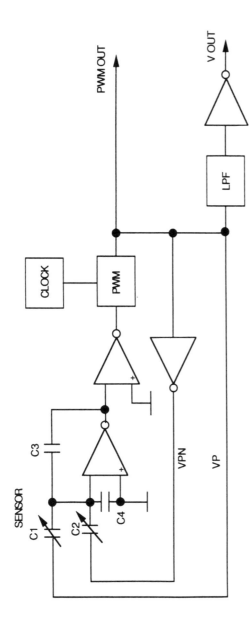

Figure 12.16 PWM block diagram.

pulsewidth modulator [14] (PWM) for use in a closed-loop system. The circuit wave-forms are shown in Figure 12.17. The difference in value between the two sensing capacitors is converted into a voltage using the same technique as in the open-loop case. The output signal from the converter is amplified and converted into a PWM signal (VP). This signal is inverted (VPN) and the true and inverted signals used to drive the top and bottom capacitor plates. With no acceleration, the output from the PWM circuit has a 50% duty cycle, so that with the seismic mass referenced to ground potential, the electrostatic forces on the center plate are equal and opposite. When an external force is applied, a change in capacitance is sensed, and the duty cycle of the driving waveform changes, increasing the duty cycle on one plate and reducing it on the other, as shown by the waveforms. The net effect is to produce an electrostatic force to counterbalance the externally applied force on the center plate. The closed-loop transfer function of the PWM system with a high-gain am-plifier is given by

$$\text{Duty cycle} = \frac{Acc \times g \times d}{C \times V^2} \tag{12.8}$$

That is, the duty cycle is proportional to the acceleration. The output from the PWM section is also fed to the output via a low-pass filter, which will convert the PWM signal into an analog signal. The output stage as shown is normally trimmed for offset and gain.

Another example of closed-loop operation is the delta modulator [15], the cir-cuit for which is shown in Figure 12.18, with the waveforms shown in Figure 12.19. This method uses digital techniques. A reference voltage is switched across the plates of capacitors $C1$ and $C2$, and when an external force is applied, the resulting charge on the center plate caused by a difference in the capacitance of $C1$ and $C2$ is con-verted into a voltage, as in the circuit shown in Figure 12.13. This voltage is then clocked into a latch, where it sets up a 1 or 0 depending upon the sign of the sensed capacitor charge. The output from the latch is then used to apply a voltage to the appropriate capacitor plate to force the center plate back to its zero position. With no external forces applied, the center plate vibrates about its center position. Clock-ing for this type of system is nominally about 2 MHz. The datastream from the latch also goes to a decimation circuit, which consists of a low-pass filter (second-order Butterworth) that is used to convert the 2MHz, 1-bit serial signal into an 8-bit word that can be clocked onto an 8-bit MPU bus at an 8 kHz clock rate. Alternatively, a low-pass filter can be used to convert the data into an analog output. The waveforms shown in Figure 12.19 compare the displacement of the seismic mass in an open-loop system and a closed-loop system. The movement with the closed-loop system is minimal. Also shown is the modulated data output from the delta modulator. Trim-ming the zero offset and sensitivity is required.

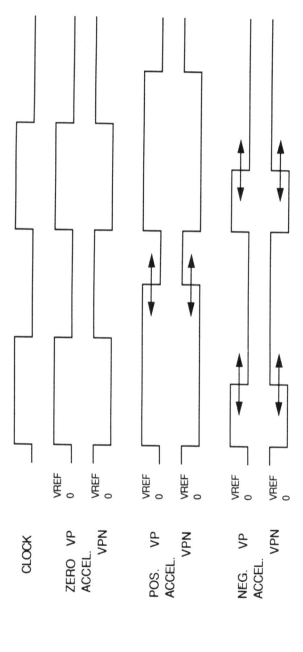

Figure 12.17 PWM waveforms.

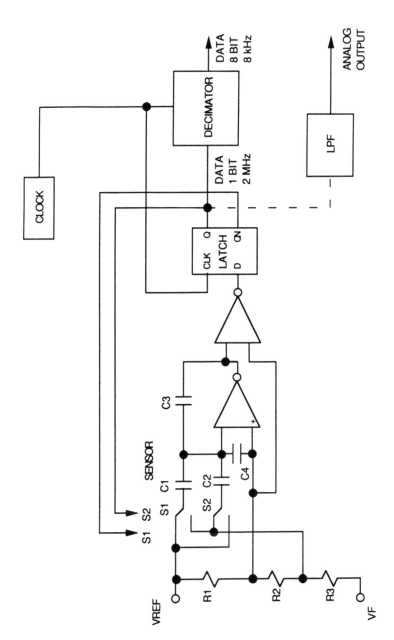

Figure 12.18 Delta modulator block diagram.

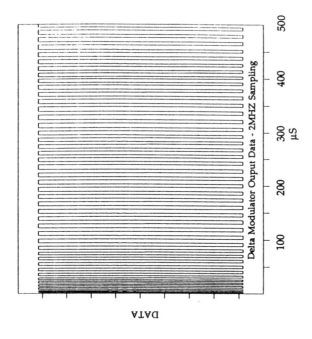

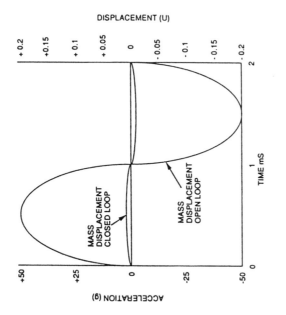

Figure 12.19 Delta modulator waveforms.

Closed-loop capacitive sensing can be used at high temperatures (>150°C), and, because the capacitors are dielectrically isolated, temperature correction is not required.

12.4 INTEGRATION

For monolithic integration of sensors and microcircuits, the processing must have compatible steps, and the integrated device has to be cost effective. Any processing steps specific for one device say, a sensor circuit, must not adversely affect the other. The starting material must be compatible to both. Integration usually provides better temperature sensing and control, reduces the silicon area, and grounds capacitances by eliminating the chip-to-chip bonding pads. Other advantages are better noise immunity, reduced lead inductance, and better reliability.

12.4.1 Integrated Pressure Sensor

Figure 12.20 shows a three-stage bipolar amplifier and its trimming network used with the MPX5100D integrated X-ducer strain gauge. The trimming network compensates for the characteristics of the strain gauge. The bipolar amplifier is trimmed using thin-film resistors. The span is compensated by resistor R_s, Figure 12.6. The temperature and offset calibration circuit uses two operational amplifiers, OA1 and OA2. Resistors $R1$ and $R2$ are used to set the operating point of the amplifiers to match that of the X-ducer. R_{to} is then set to match the shift in output voltage of the X-ducer over temperature. Amplifier OA3 is the main gain stage. The zero offset is adjusted by $R8$ and $R9$ and the gain or sensitivity is set by R_g. The sequence of trimming is critical to prevent interaction between the adjustments. Using this approach accuracies of ±1% over the temperature range −40°C to +100°C can readily be achieved.

12.4.2 Integrated Accelerometer

An integrated accelerometer was described in [5]. The device uses CMOS signal conditioning circuits and capacitive micromachined sensing structures of the type shown in Figure 12.10(b). The conditioning circuits use switched-capacitor sensing techniques in an open-loop configuration (Figure 12.13) and can operate over the temperature range −55 to +125°C. The output stage is a voltage-to-frequency converter, so that a frequency output is obtained proportional to the sensed acceleration. Trimming is done off-chip.

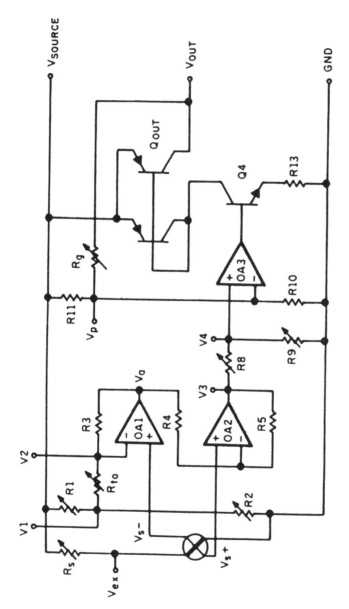

Figure 12.20 Integrated X-ducer sense amplifier.

12.4.3 Dynamic Considerations

No section on sensing and control circuits for micromachined structures would be complete without looking at dynamic characteristics. For instance, in the accelerometer air-film damping is desirable to reduce the effect of shock under high-g forces [16,17], but reduces the frequency response of the system. In the case of the cantilever device, the natural resonant frequency (f_0) of the undamped accelerometer is given by [18]

$$f_0 = \frac{(Yab^3)^{1/2}}{2\pi(4L^3m)^{1/2}} \quad (12.9)$$

A typical f_0 for the undamped cantilever shown in Figure 12.10(a) would be about 5 kHz, thus giving a high peak output at resonance. Such a structure, when critically damped [19–21] (the damping constant ∂ is 0.7), will give a response that is flat to about 2 kHz. Typical response curves for both an undamped and a critically damped cantilevers are shown in Figure 12.21. The damping depends upon the area of the seismic mass, the thickness, and the pressure and type of gas in the air film. In

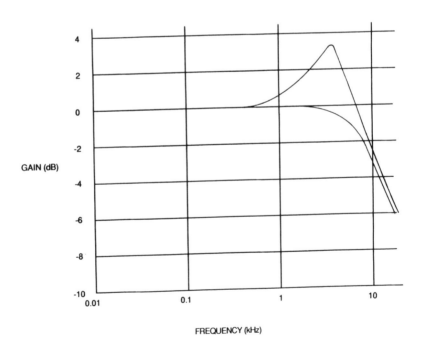

Figure 12.21 Cantilever response, undamped and critically damped.

applications where a high-frequency response is required, damping can be reduced by increasing the thickness of the air gap. On the other hand, this is undesirable, because to maintain the value of the sensing capacitance, the size of the device must be increased. The pressure of the gas in the accelerometer can be reduced, but this necessitates sealing the device. Also, holes can be put into the moving plate to reduce the effect of the air damping. Using this approach the frequency (f) versus gain is given by [19]

$$\text{Gain} = \text{Force}/\{(f_0^2 - f^2)^2 + 4\partial^2 f^2\}^{1/2} \tag{12.10}$$

where ∂ is the damping constant. With the cantilever mass as a solid plate and a 2 μm air gap, the damping at atmospheric pressure gives the ratio [13,16] $\partial/f_0 = 16$. If slits are used in the plate to reduce the effect of film damping, as shown in Figure 12.14, the ratio, $\partial/f_0 = 7.5, 3,$ and 1 for one, three, and five slits, respectively. If these numbers are put into (12.10) and plotted, a set of response curves similar to those shown in Figure 12.22 will be obtained for frequency vs. gain. As can be seen with a solid seismic mass the frequency response is flat to about 60 Hz, increasing to 700 Hz with five slits.

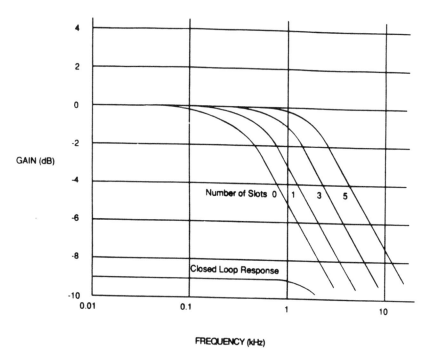

Figure 12.22 Cantilever frequency response, with and without slots.

When considering the dynamic response, the circuit configuration needs also to be taken into account. The curves plotted in Figures 12.21 and 12.22 apply to open-loop systems. In a closed-loop system, where the seismic mass is maintained in its central position, the film-damping effects are greatly reduced and will become of only second order, as shown in [14]. This enables the frequency response to be controlled by the signal-conditioning circuits. The effect is also shown in Figure 12.22, where the closed-loop curve shows the reduced gain required from the mechanical transducer, along with the corresponding increase in frequency response. For a solid plate under atmospheric pressure, the frequency response is increased from 60 Hz to 1 kHz, for the closed-loop configuration.

12.4.4 Shock Considerations

To ensure survivability a system must be designed to tolerate shock. A common criterion for micromachined devices is the ability to survive when dropped onto a concrete floor from a height of 3 ft, which is equivalent to a shock of 5,000g. The problem of survivability becomes most difficult in devices with high sensitivity, because of the increased mass and minimal strength in the suspension arms. The acceleration waveform due to shock [17] is shown in Figure 12.23, and although very

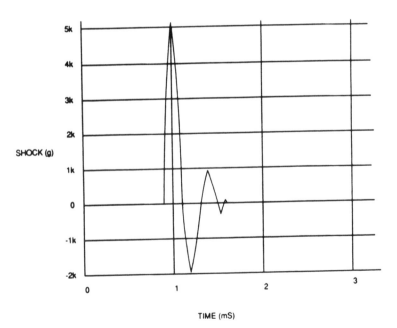

Figure 12.23 Shock waveform.

high amounts of g are involved, their duration is short (0.1 ms), so that film damping can hold the seismic mass in its central position.

An accelerometer can be considered as having three major axis of movement. These are shown in Figure 12.24. The device is most sensitive in the Z-direction by design, and in this direction film damping is used for protection. It should be noted that because of the higher damping factor that can be used with closed-loop operation, a high tolerance to shock is obtained. In the Y-direction there is no film damping, but the device is very rugged, because of the high tensile strength of silicon. The movement depends upon the size of the mass and on the width and thickness of the suspension arms. In this direction an accelerometer is normally capable of withstanding a shock of several thousand g.

In the X-direction the two parallel cantilever beams as shown in Figure 12.24, are used to give good shock tolerance, which will be much higher than the tolerance of a single-support beam. Some twisting of the plate can occur under shock, so that some film damping can occur. But the twisting during normal operation is prevented by the two suspension arms, and good cross-axis sensitivity is obtained. In this direction the device can also be designed to withstand several thousand g. It is worth mentioning that end stops can also be used to prevent excessive travel of the seismic plate in the vertical and transverse directions, a problem that could arise during extended periods of high acceleration. In addition, the end stops can be extended to support the upper capacitor plate over the moving plate, which is a configuration required for differential sensing of the movement of the mass.

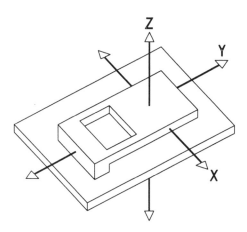

Figure 12.24 Major axis.

12.4.5 Self-Test Features

In many applications self-testing [5] is a requirement. The purpose of the self-test feature is to check that the device and the system are operating correctly. In the case of the accelerometer, this can be achieved with an additional, fixed capacitor plate, as shown in Figure 12.25. Self-testing can be used with either piezoresistive or capacitive sensing. In either case, the self-test feature uses electrostatic forces to deflect the mass and verify operation of the device. By applying a dc voltage between the plates of the test capacitor and the center plate, the mass can be deflected in a manner simulating an externally applied force. Figure 12.26 shows the relation between the plate voltage, and the equivalent g force for an accelerometer with a 2 μm spacing. The electrostatic forces are high for these spacings, and accelerations of 20g and higher can be simulated with only 5V using a test capacitor plate as small as 5,000 μm^2. The deflection can then be measured using the normal sensing circuits. Other advantages of the self-test feature are as follows:

1. The functionality of the device can be tested at the probe level.
2. Assembled devices can be checked.

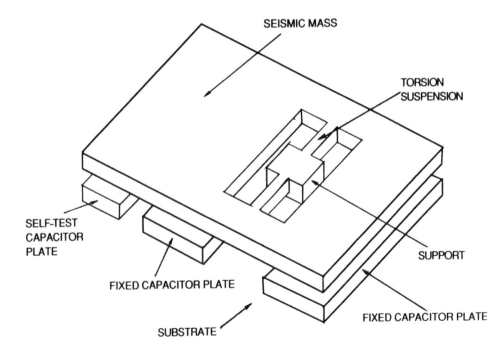

Figure 12.25 Self-test capacitor.

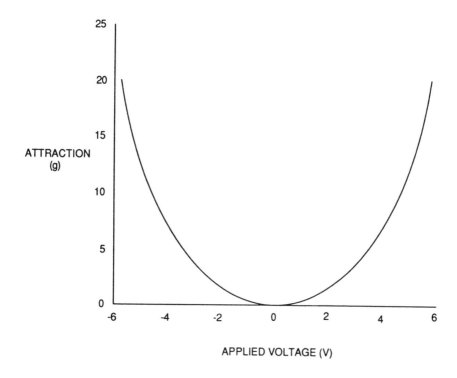

Figure 12.26 Test capacitor voltage versus g force.

3. The device can be exercised for drift and lifetime measurements without having to apply mechanical forces.
4. Operating conditions can be simulated to provide a system test.
5. Devices can be evaluated as functions of frequency and temperature.

The self-test feature can also be used for the development of a smart sensor. The self-test voltage can be trimmed, in order to give a very accurate check of the operation of an accelerometer.

12.4.6 Testing and Trimming

Mass-production testing of dynamic mechanical devices, such as accelerometers, imposes many specific problems in testing and requires special test equipment. For example, it is difficult to check for functionality by probing a wafer of accelerometers on a shaker table. Hence, the self-test capacitor is required to simulate externally applied forces at the test level. The capacitance between the plates of an accelerometer, with air as an insulator, can be designed to be as low as 0.5 pF. However, the change in capacitance can be several pF when the moving plate is pulled to within close proximity to the fixed plate. This capacitance change can be measured at a probe station, providing care is taken to minimize stray capacitance. When the capacitance is integrated with the signal-conditioning circuit, testing is simplified, although the circuit has to be trimmed for zero offset and sensitivity (and over temperature in the case of piezoresistive sensing). Although the self-test capacitor can be used to check functionality, system bandwidth, and to exercise the device for accelerated life testing, it cannot be used for absolute calibration of sensitivity, unless it is also trimmed and calibrated.

A shaker table is normally used for final testing of accelerometers. The zero offset is trimmed with the device stationary, and then the sensitivity is measured and the device calibrated during acceleration. Usually the calibration is easier at lower acceleration than at higher acceleration, because of the higher velocities and stroke required from the shaker table. Another approach is to use a centrifuge, which can be used for device evaluation but is impractical for final testing, because of the burden of loading and unloading times. During testing, the offset and sensitivity are measured and the biasing resistors trimmed to enable the device to meet the specifications. It is interesting to mention that several methods of trimming are available: NiCr or CrSi resistor biasing chains can be deposited on the die and trimmed using a laser beam to cut shorting links, which will increase the values of the biasing resistors. This method can be used to accurately trim over wide resistor ranges and high resistance values. The main disadvantage is the extra process steps required to deposit the resistors.

Alternative trimming methods using available on-chip components are zener zapping or fusible link technology, either trimming scheme can be performed during the probe operation. In either case a number of pads with biasing resistors connected between them are required. Zener zapping uses zener diodes, which can be broken down and shorted by passing a current through them, these diodes are connected between the pads to short out the appropriate biasing resistors. Fusible link technology uses of metal or polysilicon links between the pads, these links can be blown by passing a high current through them, and the appropriate resistors inserted into the bias chain. The resistors are normally binary weighted with respect to each other and are normally used for course trimming.

Finally, for fine trimming over a limited working range either zipper or polysilicon trimming can be used. With zipper trim a n^+ diffusion or resistor is partially covered with metal that can be made to migrate either way along the resistor to increase or decrease its value. This is accomplished by passing an excessive current through the n^+ and metal combination. In the case of the poly resistor, its value can be reduced by passing a high current through it, thereby changing its grain structure and hence its resistance.

12.5 SUMMARY

This chapter has been devoted to the issues dealing with both the conversion of mechanical forces into electrical signals and the signal processing of those signals, as they apply to mechanical sensors. The basic methods of sensing mechanical forces have been discussed, together with their pros and cons, and the various signal-conditioning schemes used to give an electrical output have been presented. The accelerometer has been used as an example to show how it can be done in an open- or closed-loop system configuration. Dynamic and shock considerations have been discussed, as well as self-test features. Finally, sensor integration has been discussed together with methods of trimming the signal conditioning circuits to obtain zero offset and to set the sensitivity.

REFERENCES

[1] W. C. Tang, T. C. H. Nguyen, and R. T. Howe, "Laterally Driven Polysilicon Resonant Microstructures," IEEE Micro Mechanical Systems Workshop, 1989.
[2] K. Peterson and P. Barth, "Silicon Fusion Bonding: Revolutionary New Tool for Silicon Bonding and Microstructures," Wescon, 1989, pp. 220–224.
[3] I. Basket, R. Frank, and D. Slocum, "Integrated Sensors Today and Tomorrow," Sensors, Vol. 8, No. 3, 1991, pp. 32–40.
[4] T. Tamagawa, P. Schiller, H. Yoon and D. L. Polla, "Micromachined Zinc Oxide Thin Film Sensors," J. of the Electrochemical Society, 1990, pp. 1084–1085.
[5] J. C. Cole, "A New Capacitive Technology for Low-Cost Accelerometer Applications," Sensors Expo. Int., 1989.
[6] R. P. Payne and K. A. Dinsmore, "Surface Micromachined Accelerometer: A Technology Update," SAE Int. Cong. and Exposition, Detroit, 1991, pp. 127–135.
[7] D. E. Bergfried, B. Mattes, and R. Rutz, "Electronic Crash Sensors for Restraint Systems," Proc. Int. Cong. on Transportation Electronics, Detroit, 1990, pp. 169–177.
[8] P. L. Chen et al., "Integrated Silicon Microbeam PI-FET Accelerometer," IEEE Trans. Electron Devices, Vol. ED-29, 1982, p. 27.
[9] C. Swartz, C. Derrington, and J. Gragg, "Temperature Compensation Methods for the Motorola X-ducer Pressure Sensor," Motorola Application Note AN-840, 1990.
[10] K. Watanabe and W.-S. Chung, "A Switched Capacitor Interface for Intelligent Capacitive Transducers," IEEE Trans. on Inst. and Meas., Vol. IM35, No. 4, 1986.
[11] J. T. Kung, H.-S. Lee, and R. T. Howe, "A Digital Readout Technique for Capacitive Sensor Applications," IEEE J. Solid-State Circuits, Vol. SC-23, No. 4, 1988, pp. 972–977.
[12] H. Kandler, J. Eichholz, Y. Manoli, and W. Mokwa, "CMOS Compatible Capacitive Pressure Sensor with Read-out Electronics," Micro System Technologies, Vol. 90, 1990, pp. 574–580.
[13] E. P. Popov, Mechanics of Materials, Englewood Cliffs, NJ, Prentice-Hall, 1952, pp. 580–581.
[14] S. Suzuki et al., "Semiconductor Capacitance-Type Accelerometer with PWM Electrostatic Servo Technique," Sensors and Actuators, Vols. A21–A23, 1990, pp. 316–319.
[15] H. Leuthold and F. Rudolf, "An ASIC for High-Resolution Capacitive Microaccelerometers," Sensors and Actuators, Vols. A21–A23, 1990, pp. 278–281.
[16] H. V. Allen, S. C. Terry, and D. W. De Bruin, "Accelerometer Systems with Self-Testable Features," Sensors and Actuators, Vol. A20, 1989, pp. 153–161.

[17] M. Mutoh et al., "Development of Integrated Semiconductor-Type Acceleration," 1990 IEEE Workshop on Electronic Applications in Transportation, 1990, pp. 35–38.

[18] H. F. Schlaak et al., "Micromechanical Capacitive Acceleration Sensor with Force Compensation," *Micro System Technologies*, Vol. 90, 1990, pp. 617–622.

[19] U. E. Gerlach-Meyer, "Capacitive Accelerometer Made by Silicon Micromechanics," *Micro System Technologies*, Vol. 90, 1990, pp. 623–628.

[20] *American Institute of Physics Handbook* 3d ed., McGraw-Hill Book Co., 1972, p. 9.

[21] W. T. Thomson, "Vibration," *Mark's Standard Handbook for Mechanical Engineers*, 8th ed., McGraw-Hill Book Co., 1978, pp. 5–67.

Chapter 13
Controlled Oscillators and Their Applicability to Sensors

I. M. Filanovsky

University of Alberta

13.1 INTRODUCTION

Sinusoidal oscillators and multivibrators represent a widely used group of electronic circuits. A lot of work has been done to investigate and design the circuits where the oscillation parameters (amplitude, frequency, phase, and in multivibrators, the duty cycle) provide information on the values of the passive (resistors, capacitors, and inductors) and active (current and voltage sources) elements used as parts of the oscillating circuits themselves. These values can be functions of external mesurands (mechanical force, electric or magnetic field, temperature, and so on). Then, the modification of oscillation parameters gives information about the external factor.

Many integrated sensors can be used exactly as the aforementioned elements and the oscillating circuits driven from sensors become important interface circuits [1]. Here the circuits of this kind, especially voltage and current controlled oscillators and multivibrators, are reviewed. Preference is given to oscillating circuits with linearly controlled frequency. The frequency output, being noise immune, is required in most cases.

The main attention is given to multivibrator circuits (only one circuit of a voltage-controlled sinusoidal oscillator is considered). A current trend in the realization of integrated sensors is the demand for communication with a microprocessor. The rectangular multivibrator output is easier convertible in the final digital form. Yet, the duty cycle modulation available in a sensor-driven multivibrator can be a very robust form of signal as well and is used in some interesting applications.

The chosen circuits are treated mostly on the level of operational amplifiers, comparators, multivibrators and switches considered as building blocks. Not only the operation principle of the circuits is described. The main goal of the approach is to present the control characteristics in a form allowing estimation of the imperfections (nonlinearity, instability, drift, and offset) of control characteristics. Special attention is given to the slew-rate limitation of operational amplifiers and comparators. This nonideality, which is not "visible" in the ordinary self-oscillating circuits, can be a main source of errors in sensor applications.

Voltage- and current-controlled multivibrators that can be designed for high-frequency operation are considered on the device level. Here, again, attention is paid mainly to the derivation of the control characteristics to obtain the result providing correct calculation of the oscillation frequency and also for estimation of the control characteristics imperfections.

The scope is limited by continuous-time oscillators and multivibrators. The application of switched-capacitor self-oscillating circuits to the conversion of RLC parameters into a frequency or digital signal can be found in [2].

13.2 CONTROLLED SINUSOIDAL OSCILLATORS

13.2.1 The Oscillator Circuit and Its Control Characteristics

Circuits of RC-sinusoidal oscillators with electronic control of frequency have been known for some years [3–5] and were even systematized in [6]. Their later versions [7–9] are based on multipliers. Practical applications of these circuits seem limited to laboratory experiments even though the authors [8,9] underline the useful features of these oscillators; for example, linear control of frequency.

One of the best oscillators with electronic amplitude and frequency control [8] is shown in Figure 13.1. The circuit includes the main positive feedback loop (operational amplifiers A_1, A_2, and A_4 and multipliers M_1 and M_2). The oscillation frequency is controlled by the voltage V_f. The sinusoidal voltages v_o and v_d are shifted by 90°. The amplitude control circuit includes the multipliers M_3 and M_4 operating as squarers, the error amplifier A_3, and the multiplier M_5, which controls the main loop damping. The oscillation amplitude is controlled by the voltage V_a.

The circuit can be described by a simplified two-loop block diagram (Figure 13.2), including two integrators in the main loop and the negative feedback loop around one of the integrators. From this block diagram one obtains that

$$v_o = -\frac{K_2 V_f}{T_2} \int v_d dt \qquad (13.1)$$

and

$$v_d = -\frac{1}{T_1} \int \left(\frac{R_1 K_5 V_e}{R_8} v_d - H K_1 V_f v_o \right) dt \qquad (13.2)$$

where $H = R_9/R_{10}$, $T_1 = R_1C_1$, and $T_2 = R_2C_2$. Eliminating the voltage v_o from (13.1) and (13.2), one can establish the oscillator differential equation

$$\frac{d^2v_d}{dt^2} + \frac{R_1K_5V_e}{T_1R_8}\frac{dv_d}{dt} + \frac{HK_1K_2V_f^2}{T_1T_2}v_d = 0 \tag{13.3}$$

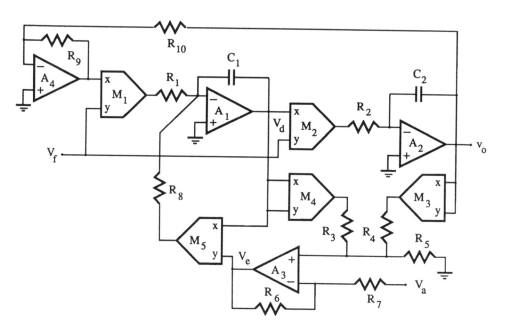

Figure 13.1 Sinusoidal oscillator with electronic amplitude and frequency controls.

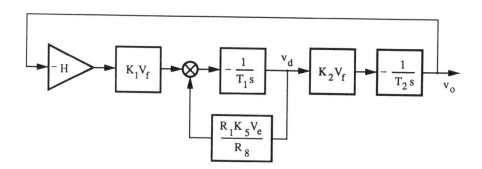

Figure 13.2 Feedback loops in the oscillator.

In the steady state oscillations $V_e = 0$ and the oscillation frequency as follows from (13.3) is

$$\omega_0 = V_f \sqrt{\frac{HK_1K_2}{T_1T_2}} \tag{13.4}$$

The resistor and the timing capacitor values in the main loop are usually chosen so that $H = 1$, $R_1 = R_2 = R_f$, $C_1 = C_2 = C_f$. Then $T_1 = T_2 = T_f$. The multiplying constants are also usually set equal. If $K_1 = K_2 = K$, then (13.4) is simplified to

$$\omega_0 = \frac{KV_f}{T_f} = KV_f\omega_f \tag{13.5}$$

Therefore, the frequency control characteristic is linear. This linear frequency control was experimentally verified. The theoretical and experimental characteristics coincide very well [10] when the circuit is realized with standard operational amplifiers (LF356) and multipliers (ICL8013). They even preserve linearity for the control voltages V_f, which are above the multiplier full voltage range of 10V.

In this oscillator it is difficult to obtain the frequency control of more than two decades for one capacitor setting. The decreasing control voltage V_f opens the main loop, and the level of nonlinear distortions, caused by the operational amplifiers offset and drift, increases. If the standard [11,12] offset compensation circuits are used both for operational amplifiers and multipliers the total harmonic distortion (THD) can be reduced to -80 dB at $V_f = 10$V. In case of offset compensation circuits in multipliers, the THD becomes about -60 dB only at $V_f = 10$V [10], it increases approximately linearly when V_f decreases, and at $V_f = 0.1$V it becomes -30 dB.

In the steady state oscillation $v_d = V_{dm} \sin \omega_0 t$, and then one obtains from (13.1) and (13.4), with the aforementioned choice of parameters in the main loop,

$$v_o = V_{om} \cos \omega_0 t = V_{dm} \cos \omega_0 t \tag{13.6}$$

so that the amplitudes V_{om} and V_{dm} become equal. Then one finds that

$$V_e = \frac{R_5(K_4R_4v_d^2 + K_3R_3v_o^2)}{(R_3R_4 + R_4R_5 + R_3R_5)}\left(1 + \frac{R_6}{R_7}\right) - \frac{R_6}{R_7}V_a \tag{13.7}$$

When $V_{om} = V_{dm}$ and, in addition, in the amplitude control circuit $R_3 = R_4 = R_a$ and $K_3 = K_4 = K$ as well, then the condition $V_e = 0$ gives the oscillation amplitude

$$V_{om} = \left\{\frac{V_a[2 + (R_a/R_5)]}{K[1 + (R_7/R_6)]}\right\}^{1/2} \tag{13.8}$$

The experimental and theoretical amplitude control characteristics for this oscillator are in a close agreement as well [10]. The square root law of (13.8) can be transformed into a linear law using an additional multiplier.

13.2.2 Oscillator Amplitude and Frequency Transients

The oscillator amplitude transients are also predicted using the model shown in Figure 13.2 and the assumption that the amplitude dependent feedback around the first integrator changes slowly within an oscillation period. Then, if the time constants in the integrators are equal and the multiplying coefficients are equal, the oscillator is described by the characteristic equation

$$s^2 + \frac{R_f K V_e}{T_f R_8} s + \frac{H K^2 V_f^2}{T_f^2} = 0 \qquad (13.9)$$

Deviations of V_e pull the complex conjugate poles of this equation to the right half or the left half of the s-plane from their position on the $j\omega$-axis, thereby producing growth or decay of the oscillation amplitude. The oscillation amplitude can then be given [13] as

$$V_{dm} = V_{dm0} \exp\left(\int_0^t \delta p_r(t) dt \right) \qquad (13.10)$$

where $\delta p_r(t)$ is the real component of the change in the $j\omega$-poles, and V_{dm0} is the initial steady state value of amplitude, when $\delta p_r(t) = 0$. It follows from the (13.9) that

$$\delta p_r = -\frac{R_f K}{2 T_f R_8} \delta V_e \qquad (13.11)$$

and, hence,

$$V_{dm} = V_{dm0} \exp\left[-\frac{K}{2 C_f R_8} \int_0^t \delta V_e(t) dt \right] \qquad (13.12)$$

The following analysis considers only one type of amplitude transient response; namely, the transient response for a step change in the reference voltage. In this case the oscillation amplitude changes from the initial value V_{dm0} to the new steady state value V_{dm1} (both of these values can be calculated from (13.7) and the condition $V_{dm} = V_{om}$) when the amplitude control voltage changes from V_{a0} to V_{a1}. Such a restriction allows integration in closed form of the amplitude control differential equation, which follows. Considering the steps of different amplitudes at different initial conditions, one can obtain adequate information on the oscillator amplitude dynamics [10,14]. As follows from (13.7) for a step change in the amplitude control voltage,

$$\delta V_e = \frac{[1 + (R_6/R_7)]}{[2 + (R_a/R_5)]} KV_{dm}^2 - \frac{R_6}{R_7} V_{a1} \tag{13.13}$$

The inversion of (13.8) gives

$$V_{a1} = \frac{[1 + (R_7/R_6)]}{[2 + (R_a/R_5)]} KV_{om1}^2 = \frac{[1 + (R_7/R_6)]}{[2 + (R_a/R_5)]} KV_{dm1}^2 \tag{13.14}$$

and the substitution of (13.13) and (13.14) into (13.12) results in

$$V_{dm} = V_{dm0} \exp\left[\int_0^t a_0(V_{dm1}^2 - V_{dm}^2)dt\right] \tag{13.15}$$

Here $a_0 = \{[1 + (R_6/R_7)]K^2\}/\{2[2 + (R_a/R_5)]R_8C_f\}$. Differentiating the (13.15) one obtains the amplitude control differential equation

$$\frac{dV_{dm}}{dt} = a_0 V_{dm}(V_{dm1}^2 - V_{dm}^2) \tag{13.16}$$

This equation can be integrated [12]. The final result of integration is

$$V_{dm} = \frac{V_{dm1}}{\{1 + [(V_{dm1}^2/V_{dm0}^2) - 1]\exp(-2a_0V_{dm1}^2 t)\}^{1/2}} \tag{13.17}$$

The calculation shows that it is easy to choose the parameters (say, for example, $R_a = R_f = R_6 = R_8 = 100$ kΩ, $R_5 = R_7 = 62$ kΩ) so that if, for example, $V_{dm0} = 3$V and $V_{dm1} = 10$V the amplitude transient is about one oscillation period. If the amplitude decreases by a step, and $V_{dm0} = 10$V and $V_{dm1} = 3$V, the transient is about three oscillation periods. The experimental transients are in a good agreement with calculations [10] and show that this oscillator has a very fast amplitude response. It is possible to create noncontrolled oscillators with fast amplitude transients [13–20] but the solutions for voltage controlled oscillators are very rare, and this criterion was decisive in the choice of the considered circuit. In any real oscillator the controls of frequency and amplitude interact (for the considered circuit this is considered later) and the frequency variation produces perturbations in the amplitude control systems. These small perturbations can cause long amplitude transients [21].

The oscillator also has a fast frequency response. Equation (13.5) shows that the new value of the oscillation frequency should be established instantly when the frequency control voltage V_f changes by a step. Experimentation [10] confirms this assumption, even, for example, for the steps of V_f from 10V to 1.25V and vice versa, resulting in an eightfold change of the oscillation frequency with no noticeable transients.

13.2.3 Interaction of the Frequency and Amplitude Controls

The frequency and amplitude controls in the considered oscillator should be independent, as follows from (13.5) and (13.8). In practice, a weak interaction exists between them. This interaction is more noticeable when the oscillator operates at high frequencies. The mechanism of this interaction deserves some attention because it is also the cause of frequency errors in some sensor-controlled multivibrators.

Assume that an operational amplifier has a slew rate S_r limitation but is ideal otherwise. Let this amplifier operate as an integrator (Figure 13.3), of which the input voltage v_i is a rectangular wave with an amplitude of V_l and frequency of $f_o = 1/T$. The output voltage v_o is a triangular wave of the amplitude $V_o = 4V_l/T$. The relationship between v_o and v_i is

$$v_o = -\frac{1}{R_i C_i} \int v_i dt \tag{13.18}$$

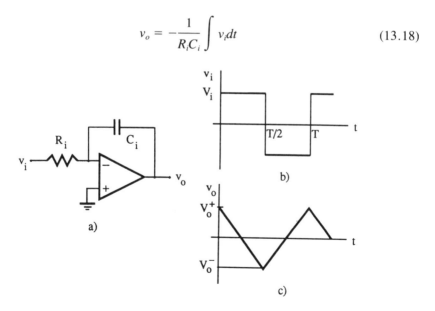

Figure 13.3 Integration of a rectangular wave: (a) integrator; (b) input voltage; (c) output voltage.

which, in this case, can be rewritten as

$$v_i = -R_iC_i \frac{dv_o}{dt}$$ (13.19)

If the input amplitude changes and achieves the value V_{IM}, such that the output wave is limited by S_r, then

$$V_{IM} = -R_iC_iS_r$$ (13.20)

The intrinsic time constant R_iC_i can be eliminated now, and one can write that

$$v_i = -\frac{V_{IM}}{S_r} \frac{dv_o}{dt}$$ (13.21)

Let the circuit shown in Figure 13.3, become an intrinsic stage, causing a slew rate limitation of another operational amplifier (Figure 13.4(a)). Then, to obtain a tri-angular output wave at the output of this more complicated operational amplifier, one has to apply a voltage v_i given by (13.21) directly to the amplifier input. How-ever, if this amplifier operates with an external feedback circuit (Figure 13.4(b)) and

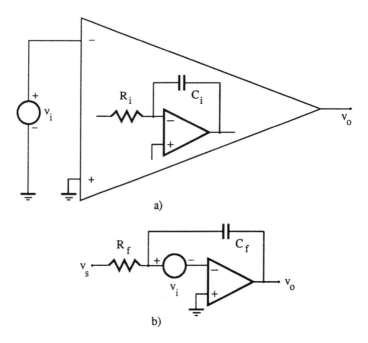

a)

b)

Figure 13.4 Slew-rate error voltage at: (a) the limiting stage input; (b) the slew-rate limited amplifier input.

one considers that the amplifier output voltage is caused by this external generator v_s, the compensating voltage v_i (note the polarity) should be introduced in the input terminal (like an offset voltage in dc analysis). Then one can write, from Figure 13.4(b), that

$$v_o = -\frac{1}{R_f C_f} \int v_s dt + v_i + \frac{1}{R_f C_f} \int v_i dt \tag{13.22}$$

Substituting (13.21) in (13.22) and differentiating one obtains that

$$\frac{dv_o}{dt} = -\frac{1}{R_f C_f} v_s - \frac{V_{IM}}{S_r} \frac{d^2 v_o}{dt^2} - \frac{1}{R_f C_f S_r} \frac{dv_o}{dt} \tag{13.23}$$

Thus, the transfer function of the integrator using an operational amplifier with a slew rate limitation is

$$\frac{v_o(s)}{v_s(s)} = -\frac{\omega_f}{s(1 + \eta\omega_f + \eta s)} \approx -\frac{\omega_f}{s(1 + \eta s)} \tag{13.24}$$

where $\omega_f = 1/(R_f C_f)$ and $\eta = V_{IM}/S_r$. The slew rate limitation introduces an additional pole located at $s = -1/(R_f C_f) - 1/\eta \approx -1/\eta$.

The circuit (Figure 13.5) that includes two such integrators and an inverter with the same limitation will oscillate. Assuming that all operational amplifiers in this circuit are identical, one can find that the characteristic equation of this circuit is

$$s^2(1 + \eta\omega_f + \eta s)^2(1 + \eta s) + \omega_f^2 = 0 \tag{13.25}$$

If η is small and the circuit oscillates at the frequency $\omega_f \ll 1/(2\eta)$ the dominant roots satisfy the equation

$$3\eta s^3 + s^2 + \omega_f^2 = 0 \tag{13.26}$$

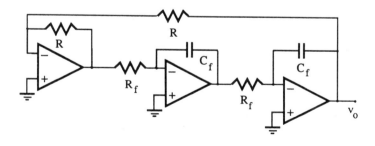

Figure 13.5 Oscillating subcircuit.

which can be rewritten as

$$(3\eta s + 1)(s^2 - 2\eta\omega_f^2 s + \omega_f^2) = 0 \tag{13.27}$$

Thus, the circuit has two roots with positive real parts located near the points $\pm j\omega_f$. The circuit will oscillate, of course, at the frequency very close to ω_f.

If the circuit of Figure 13.5 is used with the same amplitude and frequency control circuits as in Figure 13.1 the circuit characteristic equation will be

$$s^2(1 + 4\eta s) + KV_e\omega_f s(1 + 3\eta s) + (KV_f\omega_f)^2 \approx 0 \tag{13.28}$$

which can be represented as

$$(1 + 4\eta s)\{s^2 - [KV_e\omega_f - 4\eta(KV_f\omega_f)^2]s + (KV_f\omega_f)^2\} \approx 0 \tag{13.29}$$

The circuit will oscillate at the frequency $KV_f\omega_f$ if

$$V_e = 4\eta\omega_f KV_f^2 \tag{13.30}$$

Then, using (13.7) one can find that the oscillation amplitude will be

$$V_{om} = \left\{ \frac{(2 + R_a/R_5)}{K(1 + R_6/R_7)} \left(\frac{R_6}{R_7} V_a + 4\eta\omega_f kV_f^2 \right) \right\}^{1/2} \tag{13.31}$$

This result shows that if the operational amplifiers have slew-rate limitation, the oscillator will have interacting controls and the oscillation amplitude will depend on both V_a and V_f. This conclusion was verified experimentally [10]. The oscillation frequency still changes linearly with the control voltage V_f.

13.2.4 Sinusoidal Oscillators in Applications

Thus, it is feasible to design an oscillator with linear dependence of the oscillation frequency on the control voltage. The transients will be practically absent even when one has fast control voltage changes. The frequency and amplitude control interaction is weak when the oscillation frequencies are low, but at high frequencies this inter-action is noticeable. The oscillation amplitude increases, and this can introduce the measurements errors if the oscillator output is used in a phase-locked system [22]. The lower oscillation frequency is also limited by increasing nonlinear distortions. Yet, a two decade variation of frequency can be easily achieved with one capacitor value. The circuit has to include rather bulky offset compensation circuitry, which makes difficult circuit integration. These circuits are more complicated than the multivibrator-based circuits designed for frequency control. In addition, their useful properties, such as independent control of frequency and amplitude, are essentially deteriorating when they are used in sensors, that is, with low voltage power supplies and with small capacitors.

13.3 CONTROLLED MULTIVIBRATORS

13.3.1 Multivibrators versus Sinusoidal Oscillators

The sinusoidal oscillators with control of oscillation parameters represent rather complicated circuits, although they can be simplified when only one oscillation parameter (usually frequency) must be controlled. Even in that case, however, amplitude stabilization is required, otherwise the circuit will have an oscillation frequency different from that designed [23,24]. The simplest amplitude stabilization system usually includes an amplitude detector and filter, and the need for small amplitude modulation increases the filter capacitor size and reduces the amplitude stabilization speed. A perturbation of this stabilization system results in a transient with a duration of tens and even hundreds of oscillation periods. The frequency control in sinusoidal RC-oscillators usually requires two changing elements (resistors or capacitors). Even though the voltage controlled oscillators do not require tightly controlled components (as is the case of tuned RC oscillators [25]), these difficulties prevent wide use of sinusoidal oscillators in sensor applications. There are sensors where the oscillating system or the oscillator feedback circuit is a highly selective structure (mechanical or quartz resonator) [26,27], allowing one to neglect the influence of nonlinear distortions on the oscillation frequency. Unfortunately, such sensors are not compatible with microelectronic technology. The oscillators with distributed structures in active feedback [28,29] are not frequently realized in sensor applications [30]. Here the condition of oscillation is sensitive to the feedback structure parameters (which is used for sensing) and the feedback structure has high selectivity, allowing one to omit amplitude stabilization and use the power supply voltage limitation.

The amplitude stabilization by power supply voltage imposes no requirements on the multivibrator structure. The circuits of amplitude stabilization are very simple (a resistor and a temperature compensated Zener diode [8]) or even omitted (if switches and the power supply voltage are used for this purpose). The amplitude and frequency transients are very short in multivibrators as well, usually their duration is about one oscillation period. In addition, frequency control is usually achieved by modification of one element (resistor or capacitor). All these advantages of multivibrators resulted in their use in much wider sensor applications.

Frequency modulation by the sensor signal of the multivibrator output voltage or current is used more frequently than with resistor or capacitor modulation. This is because in many cases the multivibrator is preceded by a voltage or current amplifier.

The method of conversion of the frequency-modulated rectangular signal into a useful analog or digital form is highly immune to noise and interference signals. Here the phase-lock loop techniques are the most important [31,32]. They allow not only easy transformation of the signal in analog form but the frequency conversion and multiplexing as well. The signal can be put in the frequency band suitable for telecommunication transmission.

13.3.2 Multivibrators with Operational Amplifiers

A frequently used multivibrator circuit is shown in Figure 13.6(a). It includes an integrator and a Schmitt trigger. The output voltage v_o is a rectangular wave limited by the values V_o^+ and V_o^- (Figure 13.6(b)) and the integrator output voltage v_1 is a triangular wave achieving the values V_1^+ and V_1^- (Figure 13.6(c)). The transition of the multivibrator from one quasi-stable state to another is determined by the balance of voltage at the positive terminal of the Schmitt trigger operational amplifier. At the end of the interval T_1, one has

$$V_1^- \frac{R_b}{R_a + R_b} + V_o^+ \frac{R_a}{R_a + R_b} = 0 \qquad (13.32)$$

which gives $V_1^- = -V_o^+ (R_a/R_b)$. By the same token, at the end of interval T_2, one has $V_1^+ = -V_o^- (R_a/R_b)$. Thus, the capacitor voltage v_c has the swing of $\Delta V_c =$

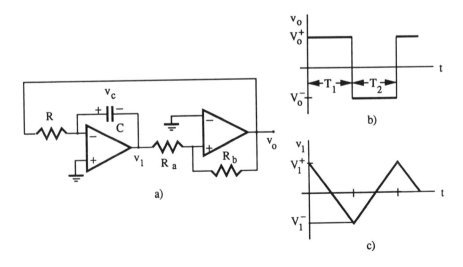

Figure 13.6 Common two-amplifier multivibrator: (a) circuit; (b) output voltage; (c) integrator output voltage.

$V_1^+ - V_1^- = (V_o^+ - V_o^-)\,(R_a/R_b)$. During the first interval capacitor C is recharged by current $I_1 = V_o^+/R$, thus, the duration

$$T_1 = RC\left(\frac{V_o^+ - V_o^-}{V_o^+}\right)\frac{R_a}{R_b} \qquad (13.33)$$

To obtain the duration T_2 it is enough to interchange the plus and minus in the superscripts of (13.33). Hence,

$$T_2 = RC\left(\frac{V_o^- - V_o^+}{V_o^-}\right)\frac{R_a}{R_b} \qquad (13.34)$$

and the oscillation period is

$$T = T_1 + T_2 = RC\,\frac{R_a(V_o^+ - V_o^-)(V_o^- - V_o^+)}{R_b V_o^+ V_o^-} \qquad (13.35)$$

If $V_o^+ = -V_o^- = V_o$ then (13.35) is simplified, and the oscillation frequency becomes

$$f_0 = \frac{1}{4RC}\left(\frac{R_b}{R_a}\right) \qquad (13.36)$$

The oscillation frequency can be modulated modifying R or C. The triangular wave amplitude can also be modified by R_a, R_b, or their ratio (the frequency, in this case, is modulated as well).

The main source of dynamic frequency errors in this multivibrator is the finite integrator slew rate. The correct frequency value can be calculated the following way. If the error voltage source

$$v_i = -\frac{V_{IM}}{S_{ri}}\frac{dv_1}{dt} \qquad (13.37)$$

is introduced in the input terminal of the integrator (Figure 13.7(a)) the integrator output voltage, say, when the multivibrator output voltage is V_o^+, will be

$$v_1 = -\frac{1}{RC}\int V_o^+ dt + v_i + \frac{1}{RC}\int v_i dt \approx -\frac{1}{RC}\int V_o^+ dt + \frac{1}{RC}\int v_i dt \qquad (13.38)$$

Differentiating (13.38) and using (13.37) one finds that during this period of time

$$\frac{dv_1}{dt} = -\frac{V_o^+}{RC + \eta_i} \qquad (13.39)$$

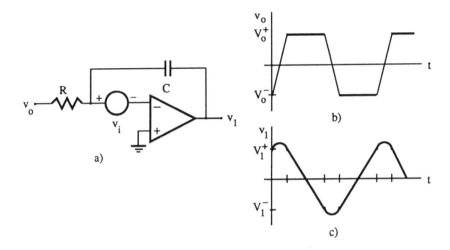

Figure 13.7 Slew-rate error voltage and multivibrator wave distortions: (a) integrator slew-rate error voltage; (b) output wave; (c) integrator output.

where $\eta_i = V_{IM}/S_{rs}$. From the other side, the capacitor voltage swing

$$\Delta V_c = \pm(V_o^+ - V_o^-)\frac{R_a}{R_b} \tag{13.40}$$

is not influenced by the presence of v_i. Thus, the correct oscillation period will be

$$T = T_1 + T_2 = (RC + \eta_i)\frac{R_a(V_o^+ - V_o^-)(V_o^- - V_o^+)}{R_b V_o^+ V_o^-} \tag{13.41}$$

that is, the source v_i introduces the error

$$\frac{\delta f}{f_0} \approx -\frac{\eta_i}{RC} = -\frac{4f_0 \eta_i R_a}{R_b} \tag{13.42}$$

which increases with the frequency.

When the oscillation frequency becomes very high it is necessary to consider the finite slew rate of the Schmitt trigger operational amplifier. The Schmitt trigger has positive feedback, and one can consider that its output voltage always changes with a constant speed of S_{rs} between multivibrator quasi-stable states (Figure 13.7(b)). The oscillation period becomes

$$T = T_1 + T_2 = (RC + \eta_i)\frac{R_a(V_o^+ - V_o^-)(V_o^- - V_o^+)}{R_b V_o^+ V_o^-} + \frac{2(V_o^+ - V_o^-)}{S_{rs}} \tag{13.43}$$

so that the frequency error is

$$\frac{\delta f}{f_0} \approx -\frac{\eta_i}{RC} - \frac{4V_0 f_0}{S_{rs}} \tag{13.44}$$

The integrator output voltage has now visibly smoothed corners (Figure 13.7(b)), and the capacitor voltage swing becomes

$$\Delta V_c = 2V_1 + \frac{V_o^2}{(RC + \eta_i)S_{rs}} \tag{13.45}$$

where the symmetric swing is assumed.

13.3.3 Switches in the Basic Multivibrator Circuits

Some useful modification occurs when additional switches are incorporated in the basic multivibrator circuit. The switches being realized by active devices of minimal geometry are not a heavy burden, either in terms of the multivibrator complexity or in chip area.

13.3.3.1 Duty-Cycle Modulation

The first useful modification is shown in Figure 13.8. The output voltage in this multivibrator controls the switches P_1 and P_2, connecting the resistors R_1 and R_2 to the integrator input. The durations T_1 and T_2 will be

$$T_1 = CR_1 \frac{(V_o^+ - V_o^-)R_a}{V_{CC}^+ R_b} \tag{13.46}$$

$$T_2 = CR_2 \frac{(V_o^- - V_o^+)R_a}{V_{CC}^+ R_b} \tag{13.47}$$

If the output wave is symmetric and, in addition, $V_{CC}^+ = -V_{CC}^- = V_{CC}$ the oscillation period becomes

$$T = 2(R_1 + R_2)C \frac{V_o R_a}{V_{CC} R_b} \tag{13.48}$$

This circuit is used when the resistor change is transformed into a duty cycle change. For this purpose it is more convenient to consider the quasi-stable state

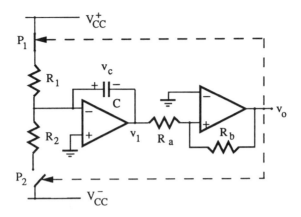

Figure 13.8 Switch-commutated resistors.

duration ratio $D = T_1/T_2$. If $R_1 = R + \delta R$ and $R_2 = R - \delta R$, then the oscillation frequency for the circuit of Figure 13.8 does not change and

$$D = \frac{T_1}{T_2} = \frac{R_1}{R_2} \approx 1 + \frac{\delta R}{R} \tag{13.49}$$

The errors caused by the slew-rate of the circuit operational amplifiers can be calculated individually for T_1 and T_2, as was outlined in the previous section. In addition, these durations should be augmented by the switching time of the switches.

13.3.3.2 Bridge-to-Frequency Converter

Using switches in the two-operational-amplifier multivibrator one can obtain the circuit where the oscillation frequency is controlled by detuning a resistive bridge. Such a circuit is shown in Figure 13.9. For normal circuit operation the bridge should be detuned (at least, the resistors R_1 and R_2 should be different).

The conditions for the circuit transition from one quasi-stable state to another are $V_1^- = -V_o^+(R_a/R_b)$ and $V_1^+ = -V_o^-(R_a/R_b)$ as in the circuit of Figure 13.6. Let the position of switches shown in Figure 13.8 correspond to the first quasi-stable state, and the output voltage is $v_o = V_o^+$ at this state. Then, at the end of this state the capacitor voltage v_c will be

$$v_c(T_1) = \frac{V_{CC}^+ R_4 + V_{CC}^- R_3}{(R_4 + R_3)} + V_o^+ \frac{R_a}{R_b} \tag{13.50}$$

The voltage v_c at the end of second quasi-stable state (equal to v_c at the beginning

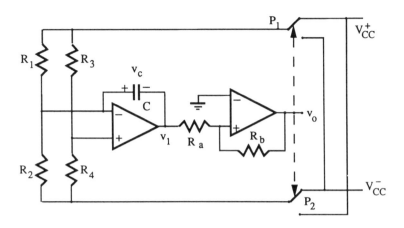

Figure 13.9 Switch-commutated bridge in the two-amplifier multivibrator.

of the first quasi-stable state) is obtained by interchanging the plus and minus in the superscripts (13.50):

$$v_c(0) = \frac{V_{CC}^- R_4 + V_{CC}^+ R_3}{(R_4 + R_3)} + V_o^- \frac{R_a}{R_b} \tag{13.51}$$

Thus, the capacitor voltage swing during the first quasi-stable state is

$$\Delta V_c = \frac{(V_{CC}^+ - V_{CC}^-)(R_4 - R_3)}{(R_4 + R_3)} + (V_o^+ - V_o^-)\frac{R_a}{R_b} \tag{13.52}$$

During this quasi-stable state the capacitor is recharged by the current

$$I_1 = \frac{(V_{CC}^+ - v^+)}{R_1} + \frac{(V_{CC}^- - v^+)}{R_2} = \frac{(V_{CC}^+ - V_{CC}^-)}{(R_3 + R_4)}\left(\frac{R_3}{R_4} - \frac{R_4}{R_2}\right) \tag{13.53}$$

One finds that the duration of the quasi-stable state

$$T_1 = T_2 = \frac{(R_4 - R_3)C}{[(R_3/R_1) - (R_4/R_2)]} + \frac{R_a(R_3 + R_4)C}{R_b[(R_3/R_1) - (R_4/R_2)]}\left(\frac{V_o^+ - V_o^-}{V_{CC}^+ - V_{CC}^-}\right) \tag{13.54}$$

and the oscillation frequency is

$$f_0 = \frac{1}{2C}\left(\frac{R_3}{R_1} - \frac{R_4}{R_2}\right)\frac{(V_{CC}^+ - V_{CC}^-)R_b}{[R_b(R_4 - R_3)(V_{CC}^+ - V_{CC}^-) + R_a(R_4 + R_3)(V_o^+ - V_o^-)]} \tag{13.55}$$

The switches allow the elimination of the feedback resistances of the Schmitt trigger (Figure 13.10) [33]. For this circuit the transition from the first quasi-stable state to the second is determined by the condition $v_1 = V_1^- = V_{CC}^-$ (the condition of the opposite transition is $v_1 = V_1^+ = V_{CC}^+$). The capacitor recharges from

$$v_c(0) = \frac{V_{CC}^- R_4 + V_{CC}^+ R_3}{(R_4 + R_3)} - V_{CC}^+ \tag{13.56}$$

to

$$v_c(T_1) = \frac{V_{CC}^+ R_4 + V_{CC}^- R_3}{(R_4 + R_3)} - V_{CC}^- \tag{13.57}$$

so that

$$\Delta V_c = 2(V_{CC}^+ - V_{CC}^-) \frac{R_4}{(R_3 + R_4)} \tag{13.58}$$

The recharge current is given by (13.53) again and one obtains that

$$T_1 = T_2 = \frac{2CR_4}{[(R_3/R_1) - (R_4/R_2)]} \tag{13.59}$$

and the oscillation frequency is

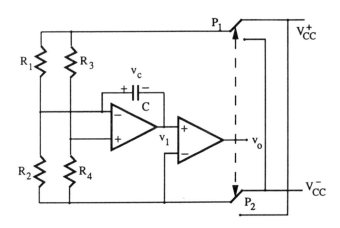

Figure 13.10 Switch-commutated bridge in the amplifier-comparator multivibrator.

$$f_0 = \frac{1}{4CR_4}\left(\frac{R_3}{R_1} - \frac{R_4}{R_2}\right) \tag{13.60}$$

The finite slew rate of the integrator can be a main source of frequency error. For error analysis it is convenient, for this case, to use two separate error voltages V_I^+ and V_I^- for each quasi-stable state and to represent them as

$$V_I^+ = -V_I^- = \frac{V_{IM}}{S_{ri}}\left(\frac{V_1^+ - V_1^-}{T/2}\right) = \frac{2f_0 V_{IM}(V_o^+ - V_o^-)R_a}{R_b S_{ri}} \tag{13.61}$$

Figure 13.11 shows the integrator of the multivibrator with this source of error during the first quasi-stable state. Then one can write that

$$v_c(T_1) = \frac{V_{CC}^+ R_4 + V_{CC}^- R_3}{(R_3 + R_4)} + V_I^+ + V_o^+\frac{R_a}{R_b} \tag{13.62}$$

Again, the initial value of the capacitor voltage can be obtained by interchanging the signs in the superscripts. This gives

$$v_c(0) = \frac{V_{CC}^- R_4 + V_{CC}^+ R_3}{(R_3 + R_4)} + V_I^- + V_o^-\frac{R_a}{R_b} \tag{13.63}$$

so that

$$\Delta V_c = (V_{CC}^+ - V_{CC}^-)\left(\frac{R_4 - R_3}{R_4 + R_3}\right) + (V_I^+ - V_I^-) + (V_o^+ - V_o^-)\frac{R_a}{R_b} \tag{13.64}$$

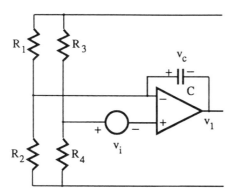

Figure 13.11 Slew-rate error voltage in the integrator with a bridge.

The recharge current in this case will be

$$I_1 = \left(\frac{V_{CC}^+ - V_{CC}^-}{R_4 + R_3}\right)\left(\frac{R_3}{R_1} - \frac{R_4}{R_3}\right) - V_I^+\left(\frac{1}{R_1} + \frac{1}{R_2}\right) \tag{13.65}$$

Using (13.64) and (13.65) one can find the duration of T_1. Then, changing the signs of the subscripts in T_1, the duration of T_2 will be found. After this, one can find the oscillation frequency. These cumbersome expressions are omitted here. In the case of $V_{CC}^+ = -V_{CC}^- = V_{CC}$, $V_o^+ = -V_o^- = V_o$, and $V_I^+ = -V_I^- = V_I$ the result will be

$$f = f_0 + \delta f = \cfrac{\cfrac{2V_{CC}}{(R_3 + R_4)}\left(\cfrac{R_3}{R_1} - \cfrac{R_4}{R_2}\right) - V_I\left(\cfrac{1}{R_1} + \cfrac{1}{R_2}\right)}{4C\left[V_{CC}\left(\cfrac{R_4 - R_3}{R_3 + R_4}\right) + V_I + V_o\cfrac{R_a}{R_b}\right]} \tag{13.66}$$

Neglecting the term V_I in the denominator of (13.66) one finds that

$$\frac{\delta f}{f_0} \approx -\frac{V_I}{2V_{CC}}F(R_B) = -\frac{f_0}{S_{ri}}\frac{V_{IM}V_o}{V_{CC}}\frac{R_a}{R_b}F(R_B) \tag{13.67}$$

where $F(R_B) = [(R_1 + R_2)(R_3 + R_4)]/(R_2R_3 - R_1R_4)$.

It is easy to verify that, for the circuit of Figure 13.10, the frequency error is [33]

$$\frac{\delta f}{f_0} \approx \frac{f_0 V_{IM}}{S_{ri}}F(R_B) \tag{13.68}$$

When the oscillation frequency increases the slew rate of the output comparator in Figure 13.9 or Figure 13.10, t increases the durations of T_1 and T_2, as in the circuit of Figure 13.6. In addition, the switching time of P_1 and P_2 should be added to T_1 and T_2 as well.

13.3.4 One-Amplifier Multivibrator

The circuit of a one-amplifier multivibrator (Figure 13.12(a)) is well known and can be used in many sensor applications. At low oscillation frequency the output voltage v_o is a rectangular wave swinging from V_o^+ to V_o^- (Figure 13.12(b)) and the capacitor voltage v_c, which consists of the exponent pieces, changes from $V_o^+\rho$ to $V_o^-\rho$ (Figure

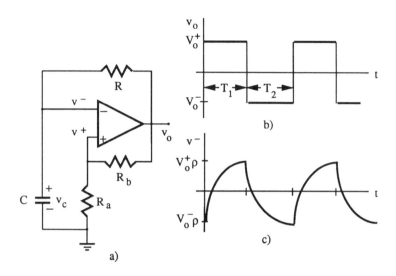

Figure 13.12 One-amplifier (comparator) multivibrator: (a) circuit; (b) output voltage; (c) OpAmp inverting input voltage.

13.12(c)). Here $\rho = R_a/(R_a + R_b)$. At high oscillation frequencies one has to take into consideration the operational amplifier finite slew rate. The output voltage can be approximated by a trapezoidal waveform (Figure 13.13(a)). The capacitor voltage amplitude increases (Figure 13.13(b)).

Assume that the voltage v_c has achieved the value of $V_o^+\rho$ and the voltage v_o starts to decrease. Then, for $0 < t < T_{1s} = (V_o^+ - V_o^-)/S_{rs}$, this voltage is

$$v_o(t) = V_o^+ - S_{rs}t \tag{13.69}$$

Solving the differential equation

$$v_c + \tau \frac{dv_c}{dt} = v_o \tag{13.70}$$

where $\tau = RC$. Using the initial condition $v_c(0) = V_o^+\rho$ one can find that

$$v_c = V_o^+[1 - \exp(-t/\tau)] + V_o^+\rho \exp(-t/\tau) - S_{rs}t + \tau S_{rs}[1 - \exp(-t/\tau)] \tag{13.71}$$

This result, for $0 < t < T_{1s}$, can be approximated as

$$v_c \approx V_o^+\rho + V_o^+(1 - \rho)\frac{t}{\tau} \tag{13.72}$$

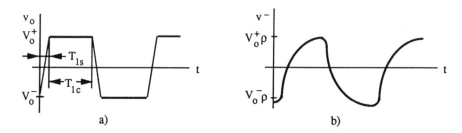

Figure 13.13 Slew-rate wave distortions in one comparator multivibrator: (a) slew-rate limited output voltage; (b) OpAmp inverting input voltage.

so that $v_c(T_{1s}) = V_o^+ \rho + [V_o^+ (1 - \rho)T_{1s}/\tau]$. And for $t > T_{1s}$ the output voltage is $v_o = V_o^-$.

Solving (13.70) again, with a new right side and new initial condition, one finds that

$$v_c = v_c(T_{1s}) + [V_0^- - v_c(T_{1s})]\{1 - \exp[-(t - T_{1s})/\tau]\} \tag{13.73}$$

From (13.73), using the final condition $v_c(T_{1s} + T_{1c}) = V_o^-$, one finds

$$T_{1c} = \tau \ln\left[\frac{V_o^- - v_c(T_{1s})}{V_o^-(1 - \rho)}\right] \tag{13.74}$$

Finally, for the symmetric conditions $V_o^+ = -V_o^- = V_o$, one finds that

$$T = 2(T_{1s} + T_{1c}) = 2T_{1s} + 2\tau \ln\left(\frac{1 + \rho}{1 - \rho} + \frac{T_{1s}}{\tau}\right) \approx \frac{4T_{1s}}{1 + \rho} + 2\tau \ln\left(\frac{1 + \rho}{1 - \rho}\right) \tag{13.75}$$

If the first term in (13.75) is omitted (as is usual for low oscillation frequencies), then

$$f_0 = \frac{1}{2\tau \ln\left(\dfrac{1 + \rho}{1 - \rho}\right)} \tag{13.76}$$

and, hence, the relative error introduced by the operational amplifier slew-rate limitation is

$$\frac{\delta f}{f_0} \approx -\frac{4f_0 T_{1s}}{(1 + \rho)} = -\frac{8V_o f_0}{S_{rs}(1 + \rho)} \tag{13.77}$$

13.3.5 Voltage-to-Frequency Converters

In the multivibrators considered so far it was assumed that the ratio $D = T_1/T_2$ is about 0.5. The factors that could cause D to differ from this value (like asymmetric output of the operational amplifiers) are considered parasitic. If D is modified on purpose to be strongly different from 0.5, the circuit can be adapted to be a voltage-to-frequency converter. The output voltage of these multivibrators is a train of narrow pulses having the frequency controllable by the voltage. The circuit is usually designed so that the width of these pulses is as narrow as possible; in this case a better linearity of the control characteristic is provided.

13.3.5.1 VFC with an Active Integrator

Such modifications can be easily done, for example, in the circuit of Figure 13.8. The modified circuit [34] is redrawn here in Figure 13.14. During the long part T_1 of the oscillation period, the switch P is open and the timing capacitor C is charged from the input voltage V_i. During the short part T_2 the switch P is closed and the timing capacitor is recharged from the voltage V_d^-. The integrator output voltage v_1 changes from $V_1^+ = V_o^+\rho$ to $V_1^- = V_o^-\rho$, where $\rho = R_a/(R_a + R_b)$. The duration T_1 can be found from the charge balance equation:

$$T_1 \frac{V_i}{R} = C(V_1^+ - V_1^-) \tag{13.78}$$

The duration T_2 is calculated from the equation

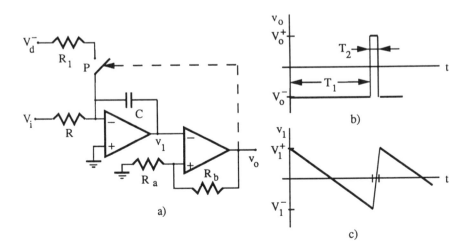

a)

b)

c)

Figure 13.14 VFC with a timing capacitor recharge: (a) circuit; (b) integrator output; (c) VFC output.

$$T_2\left(\frac{V_i}{R} + \frac{V_d^-}{R_1}\right) = C(V_1^- - V_1^+) \tag{13.79}$$

Then the oscillation frequency f can be found as

$$f = \frac{1}{T_1 + T_2} = \frac{V_i}{RC(V_1^+ - V_i^-)}\left[1 + \frac{V_iR_1}{V_d^-R}\right] \tag{13.80}$$

If the second term in the brackets of (13.80) is small and can be neglected, then the oscillation frequency

$$f_0 = \frac{V_i}{RC(V_1^+ - V_1^-)} \tag{13.81}$$

is proportional to the input voltage. When the input voltage increases, this term increases as well and determines the relative frequency error

$$\frac{\delta f}{f_0} = \frac{V_iR_1}{V_d^-R} \tag{13.82}$$

The additional errors caused by the slew-rate limitation and finite switching time are added (at the upper end of the frequency range) in the same fashion as in the circuit of Figure 13.8.

13.3.5.2 VFC with a Passive RC-Circuit and Full Discharge of the Timing Capacitor

The minimal value of the error term (13.82) is limited by the integrator maximal current. In addition, the auxilliary voltage source V_d^- should provide a high current for the capacitor recharge. These deficiencies are eliminated in the circuit shown in the Figure 13.15(a) [35–37]. Here the active integrator is replaced by a passive RC-circuit. A buffer is introduced between the integrator and the dynamic Schmitt trigger.

The circuit operation can be understood tracking the voltage v^+ (Figure 13.15(b)). During the time interval T_1 the switch P is open and the capacitor C is charging from the input signal V_i. The voltage v_1 follows the voltage at the capacitor C, and the values of R_a, R_b, and C_d are such that the voltage v^+ at the Schmitt trigger positive terminal follows v_1 as well. The capacitor C_d is small, and the initial transient for v^+ is short and even can be completely eliminated (the required parameter relationship is given later). When v^+ attains the value of V_{Ref} the output voltage v_o jumps to V_o^+ and the switch P will be closed. The voltage v_1 drops to 0, yet the voltage

v^+ becomes much higher than V_{Ref} because the voltage at C_d and V_o^+ are operating in series. Then the capacitor C_d starts to recharge and v^+ quickly diminishes. When v^+ drops to the value of V_{Ref}, the output voltage will drop to V_o^-, the switch will be open again, and the cycle will be repeated.

The durations T_1 and T_2 are calculated considering the voltages in the Schmitt trigger feedback circuit. When the switch P is open (Figure 13.15(d)) the buffer output voltage is

$$v_1 = V_i[1 - \exp(-t/RC)] \approx (V_i t)/(RC) \tag{13.83}$$

If C_d is small then

$$i_d = C_d \frac{dv_1}{dt} \approx \frac{V_i C_d}{RC} \tag{13.84}$$

At the end of the first quasi-stable state $v_1(T_1) = V_{\text{Ref}} + i_d R_a$; that is,

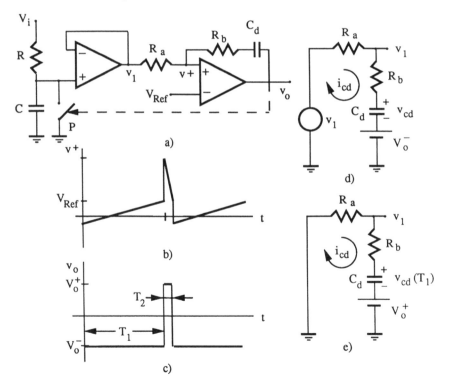

Figure 13.15 VFC with full discharge of the timing capacitor: (a) circuit; (b) comparator input; (c) VFC output; (d) C_d charge; (e) C_d discharge.

$$V_i[1 - \exp(-t/RC)] = V_{\text{Ref}} + V_i(R_aC_d/RC) \tag{13.85}$$

and from (13.85) one finds that

$$T_1 = RC \ln\left\{\left[1 - \left(\frac{V_{\text{Ref}}}{V_i} + \frac{R_aC_d}{RC}\right)\right]^{-1}\right\} \approx RC\frac{V_{\text{Ref}}}{V_i} + R_aC_d \tag{13.86}$$

At the end of this part of the oscillation period, the voltage v_{cd} at the feedback capacitor becomes $v_{cd}(T_1) = V_{\text{Ref}} - V_o^- - (V_iR_bC_d/RC)$.

When the circuit switches in the second quasi-stable state, the voltage v^+ can be calculated from the circuit of Figure 13.15(e). First of all, it is seen that the condition

$$[V_o^+ + v_{cd}(T_1)]\rho > V_{\text{Ref}} \tag{13.87}$$

should be satisfied to obtain this quasi-stable state. Here $\rho = R_a/(R_a + R_b)$ as usual. Then one can write that

$$v^+ = [V_o^+ + v_{cd}(T_1)]\rho \exp\{-(t - T_1)/[C_d(R_a + R_b)]\} \tag{13.88}$$

for $T_1 \le t \le (T_1 + T_2)$, and the condition $v^+ (T_1 + T_2) = V_{\text{Ref}}$ gives

$$T_2 \approx C_d(R_a + R_b)\left\{1 - \frac{V_{\text{Ref}}/\rho}{[V_o^+ - V_o^- + V_{\text{Ref}} - (V_iR_bC_d)/(RC)]}\right\} \tag{13.89}$$

The oscillation frequency that is

$$f = \frac{1}{T_1 + T_2} = \frac{V_i}{RCV_{\text{Ref}} + V_i(R_aC_d + T_2)} \tag{13.90}$$

includes, for this circuit, the error term increasing with V_i. Indeed, if one denotes

$$f_0 = \frac{V_i}{RCV_{\text{Ref}}} \tag{13.91}$$

then, as follows from (13.90), the relative frequency error is

$$\frac{\delta f}{f_0} \approx -f_0(R_aC_d + T_2) \tag{13.92}$$

Finally, at the end of the second quasi-stable state $v_{cd}(T_1 + T_2) = v_{cd}(0) = V_{Ref}[1 + (R_b)/(R_a)] - V_o^+$ and, if the condition $V_{Ref} = V_o^+ \rho$ is satisfied, the voltage v^+ follows the voltage at the timing capacitor without a transient in the first quasi-stable state.

13.3.5.3 VFC with a Passive RC-Circuit and Partial Discharge of the Timing Capacitor

The operation of the circuit [36] shown in Figure 13.16(a) is as follows. Here the voltage V_{Ref} is very small (it is necessary only to prevent the circuit negative latch-up) and the timing capacitor is recharged to a negative voltage when the switch P is closed for a short time by the Schmitt trigger feedback circuit. When this happens the voltage v^- at the negative terminal drops to

$$v^-(0) = V_{Ref} \frac{C}{(C + C_r)} - (V_o^+ - V_o^-)\frac{C_r}{(C + C_r)} \tag{13.93}$$

At this time the voltage at the feedback capacitor is $v_{cd}(0) = V_o^+ - V_{Ref}$. After the initial drop the timing capacitor C charges up and v^- slowly increases as

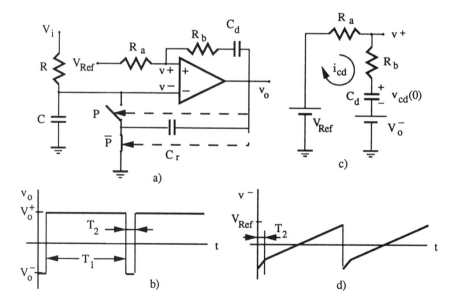

Figure 13.16 VFC with partial discharge of the timing capacitor: (a) circuit; (b) VFC output; (c) C_d charge; (d) comparator input.

$$v^- = V_i + [v^-(0) - V_i] \exp\{-t/[R(C + C_r)]\} \tag{13.94}$$

Simultaneously, the voltage at the positive terminal (which is calculated considering the feedback circuit of Figure 13.16(d)) decreases as

$$v^+ = [V_{Ref} - (V_o^+ - V_o^-)\rho] \exp\{-t/[C_d(R_a + R_b)]\} \tag{13.95}$$

When $v^+(T_2) = v^-(T_2)$, part T_2 of the oscillation period is finished. This gives

$$T_2 \approx (R_a + R_b)C_d\left[1 - \frac{C_r}{\rho C}\right] \tag{13.96}$$

Assuming that T_2 is small and using (13.94) with C instead of $C + C_r$ (C_r is not connected now) and the conditions $v^-(T_2) \approx v^-(0)$ and $v^-(T_1 + T_2) = V_{Ref}$, one obtains that

$$T_1 \approx \frac{(V_o^+ - V_o^-)RCC_r}{(V_o^+ - V_o^-)C_r + (V_i - V_{Ref})C} \tag{13.97}$$

Then the oscillation frequency

$$f = \frac{1}{T_1 + T_2} \approx \frac{(V_o^+ - V_o^-)C_r + (V_i - V_{Ref})C}{(V_o^+ - V_o^-)RCC_r + T_2(V_i - V_{Ref})C} \tag{13.98}$$

if the condition $RC \gg T_2$ is satisfied. If

$$f_0 = \frac{(V_i - V_{Ref})}{RC_r(V_o^+ - V_o^-)} + \frac{1}{RC} \tag{13.99}$$

then the relative frequency error will be

$$\frac{\delta f}{f_0} \approx -f_0 T_2 \tag{13.100}$$

13.3.6 Current-to-Frequency Converters

The previously described VFCs could be transformed into current-to-frequency converters (CFCs) if the voltage source V_i and resistor R were replaced by a current source charging the timing capacitor C. The capacitor discharge was realized via a circuit of noncritical design, providing fast operation. When the oscillation frequency increases, the discharge time becomes comparable with the charge time and results in the nonlinearity of the control characteristic.

The CFCs considered in this section are based on a different approach. It is assumed here that the timing capacitor is periodically charged and discharged from two tightly matched current sources. The rest of the multivibrator circuit provides redirection of these currents. A transistor current source has wide current and frequency ranges [22,25,39], and redirection is realized by switches that can be very fast. The approach requires more complicated circuits but results in the circuits operating at high frequencies and having very linear control characteristics. The ratio $D = T_1/T_2$ of these multivibrators is normally about 1, making them highly suitable for phase-locked loop applications [22,31,32].

These multivibrators can be divided into two types: CFCs with a grounded timing capacitor and CFCs with a floating timing capacitor. Both types are characterized by a very broad range of control (four decades of the control current) and provide approximately the same results in terms of nonlinearity and stability of control characteristic.

13.3.6.1 Comparator Current-to-Frequency Converters

The floating timing capacitor CFCs that can be realized using comparators and complementary switches are shown in Figure 13.17. The condition of switches shown corresponds to the recharge of the timing capacitor from the polarity shown in parentheses to the polarity shown without parentheses.

At this time the output voltage of the comparator A_1 in Figure 13.17(a) is low (its negative input terminal has a potential higher than that of the positive terminal). The output voltage of A_2 is high. The input terminals of this comparator have constant voltage difference of V_c between them during all this recharging. The left voltage source absorbs both currents I and the capacitor C is recharging via the right current source. When the voltage v_c becomes equal to V_c (with the polarity shown without parentheses) the output voltage of A_1 becomes high and the condition of switches P_1, P_1 changes. Even if P_2 is not yet open the output voltage of A_1 is preserved as high (especially if the resistor is chosen so that $2IR < V_c$) because the timing capacitor will continue to recharge until the switch P_1 is closed. When P_1 is, finally, closed, the voltage v_c becomes applied to the input terminals of the comparator A_2, with the polarity providing an abrupt change of the output voltage of this comparator. As a result the switches P_2, P_2 change their condition. The switching is finished when the switch P_2 is closed (thus, the correct sequence of switch condition changing is P_1, P_1, P_2, P_2). Now one has to consider the second, completely symmetric, part of the oscillation cycle.

In ideal comparators and ideal switches, the capacitor is recharged from $+V_c$ to $-V_c$ by current I. Hence, the duration of a semiperiod will be $T/2 = (2V_c)/I$, and the oscillation frequency is

$$f_0 = \frac{I}{4CV_c} \qquad (13.101)$$

and can be linearly controlled by the matched current sources.

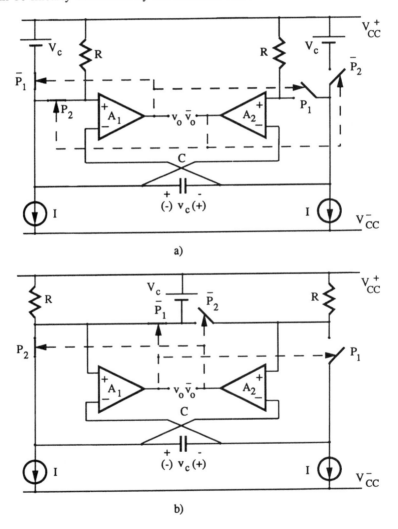

a)

b)

Figure 13.17 Comparator CFCs with: (a) two reference voltages; (b) one reference voltage.

In normal operation of this CFC both voltage sources V_c are never used at the same time. This allows one to modify this circuit into the circuit shown in Figure 13.17(b). The operation of this circuit follows directly from the previous description and the oscillation frequency is given by the (13.101) as well.

The main source of the frequency error in both circuits is the slew-rate limitation of A_1 when P_1 is open and P_1 is not closed yet, and an additional charge is deposited on the timing capacitor and the slew-rate limitation of A_2 when P_2 is still closed and P_2 is still open, and the timing capacitor is discharged via both resistors but not by a current source. If the combined delay introduced by the comparator and two connected complementary switches is denoted T_s the relative frequency error will be

$$\frac{\delta f}{f_0} \approx -2f_0 T_s \qquad (13.102)$$

The additional increase of the v_c voltage swing can be important for some low-voltage circuits.

13.3.6.2 Emitter- and Source-Coupled Multivibrators

The emitter-coupled multivibrator [22,25,40] shown in Figure 13.18(a) is, probably, the first circuit used as a current-to-frequency converter with a wide range (at least, four decades) of linear frequency control. It is similar to the circuit of Figure 13.17. Indeed, the diodes Q_5 and Q_6 play the same role as the voltage sources in that circuit. In addition, they are self-switching, which allows one to have only two additional switches, so that Q_1 and Q_2 are sufficient.

Referring to the circuit of Figure 13.18, the frequency of oscillations can be determined [22] as

$$f_0 \approx \frac{I}{4CV_{be(ON)}} \qquad (13.103)$$

Here this result is denoted as an approximate one. Indeed, during the switching from one quasi-stable state to another the circuit has a positive feedback loop and the loop gain depends on the current I. The voltage drop at Q_5 and Q_6 changes as well. The approximate values of $V_c(0) = -V_c(T_1) = V_{be(ON)} = 0.7V$, which, used in the (13.103), results in a large error for f_0. A more exact result without the overcomplication of the calculations requires the following assumptions [40].

- The exact values $V_c(0) = -V_c(T_1)$ of v_c at the instant of transition from one quasi-stable state to another are most important in the calculation of the oscillation frequency.

- One can consider that recharging of the timing capacitor C from $V_c(0)$ to $V_c(T_1)$ and vice versa is obtained with a constant current I.
- One can assume that the transition from one quasi-stable state to another is an instantaneous process.

To obtain the exact values of $V_c(0)$ and $V_c(T_1)$ the multivibrator of Figure 13.18, during a short period of time near the transition from one quasi-stable state to another, is considered a linear circuit. Then, the loop transfer function for this linear small-signal equivalent circuit is calculated. Equating this transfer function to unity one finds the exact values of the collector current I_{c1} and I_{c2} at the instant of transition. Finally, using these currents one calculates $V_c(0)$ and $V_c(T_1)$.

The simplified small signal equivalent linear circuit describing the multivibrator during the transition is shown in Figure 13.18(b). The duration of a quasi-stable state is much longer than that of a transition, so the voltage on the coupling capacitor practically does not change during this transition, and the capacitor is represented

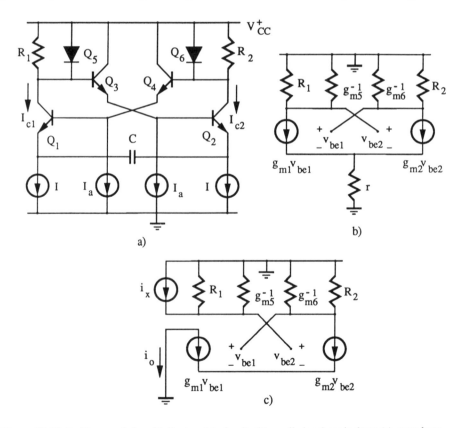

Figure 13.18 Emitter-coupled multivibrator: (a) circuit; (b) small-signal equivalent; (c) open loop.

as a short circuit in the model. Transistors Q_3 and Q_4 of the Figure 13.18(a) are linear buffers and they are not included in this equivalent circuit. The circuit with $R_1 = R_2 = R$ will be considered but the notations R_1 and R_2 are preserved for convenience. The resistor r is introduced to avoid infinite nodal impedance (it can be considered as representing the output resistance of two current sources in the emitter circuits of Q_1 and Q_2 connected in parallel by the shorting coupling capacitor and the assumption $r \to \infty$ in the final result does not introduce any noticeable error).

The loop transfer function $A_L = i_o/i_x$ can be easily obtained from the circuit shown in Figure 13.18(c) (r is omitted). It is

$$A_L = \frac{g_{m1}g_{m2}R_1}{(1 + g_{m5}R_1)[g_{m1} + g_{m2} - g_{m1}g_{m2}R_2(1 + g_{m6}R_2)^{-1}]} \tag{13.104}$$

At the instant of transition $A_L = 1$, and this gives the condition of transition

$$g_{m1} + g_{m2} = g_{m1}g_{m2}[R_1(1 + R_1g_{m5})^{-1} + R_2(1 + R_2g_{m6})^{-1}] \tag{13.105}$$

In the considered transition, Q_1 is on and changes its state to off. Correspondingly, Q_2 is off and turns on (the currents in the opposite transition are obtained by interchanging the I_{c1} and I_{c2} values). The transition starts when the collector current of Q_1 decreases from the value of $2I$ to that of $2I - \Delta I$ and the collector current of Q_2 increases from 0 to ΔI. At this instant the transconductances will be $g_{m1} = I_{c1}/V_T = (2I - \Delta I)/V_T$ and $g_{m2} = I_{c2}/V_T = \Delta I/V_T$ (here $V_T = 26$ mV at the temperature 300K). These values should be substituted in (13.105) to obtain the equation for calculation of ΔI.

The diodes Q_5 and Q_6 are clamping the base of Q_3 and Q_4. Then $(2I - \Delta I) > V_{be(ON)}$ and the currents in Q_5 and R_1 are comparable. In this case

$$g_{m5} = \frac{2I - \Delta I - [V_{be(ON)}/R_1]}{V_T} \tag{13.106}$$

When Q_1 is on, the normal circuit operation requires that the condition $V_{R2} < V_{be(ON)}$ be satisfied up to and including the instant of transition (that is, the diode Q_6 is off and one has to set $g_{m6} = 0$ in (13.105)). If all these transconductance values are substituted in (13.107), one obtains

$$\frac{2IV_T}{R} = (2I - \Delta I)\Delta I \frac{[2V_T + R(2I - \Delta I) - V_{be(ON)}]}{[V_T + R(2I - \Delta I) - V_{be(ON)}]} \tag{13.107}$$

Solving (13.107) numerically one can obtain the value of ΔI. If the term $(\Delta I)^3$ is neglected, a quadratic equation gives the approximate value of

$$\Delta I \approx \frac{V_T}{R}\left[1 - \frac{2V_T}{2IR - V_{be(ON)}}\right] \tag{13.108}$$

Then the coupling capacitor voltage V_c can be found from

$$V_{R1} + V_{be3} + V_{be2} - V_c - V_{be1} - V_{be4} - V_{R2} = 0 \tag{13.109}$$

Transistors Q_3 and Q_4 have equal emitter currents, and $V_{be3} = V_{be4}$. Also, in the considered transition point $V_{R2} = I_{c2}R_2 = \Delta IR_2$ and $V_{R1} = I_{c1}R_1 = (2I - \Delta I)R_1$. Then one finds from (13.109) that

$$V_c = (2I - \Delta I)R_1 - \Delta IR_2 - V_T \ln\left(\frac{2I}{\Delta I} - 1\right) \tag{13.110}$$

In the opposite transition (Q_2 is *on* and turns *off*, Q_1 is *off* and turns *on*) the capacitor C will have the voltage $-V_c$ on it (with $R_1 = R_2 = R$) and so the total voltage swing will be $\Delta V_c = 2V_c$. The recharge time of C, of course, increases when the current is deflecting into Q_2 during the transition. But the calculations show that the maximum value of deflected current ΔI is so small that this increase can be neglected and the charge balance equation can be written as $\Delta V_c C = (IT)/2$, where T is the oscillation period. Then the oscillation frequency is

$$f = \frac{I}{4C\left\{V_{be(ON)} - V_T\left[\ln\left(\frac{2IR}{V_T} + \frac{4IR}{2IR - V_{be(ON)}}\right) + 1 - \frac{2V_T}{2IR - V_{be(ON)}}\right]\right\}} \tag{13.111}$$

This is the required equation of the frequency control characteristic. It shows that

$$\frac{\delta f}{f_0} \approx \frac{V_T}{V_{be(ON)}}\left[\ln\left(\frac{2IR}{V_T}\right) + 1\right] \tag{13.112}$$

which varies very slowly when I increases. Equation (13.112) allows one to predict the nonlinearity for a given range of the control current I. It also can be used to evaluate the thermostability of the characteristic and the required compensating means in the bias circuit.

The diode clamping is essential for this circuit to be a current-to-frequency converter. If the transistors Q_5 and Q_6 are absent, the calculation procedure is the same except one has to use $g_{m5} = g_{m6} = 0$ from the onset of calculations [40]. The result will be

$$f = \cfrac{1}{8RC\left\{1 - \cfrac{V_T}{2IR}\left[\ln\left(\cfrac{4IR}{V_T}\right) + 1\right]\right\}} \qquad (13.113)$$

That is, the circuit becomes a fixed-frequency multivibrator (the variation of the second term in the denominator is small).

The circuit of Figure 13.17(b) can be also used as a prototype to develop an emitter-coupled multivibrator [41]. The calculation of the oscillation frequency follows the same steps. One of the most precise circuits [42] based on this model is shown in Figure 13.19. In this circuit the voltage source V_c is removed from the path of currents (this path is closed via Q_5 and Q_6). The design key point in this circuit is four matched current sources operating in the emitter and collector circuits of Q_1 and Q_2 and six matched transistors (Q_5, Q_4, and the transistors realizing D_1 to D_4). If the last requirement is satisfied, the voltage appearing between the collectors of Q_1 and Q_2 does not depend on the current I, the temperature, the exact value of $V_{be(ON)}$, and the power supply voltage. The voltage at the timing capacitor changes from $-V_c$ to $+V_c$ during one half-cycle and the oscillation frequency is given by (13.101) again. The oscillation frequency stability becomes dependent on the V_c realization and its matching with charging current sources.

The source-coupled multivibrator [44] shown in Figure 13.20(a) represents the CMOS version of the previously considered multivibrator. The frequency calculation

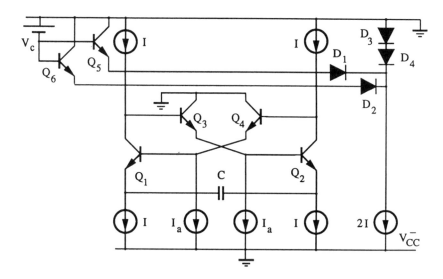

Figure 13.19 Modified emitter-coupled multivibrator.

for this circuit is based on the same assumptions. Yet a profound difference exists between the CMOS multivibrator and its bipolar prototype. This difference is due mainly to the body effect of n-channel transistors. As a result the frequency versus the control current I of the multivibrator becomes nonlinear.

During the short period just before the discontinuous change of the drain currents, the multivibrator can be described by the simplified small-signal equivalent circuit shown in Figure 13.20(b). Investigation of the loop transfer function near a transition point brings up the condition of transition

$$g_{m1} + g_{m2} = g_{m1} g_{m2} R_{oe} \tag{13.114}$$

where $R_{oe} = [R(2 + Rg_{m3})]/(1 + Rg_{m3})$. The condition (13.114) is used to find the drain currents I_{d1} and I_{d2} at this instant. If the transition in which M_1 is on and turns off is considered, then I_{d2} is small and the current through the parallel connection of R and M_3 changes very little up to the instant of discontinuous change ($I_{d1} \approx 2I$). The assumption that R_{oe} is constant can be used to simplify the calculations. To calculate R_{oe} itself, one has to find

$$g_{m3} = 2\beta_p(V_{DD} - V_{g2} - |V_{tp0}|) \tag{13.115}$$

where

$$V_{g2} \approx V_{DD} - |V_{tp0}| - \frac{\sqrt{1 + 4\beta_p R(2IR - |V_{tp0}|)} - 1}{2\beta_p R} \tag{13.116}$$

Here $|V_{tp0}|$ is the threshold voltage for p-channel transistors with the source and substrate connected together. Equation (13.115) assumes that the transistors are described by the square law characteristic, where β_p (or β_n for n-channel transistors) is the proportionality coefficient. Further discussion of β_n and β_p, which should be used for calculations, will be given later. It is also assumed that the transistors are matched in pairs (thus, $\beta_{p3} = \beta_{p4} = \beta_p$), and the voltage swing at the capacitor C is small enough to neglect the difference in the source voltages of M_1 and M_2 (thus, $\beta_{n1} \approx \beta_{n2} \approx \beta_n$). The last assumption allows us to write that $g_{m1} = 2\sqrt{(\beta_n I_{d1})}$ and $g_{m2} = 2\sqrt{(\beta_n I_{d2})}$. If these values are substituted in (13.114) one finds that

$$I_{d1} + I_{d2} = 2R_{oe}\sqrt{\beta_n I_{d1} I_{d2}} \tag{13.117}$$

This equation should be solved together with the condition $I_{d1} + I_{d2} = 2I$, which follows from the circuit topology.

Once the currents I_{d1} and I_{d2} are found, one can calculate the gate voltages

$$V_{g1} = V_{d2} = V_{DD} - I_{d2}R \tag{13.118}$$

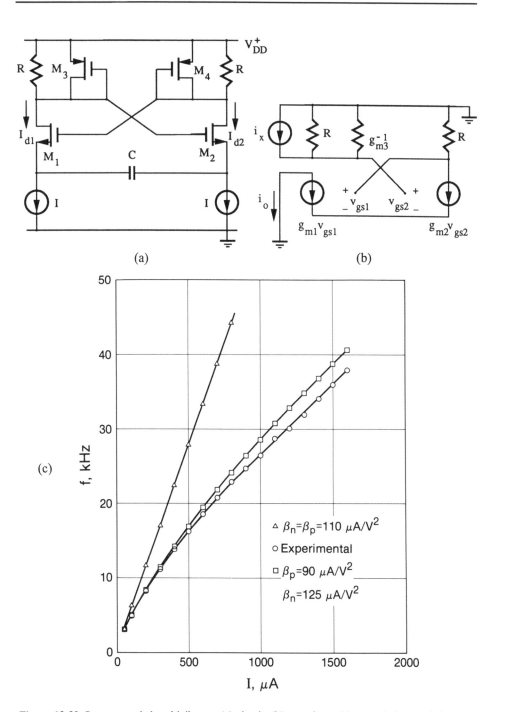

Figure 13.20 Source-coupled multivibrator: (a) circuit; (b) open loop; (c) control characteristic.

$$V_{g2} = V_{d1} = V_{DD} - |V_{tp0}| - \frac{\sqrt{1 + 4\beta_p R(RI_{d1} - |V_{tp0}|)} - 1}{2\beta_p R} \qquad (13.119)$$

Then the source voltages are obtained from the equation

$$V_{si} = \frac{1}{2}(2V_{ni} + \gamma^2 - \gamma\sqrt{\gamma^2 + 4V_{ni} + 8|\phi_f|}) \qquad (13.120)$$

where $V_{ni} = V_{gi} + 2|\phi_f| - V_{tn0} - \sqrt{(I_{di}/\beta_n)}$ and $i = 1, 2$.

In (13.119) and (13.120), γ is the body-effect coefficient, $|\phi_f|$ is the Fermi voltage, and V_{tn0} is the threshold voltage for n-channel transistors with the source and substrate connected together. The results (13.119) and (13.120) follow directly from the basic relationships of the MOS transistor [45,46] rewritten as

$$I_{di} = \beta_n(V_{gi} - V_{si} - V_{tni})^2$$

(hence, here the coefficients β_n and β_p are twice larger than in the frequently used notation [46]) and

$$V_{tni} = V_{tn0} + \gamma\,[\sqrt{(2|\phi_f| - V_{si})} - \sqrt{(2|\phi_f|)}]$$

for each transistor.

When V_{s1} and V_{s2} are found and the pinch-off operation of the devices is confirmed ($V_{dsi} \geq V_{gsi} - V_{tni}$, $i = 1, 2$), the voltage amplitude at the capacitor is calculated as

$$V_c = V_{s1} - V_{s2} \qquad (13.121)$$

Then the frequency of oscillations is $f = I/(4CV_c)$ as usual, and the linear control characteristic will be obtained if V_c is constant.

There is no close formula, and the frequency control characteristic can be calculated point by point only. Yet it can be approximately analyzed [44]. This analysis shows that

$$V_c \approx |V_{tp0}| + \sqrt{\frac{I_{d2}}{\beta_n}} + \sqrt{2I}\left(\frac{1}{\sqrt{\beta_p}} - \frac{1}{\sqrt{\beta_n}}\right) \qquad (13.122)$$

The second term is usually small and can be neglected. It is seen from (13.122) that V_c can be nearly constant if one can match the transistor parameters so that β_p and β_n are equal. The dependence (even though weak) of β_p and β_n on the level of the operating current I and source-substrate voltage makes the matching impossible over a wide range of operating current.

Figure 13.20(c) shows the results of the frequency calculations when the constant values of $\beta_n = \beta_p = 110$ mA/V^2 are used (MO415 CMOS transistor array [47], the measurements at the drain current of 50 mA and $V_{sb} = 0$ for each device). The experimental results show strong divergence with the calculated ones, as the operating current increases. As was mentioned already, the main cause of this is the voltage and current dependence of the MOS transistor gain. In a considered circuit this effect is most pronounced for n-channel transistors. If one relies on the average value of β_n, which is in this case about 125 mA/V^2, the error between the experimental and calculated results is not more than 10% (Figure 13.20(c)). This chosen average value provides the coincidence of all dc levels within the quasi-stable states and the slope of the V_c versus $\sqrt{I}$ characteristics (the calculated slope is 22 V/mA$^{0.5}$, the experimental one is 20 V/mA$^{0.5}$ for the considered devices) within the same error band. Yet, the control characteristic is, of course, nonlinear.

13.3.6.3 Schmitt-Trigger-Type Current-to-Frequency Converters

The previously considered CFCs used a floating timing capacitor and two matched current sources. In some integrated circuits it is highly desirable to have one plate of the capacitor grounded. This requirement resulted in the CFCs using the Schmitt trigger and different switching circuits providing redirection of the charging current. The circuits can be divided into three groups.

In the first group of circuits (Figure 13.21(a)) the timing capacitor is charged by the current source I from the top, and when the voltage v_c attains the value of V_{cH} the trigger turns off the current source $\bar{I}$, opens the switch P, closes the switch $\bar{P}$ and turns on the complementary current source I. Now the timing capacitor is discharged, and when v_c attains the value of V_{cL}, the trigger turns off the current source I and the switching operations are repeated in the inverse order.

In the second group (Figure 13.21(b)) the current source I permanently charges the timing capacitor from the top and the trigger redirects the current source $2I$. During the timing capacitor discharge both current sources are connected to the timing capacitor, and it is discharged by the difference of two currents.

Finally, the third group (Figure 13.21(c)) has only one current source, which is redirected by the Schmitt trigger and two complementary switches. When the switch $\bar{P}$ is on, the current is directed into current mirror and charges the timing capacitor from the top; when the switch is on, the current source is directly connected to the capacitor and discharges it.

The oscillation frequency is

$$f_0 = \frac{I}{2C(V_{cH} - V_{cL})} \tag{13.123}$$

for all three groups of circuits.

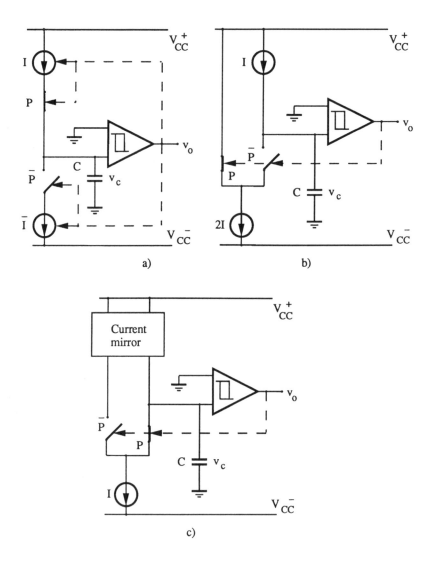

Figure 13.21 CFCs with a Schmitt trigger: (a) complementary sources; (b) I and $2I$ sources; (c) I and a current mirror.

In the circuits of the first group, the switching and trigger subcircuits are operating in series. These circuits are suitable for applications in sensors measuring slow variables (temperature, static fields) [48,49]. The trigger can be realized using a comparator and two-resistor positive feedback. When the currents I and $\bar{I}$ are changing simultaneously one obtains a CFC. The relative frequency error for these converters can be estimated using (13.102), where T_s is the combined delay introduced by the comparator and the switch. If I and $\bar{I}$ are changing separately but their sum is preserved, one obtains a duty-cycle modulation.

In the second and third groups the switching and trigger subcircuits are driven in parallel. This results in faster circuits, capable of operating at higher frequencies. In addition, these CFCs are structurally divided into two independent subcircuits; namely, the threshold circuit and the charge-discharge circuit. This makes the design more flexible and provides an abundance of circuits adapted for different design conditions. The charge-discharge circuit includes two complementary switches that only redirect the current ($2I$ or I). Nearly all designs use a differential pair, which is able to provide fast switching of its tail current. The Schmitt trigger can be realized, using an operational amplifier in this case as well. The fast converters rely on more simple structures of the Schmitt triggers, also based on switching a differential pair. The whole circuit becomes a nonsaturated multivibrator. The circuits corresponding to the second group allow the widest range of the control current because they can be realized (in bipolar technology) using n-p-n transistors only [50]. The circuits of the third group are slower (the bipolar p-n-p integrated transistors have inferior frequency properties), but their bipolar design can be topologically transferred into CMOS technology (the additional limitations are minor).

The frequency calculation for the circuits of the second and third groups is very similar to that for emitter-coupled multivibrators. In accordance with the first assumption, the upper V_{cH} and lower V_{cL} thresholds of the Schmitt trigger have to be calculated precisely via the currents in the trigger circuit at the instants of transitions. Then one can neglect small variations in the charge and discharge current before the transition from one quasi-stable state to another and consider the transition as an instantaneous process.

As a representative example the circuit with a current mirror feedback Schmitt trigger (Figure 13.22(a)) is considered. The timing capacitor C is charged and discharged by the current source I (the control current), which is redirected by the trigger output voltage v_o via the switching differential pair Q_3, Q_4, and the current mirror Q_5, Q_6. The trigger circuit includes the differential amplifier Q_1, Q_2 with positive feedback via the current mirror Q_7, Q_8 and the divider R_1, R_2.

As in all of the previously considered circuits, before a jump from one quasi-stable state to another this CFC operates as a linear circuit (when both Q_1 and Q_2 are conducting) with positive feedback. The equivalent small-signal circuit for this part of operation is shown in Figure 13.22(b). Investigating the loop transfer function of the circuit at a transition point, one obtains the condition of jump

$$g_{m2} = r_{\pi 2}^{-1} + \frac{1}{R_{oe}} \left(1 + \frac{r_{\pi 2}^{-1} + g_{m2}}{r_{\pi 1}^{-1} + g_{m1}} \right) \tag{13.124}$$

where $R_{oe} = R_1 R_2/(R_1 + R_2)$. Taking into consideration [25] that $r_\pi \approx \beta/g_m$ and, hence, $r_\pi^{-1} \ll g_m$ the condition (13.128) can be simplified to

$$g_{m1}^{-1} + g_{m2}^{-1} \approx R_{oe} \qquad (13.125)$$

Substituting $g_{m1} = I_{c1}/V_T$ and $g_{m2} = I_{c2}/V_T$ into (13.129), one finds that

$$I_{c1}^{-1} + I_{c2}^{-1} = R_{oe}V_T^{-1} \qquad (13.126)$$

The condition $I_{c1} + I_{c2} = I_a$ is also valid. Finding I_{c1} and I_{c2} from (13.126) and this last condition and using [25] $V_{bei} = V_T \ln (I_{ci}/I_s)$ for $i = 1, 2$, one finds further that the voltage between bases of Q_1 and Q_2 at the instant of switching is

$$\Delta V = V_{be1} - V_{be2} = \pm V_T \ln m \qquad (13.127)$$

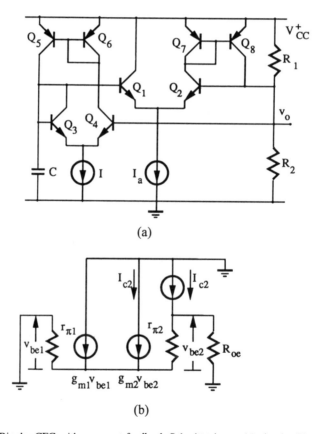

(a)

(b)

Figure 13.22 Bipolar CFC with a current-feedback Schmitt trigger: (a) circuit; (b) small-signal equivalent; (c) control characteristics.

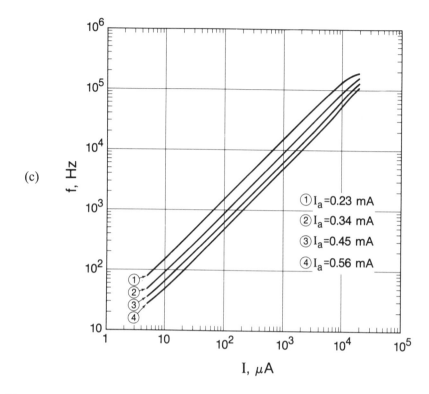

Figure 13.22 continued.

where $m = \{1 + \sqrt{[1 - (4V_T)/(I_aR_{oe})]}\}/\{1 - \sqrt{[1 - (4V_T)/(I_aR_{oe})]}\}$. Now one can calculate

$$V_{cL} = \frac{V_{CC}R_2}{R_1 + R_2} + \Delta V + I_{c2}R_{oe} \tag{13.128}$$

Substituting the value of ΔV with upper sign one finds that

$$V_{cL} \approx \frac{V_{CC}R_2}{R_1 + R_2} + V_T(1 + \ln m) \tag{13.129}$$

where one takes into consideration that m is usually large. Then one finds

$$V_{cH} = \frac{V_{CC}R_2}{R_1 + R_2} + \Delta V + I_{c2}R_{oe} - R_{oe}\frac{I_a}{\beta_{F2}} \tag{13.130}$$

where ΔV has to be taken from (13.127) with the lower sign. Here the term representing the base current of transistor Q_2 is included. Indeed, at high values of I_a, this base current reduces the value of the voltage v_o and, correspondingly, V_{cH}. The calculations gives

$$V_{cH} \approx \frac{V_{CC}R_2}{R_1 + R_2} + I_a R_{oe} - V_T(1 + \ln m) - R_{oe}\frac{I}{\beta_4} \qquad (13.131)$$

Now the peak-to-peak voltage at the timing capacitor can be calculated as

$$V_c = V_{cH} - V_{cL} = I_a R_{oe} - 2V_T(1 + \ln m) - R_{oe}\frac{I}{\beta_4} \qquad (13.132)$$

If the effect of variation in the charging current during the transition periods from one quasi-stable state to another is neglected, then the timing capacitor is charged, in accordance with this assumption, by the current I so that the charge time is $T_1 = CV_c/I$. The discharging current is $I + I_a/\beta_1$ and the discharge time is $T_2 = CV_c/(I + I_a/\beta_1)$, where the effect of the base current of transistor Q_1 is included. Thus, the oscillation frequency

$$f \approx \frac{I}{C\left[I_a R_{oe} - 2V_T(1 + \ln m) - R_{oe}\dfrac{I}{\beta_4}\right]\left(2 - \dfrac{I_a}{I\beta_1}\right)} \qquad (13.133)$$

If one denotes

$$f_0 = \frac{I}{2CR_{oe}I_a} \qquad (13.134)$$

which shows that this CFC has linear control, then the relative frequency error will be

$$\frac{\delta f}{f_0} \approx \frac{2V_T(1 + \ln m)}{I_a R_{oe}} + \frac{I}{I_a \beta_4} + \frac{I_a}{2I\beta_1} \qquad (13.135)$$

Using standard bipolar technology one can obtain the circuit with linear current control characteristic (Figure 13.22(c)) embracing three decades of control current ($R_1 = 5.09$ kΩ, $R_2 = 3.89$ kΩ, $V_{CC} = 3$V). The full-scale nonlinearity is less than 2% if the trigger current I_a is chosen to optimize the performance. At high values of the control current, the circuit performance deteriorates due to a low f_T value (about 3 MHz) for integrated *p-n-p* transistors in standard technologies.

The CMOS multivibrator with a current feedback CMOS Schmitt trigger (Figure 13.23) repeats the structure of its bipolar prototype. The analysis, with minor modifications that are clear from the previous section, follows the same steps. The instant of switching is characterized [52] by the voltage

$$\Delta V = V_{gs1} - V_{gs2} = \pm \sqrt{\frac{I_a}{\beta_{nt}} (1 - q)} \qquad (13.136)$$

Here $\beta_{nt} = (\mu C_{ox} W_1)/(2L_1) = (\mu C_{ox} W_2)/(2L_2)$ (that is, it is assumed that M_1 and M_2 are matched), and $q = [\sqrt{(1 + 4\beta_{nt} I_a R_{oe})} - 1]^{-1}$ and is small.

The timing capacitor voltage v_c changes between its highest value of

$$V_{cH} \approx \frac{V_{DD} R_2}{R_1 + R_2} + I_a R_{oe} + (1 - q) \sqrt{\frac{I_a}{\beta_{nt}}} \qquad (13.137)$$

and its lowest value of

$$V_{cL} \approx \frac{V_{DD} R_2}{R_1 + R_2} + (1 - q) \sqrt{\frac{I_a}{\beta_{nt}}} \qquad (13.138)$$

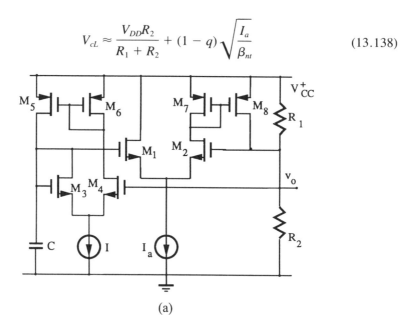

(a)

Figure 13.23 CMOS CFC with a current feedback Schmitt trigger: (a) circuit; (b) control characteristics.

(b)

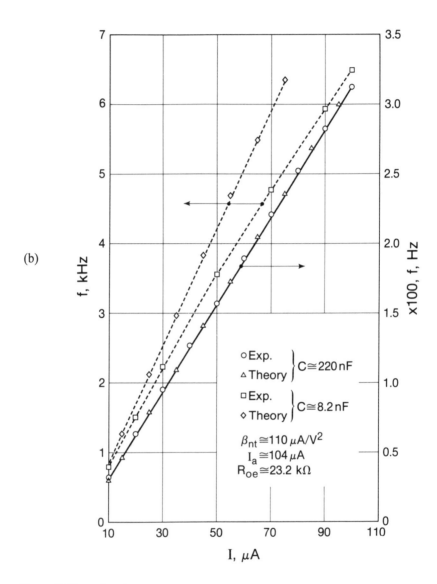

Figure 13.23 continued.

so that the voltage at the timing capacitor has the peak-to-peak value of

$$V_c \approx I_a R_{oe} - 2(1 - q)\sqrt{\frac{I_a}{\beta_{nt}}} \tag{13.139}$$

This voltage, contrary to the source-coupled multivibrator case, does not depend on the control current I or parameters of the devices through which this current flows. The timing capacitor is charged and discharged by the current I during both semiperiods. The oscillation frequency in this case (no terms in (13.139) depend on I) is

$$f_0 = \frac{I}{2CV_c} \approx \frac{I}{2C\left[I_a R_{oe} - 2(1 - q)\sqrt{\frac{I_a}{\beta_{nt}}}\right]} \tag{13.140}$$

Equation (13.140) shows that this multivibrator has linear control characteristic with I as a control current. The nonlinearity of the control characteristic is a result of switching delays. If the switching delays are constant, they can be taken into consideration by a simple correction of the result (13.140); namely, in this case

$$f = \left[\frac{2CV_c}{I} + T_s\right]^{-1} \tag{13.141}$$

where T_s is the total switching delay. Then

$$\frac{\delta f}{f_0} = -f_0 T_s \tag{13.142}$$

This result was confirmed [53] experimentally (breadboarding, MO415 CMOS transistor array devices), where the frequency was measured for the range of I from 10 to 100 μA. Two values of timing capacitor C were used: 220 and 8.2 nF. The test with the large capacitor (right scale) confirms that the theoretical and experimental results correspond and show a linear relationship between frequency and current, as expected. For CFC with the small capacitor (left scale, a higher overall frequency range) the theoretical and experimental results diverge and the experimental characteristic becomes nonlinear. The total switching delay of the circuit was measured to be about 30 μs. Subtracting this delay from the frequency characteristic returns it to a straight line.

The lower limit for the current is set by the level that could be accurately measured using the available test equipment (in fact, the circuit can operate at lower currents as well, which was confirmed for I_c down to 1 μA), and the upper limit is set by the condition

$$I < \frac{\beta_{nc}}{\beta_{nt}} \qquad (13.143)$$

This condition, specific for CMOS integrated circuits, provides that the charge current is entirely switched by the threshold circuit. In (13.143) $\beta_{nc} = (\mu C_{ox} W_3)/(2L_3)$ $= (\mu C_{ox} W_4)/(2L_4)$. If $\beta_{nc} = \beta_{nt}$, the condition (13.143) limits I to I_a.

Using the extended control range configuration [53,54], where a buffer stage is added to the basic configuration of Figure 13.23, allows CFC to be obtained with the control characteristic over a current range of 3 to 1,000 mA with the integral nonlinearity of 1.5%.

13.3.7 Duty-Cycle Modulation

A multivibrator with duty-cycle modulation was mentioned in the previous sections. The ratio T_1/T_2 is determined by the ratio of two elements (as it was in (13.49)), which is an attractive feature allowing wide element tolerance.

As was mentioned previously the frequency in the current-controlled multivibrators can be calculated with sufficient precision if the exact swing of the voltage at the timing capacitor is known. In the emitter-coupled multivibrator this voltage changes with control current, introducing a nonlinearity in frequency control characteristic. A multivibrator with duty-cycle modulation represents an attempt to use a controllable oscillation parameter that is independent of the V_c values.

The circuit of an emitter-coupled multivibrator with nonidentical current sources in the emitters of Q_1 and Q_2 (Figure 13.24(a)) oscillates with a duty cycle different from 50%. Indeed, using the model shown in Figure 13.18 with $g_{m5} = g_{m6} = 0$ and following the same calculation steps, one can find the voltage V_c at the timing capacitor at the instant of switching. The oscillation period consists of two parts, $T_1 = (2V_c)/I_1$ and $T_2 = (2V_c)/I_2$; and their ratio $T_1/T_2 = I_2/I_1$, indeed, does not depend on V_c.

The exact control characteristic, hence, depends on the circuit that controls the ratio I_2/I_1. If, for example [55], these currents are the output currents of a differential pair driven from a bridge (Figure 13.24(b)), then

$$\frac{T_1}{T_2} = \exp\left(\frac{v_2 - v_1}{V_T}\right) \approx 1 + \frac{2V_{CC}\delta R}{V_T(R_0 + 2R)} \qquad (13.144)$$

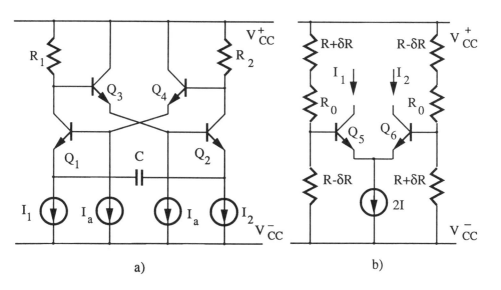

Figure 13.24 Emitter-coupled multivibrator with a controlled duty cycle: (a) multivibrator; (b) control of I_1 and I_2.

is linearly dependent on the bridge detuning when the differential signal $v_2 - v_1$ is small (which is usually the case). One can estimate the control characteristic non-linearity using the additional term of the exponent expansion.

The oscillation frequency depends on the ratio I_2/I_1 as well, but

$$f = \frac{I}{4V_c}\left[1 - \left(\frac{v_2 - v_1}{2V_T}\right)^2\right]$$
(13.145)

$$\frac{\delta f}{f_0} = -\left(\frac{v_2 - v_1}{2V_T}\right)^2$$
(13.146)

is by the order of magnitude less than the change of T_2/T_1 (the voltage is constant if the sum $I_1 + I_2 = 2I$ and constant). Here $f_0 = I/(4V_c)$.

Cascoding of an emitter-coupled multivibrator with a voltage-controlled differential pair reduces the circuit dynamic range. The reduction can be avoided by using the parallel connection of a Schmitt-trigger-type multivibrator and a voltage-controlled current source (Figure 13.25).

The multivibrator of the circuit is shown in Figure 13.25(a). Ignoring for a while the current source δI, assume that the power supply is turned *on*. The capacitor C is initially discharged and the transistor Q_1 will be *on* and Q_3 will turn *off*. The

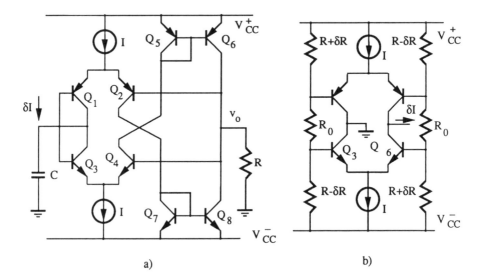

Figure 13.25 Schmitt-trigger multivibrator with a controlled duty cycle: (a) multivibrator; (b) bridge control of δI.

current I from the bottom current source will be taken by Q_4 and repeated by the current mirror Q_5, Q_6. The voltage v_o at the resistor R will jump to its positive value of $V_{oH} = IR$. The current from the top current source will charge, via Q_1, the capacitor C. When the voltage v_c attains the value of $V_{cH} \approx IR - 120$ mV, the transistor Q_1 starts to turn *off*, and Q_3 starts to turn *on*. Positive feedback will accelerate this switching, and the resistor voltage will jump to its negative value of $V_{oL} = -IR$. Now the multivibrator in its steady-state oscillation, the voltage v_c changes from V_{cH} to $V_{cL} \approx -IR + 120$ mV and vice versa.

The detailed calculation that follows the same steps as in the multivibrator with a noncomplementary trigger just considered shows that

$$V_c = V_{cH} - V_{cL} = 2IR - 2V_T(1 + \ln m) \tag{13.147}$$

where $m = \{1 + \sqrt{[1 - (4V_T)/(IR)]}\}/\{1 - \sqrt{[1 - (4V_T)/(IR)]}\}$. The oscillation frequency, hence, is $f_0 = I/(2CV_c)$.

If the parallel current source δI is in operation, the oscillation period becomes $T = T_1 + T_2$, where $T_1 = (CV_c)/(I + \delta I)$ and $T_2 = (CV_c)/(I - \delta I)$. The ratio

$$\frac{T_1}{T_2} = \frac{I + \delta I}{I - \delta I} \approx 1 - \frac{2\delta I}{I} \tag{13.148}$$

and also does not depend on V_c. The oscillation frequency becomes $f = f_0[1 - (\delta I/I)^2]$. Thus, for small values of $\delta I/I$, the ratio T_1/T_2 changes as a linear function of $\delta I/I$, and the frequency variation $\delta f/f_0 = -(\delta I/I)^2$ is one order of magnitude less.

This parallel source δI can be arranged, as in the previous circuit, as the bridge-controlled current source (Figure 13.25(b)). If the base current is neglected, one can obtain $\delta I = (8V_{CC}g_m\delta R)/(R_0 + 2R)$. Here $g_m = I_a/(2V_T)$ is the transconductance of the transistors Q_9 to Q_{12}. Then

$$\frac{T_1}{T_2} = 1 - \left[\frac{8V_{CC}\delta R}{V_T(R_0 + 2R)}\right]\frac{I_a}{I} \tag{13.149}$$

Comparing (13.148) and (13.149) one can see that in this circuit the sensitivity to the bridge detuning can be additionally controlled by the ratio I_a/I. The circuit of Figure 13.25(b) can be, with some modifications, used as a duty-cycle detector [56].

13.3.8 Controlled Multivibrators in Applications

It is feasible to design a multivibrator with a linear dependence of the oscillation frequency on the control voltage. The three decade (four decades in bipolar case) variation of frequency can be achieved with one capacitor value. The offset compensation circuits are simple. The multivibrators do not require amplitude control-stabilization circuits; hence, they are less complicated than the sinusoidal-oscillator-based circuits designed for the frequency control. The circuit integration is less difficult. The factors limiting the linearity of control characteristics are or the same (finite slew rate of the operational amplifiers) or become important at higher frequencies (switching delays).

13.4 SUMMARY

The considered circuits show that there are many ways of conversion from analog signal to frequency, pulse rate, and duty cycle. The conversion of an analog signal in the frequency of a sinusoidal oscillator is used mainly when the amplitude control system can be essentially simplified (a rectifier with a filter of a long time constant) or completely omitted. As a result the sinusoidal oscillators are combined with sensors of slow changing variables (temperature, humidity). The development of CMOS multiplier circuits (which are now in the stage of intensive investigation) with small offsets is vital for this method of conversion. Analog-to-frequency conversion using multivibrators is very flexible. The frequency can be controlled by one element (passive or active), which makes these circuits suitable for applications with many types of sensors. A resistive bridge can be also a part of multivibrator structure, which makes the multipliers even more flexible in sensor applications. These properties are preserved for multivibrators with duty cycle control. Also, in multivibrators the frequency or pulse output can be easily encoded into digital form.

REFERENCES

[1] S. Middelhoek, P. J. French, J. H. Huising, and W. J. Lian, "Sensors with Digital or Frequency Output," *Sensors and Actuators*, Vol. A15, 1988, pp. 119–133.

[2] A. Cichocki and R. Unbehauen, "Application of Switched-Capacitor Self-Oscillating Circuits to the Conversion of RLC Parameters into a Frequency or Digital Signal," *Sensors and Actuators*, Vol. A24, 1990, pp. 129–137.

[3] A. G. J. Holt and M. R. Lee, "A Class of RC-Oscillators," *Proc. IEEE*, Vol. 55, 1967, p. 1119.

[4] W. G. Howard and D. O. Pederson, "Integrated Voltage-Controlled Oscillators," *Proc. Nat. Electron. Conf.*, Vol. 23, 1967, pp. 279–284.

[5] A. B. Grebene, "A High-Frequency Voltage-Controlled Oscillator for Integrated Circuits," *Proc. Nat. Electron. Conf.*, Vol. 24, 1968, pp. 216–220.

[6] Y. Sun, "Generation of Sinusoidal Voltage (Current) Controlled Oscillators for Integrated Circuits," *IEEE Trans. Circuit Theory*, Vol. CT-19, 1972, pp. 137–141.

[7] J. G. Graeme, *Applications of Operational Amplifiers*, New York, McGraw-Hill, 1973.

[8] M. Herpy, *Analog Integrated Circuits*, New York, John Wiley & Sons, 1980.

[9] V. P. Singh and S. K. Saha, "Linear Sinusoidal VCO," *Int. J. of Electronics*, Vol. 65, 1988, pp. 243–247.

[10] I. M. Filanovsky, S.-S. Qiu, and G. Kothapalli, "Sinusoidal Oscillator with Voltage Controlled Frequency and Amplitude," *Int. J. of Electronics*, Vol. 68, 1990, pp. 95–112.

[11] *Intersil Data Book*, 1981.

[12] *Linear Databook*, National Semiconductor Corporation, 1982.

[13] D. Meyer-Ebrecht, "Schnelle amplitudenregelung harmonischer Oscillatoren," *Philips Research Rep. Supplement*, Vol. 6, 1974, pp. 1–85.

[14] I. M. Filanovsky, G. J. Fortier, and L. F. Taylor, "Amplitude Transients in the Twin-T Bridge RC-Oscillator with a Multiplier Amplitude Control System," *Int. J. Electronics*, Vol. 64, 1988, pp. 547–561.

[15] E. Vannerson and K. C. Smith, "A Low Distortion Oscillator with Fast Amplitude Stabilization," *Int. J. Electronics*, Vol. 39, 1975, pp. 465–472.

[16] E. Vannerson and K. C. Smith, "Fast Amplitude Stabilization of an RC-Oscillator," *IEEE J. Solid-State Circuits*, Vol. SC-9, 1974, pp. 176–179.

[17] J. Pahor, J. Fettich, and M. Tavzes, "A Harmonic Oscillator with Low Harmonic Distortion and Stable Amplitude," *Int. J. Electronics*, Vol. 37, 1974, pp. 765–768.

[18] N. Vannai and L. Pap, "RC Oscillators with Extremely Low Harmonic Distortion," *Periodica Polytechn., Elect. Eng.* (Hungary), Vol. 24, No. 1–2, 1980, pp. 59–65.

[19] I. M. Filanovsky, "A Wien-Bridge RC-Oscillators with Fast Amplitude Control," *Int. J. Electronics*, Vol. 58, 1985, pp. 817–826.

[20] I. M. Filanovsky, "Oscillators with Amplitude Control by Restoration of Capacitor Initial Conditions," *IEEE Proc.*, Vol. 134, No. 1, 1987, pp. 31–37.

[21] D. Meyer-Ebrecht, "Fast Amplitude Control of a Harmonic Oscillator," *Proc. IEEE*, Vol. 60, No. 6, 1972, p. 736.

[22] P. R. Gray and R. G. Meyer, *Analysis and Design of Analog Integrated Circuits*, 2d ed., New York, J. Wiley & Sons, 1984.

[23] B. Van der Pol, "The Nonlinear Theory of Electric Oscillations," *Proc. IRE*, Vol. 22, 1934, pp. 1051–1086.

[24] H. S. Tan and K. W. H. Foulds, "Analytical Determination of the Waveforms of Nonlinear Oscillators," *Electronic Circuits and Systems*, Vol. 1, 1977, pp. 99–102.

[25] A. Grebene, *Bipolar and MOS Analog Integrated Circuit Design*, New York, J. Wiley & Sons, 1984.

[26] A. Kindlund, H. Sundgren, and I. Lundstrøm, "Quartz Crystal Gas Monitor with a Gas Concentrating Stage," *Sensors and Actuators*, Vol. A6, 1984, pp. 1–17.

[27] R. M. Landon, "Resonator Sensors," *J. Phys. E: Sci. Instrum.*, Vol. 18, 1985, pp. 103–115.

[28] J. D. Maines, E. G. S. Paige, A. F. Saunders and A. S. Young, "Simple Technique for the Accurate Determination of Delay-Time Variations in Acoustic-Surface-Wave Structures," *Electron. Letters*, Vol. 5, No. 26, 1969, pp. 678–680.

[29] J. Crabb, M. F. Lewis, and J. D. Maines, "Surface-Acoustic-Wave Oscillators: Mode Selection and Frequency Modulation," *Electron. Letters*, Vol. 9, No. 10, 1973, pp. 195–197.

[30] A. Venema, E. Nieuwkoop, M. J. Vellekoop, M. S. Nieuwenhuizen, and A. W. Barendsz, "Design Aspects of SAW Gas Sensors," *Sensors and Actuators*, Vol. A10, 1986, pp. 47–64.

[31] R. Best, *Phase-Locked Loops*, New York, McGraw-Hill, 1984.

[32] D. H. Sheingold, ed., *Transducer Interfacing Handbook*, Norwood, MA, Analog Devices, 1980.

[33] J. H. Huising, G. A. Van Rossum and M. Van der Lee, "Two-Wire Bridge-to-Frequency Converter," *IEEE J. Solid-State Circuits*, Vol. SC-22, 1987, pp. 343–349.

[34] G. B. Clayton, *Operational Amplifiers*, 2d ed., London, Newnes-Butterworths, 1979.

[35] I. M. Filanovsky, V. A. Piskarev and K. A. Stromsmoe, "Nonsymmetric Multivibrators with an Auxiliary RC-Circuit," *IEE Proc.*, Vol. 131, No. 4, 1984, pp. 141–146.

[36] I. M. Filanovsky, "A Simple Voltage-to-Frequency Converter," *IEEE Circuits and Devices*, Vol. 6, No. 5, 1990, p. 47.

[37] I. M. Filanovsky and V. A. Piskarev, "A Simple Bridge RC Multivibrator," *Int. J. Electronics*, Vol. 57, No. 2, 1984, pp. 217–226.

[38] T. M. Frederiksen, *Intuitive Operational Amplifiers*, New York, McGraw-Hill, 1988.

[39] T. J. Van Kessel and R. J. Van de Plassche, "Integrated Linear Basic Circuits," *Philips Tech. Review*, Vol. 32, No. 1, 1971, pp. 1–12.

[40] I. M. Filanovsky, "Remarks on Design of Emitter-Coupled Multivibrators," *IEEE Trans. Circ. Syst.*, Vol. CAS-35, 1988, pp. 751–755.

[41] R. R. Cordell and W. G. Garrett, "A Highly Stable VCO for Application in Monolithic Phase-Locked Loops," *IEEE J. Solid-State Circ.*, Vol. SC-10, No. 10, 1975, pp. 480–485.

[42] B. Gilbert, "A Versatile Monolithic Voltage-to-Frequency Converter," *IEEE J. Solid-State Circuits*, Vol. SC-11, No. 6, 1976, pp. 852–864.

[43] F. V. J. Sleeckx and W. M. C. Sansen, "A Wide-Band Current-Controlled Oscillator Using Bipolar-FET Technology," *IEEE J. Solid-State Circuits*, Vol. SC-15, No. 5, 1980, pp. 875–881.

[44] I. G. Finvers and I. M. Filanovsky, "Analysis of a Source-Coupled CMOS Multivibrator," *IEEE Trans. Circ. Syst.*, Vol. CAS-35, 1988, pp. 1182–1185.

[45] R. Gregorian and G. C. Temes, *"Analog MOS Integrated Circuits for Signal Processing,"* New York, J. Wiley & Sons, 1986.

[46] P. E. Allen and D. R. Holberg, *CMOS Analog Circuit Design*, New York, Holt, Rinehart and Winston, 1987.

[47] *Analog MOS Integrated Circuit Design Manual*, Scotts Valley, CA, Interdesign Corp., 1983.

[48] A. Nathan, I. A. McKay, I. M. Filanovsky, and H. P. Baltes, "Design of a CMOS Oscillator with Magnetic Field Frequency Modulation," *IEEE J. Solid-State Circuits*, Vol. SC-21, 1987, pp. 230–232.

[49] C. C. M. Meijer, "A Three-Terminal Wide-Range Temperature Transducer with Microcomputer Interface," European Solid-State Circuits Conf., Delft, The Netherlands, 1986.

[50] J. F. Kukielka and R. G. Meyer, "A High-Frequency Temperature Stable Monolithic VCO," *IEEE J. Solid-State Circuits*, Vol. SC-16, 1981, pp. 639–647.

[51] I. M. Filanovsky, "A Current Controlled Multivibrator for Low Voltage Power Supply," *Int. J. Electronics*, Vol. 65, No. 1, 1988, pp. 37–43.

[52] I. M. Filanovsky and I. G. Finvers, "A Simple Nonsaturated CMOS Multivibrator," *IEEE J. Solid-State Circuits*, Vol. SC-23, No. 1, 1988, pp. 290–292.

[53] I. G. Finvers, "CMOS Current Controlled Oscillators," MSc thesis, University of Alberta, Edmonton, Canada, 1988.

[54] I. M. Filanovsky, I. G. Finvers, Lj. Ristic, and H. P. Baltes, "Multivibrators with Frequency Control for Application in Integrated Sensors," Proc. *First Int. Forum on ASIC and Transducer Technology* (ASICT'88), Honolulu, 1988, p. 65.

[55] R. R. Spencer and J. B. Angell, "A Voltage-Controlled Duty-Cycle Oscillator," *IEEE J. Solid-State Circuits*, Vol. SC-25, No. 1, 1990, pp. 274–281.

[56] I. M. Filanovsky and V. A. Piskarev, "Duty Cycle Detector," *IEEE Circuits and Devices*, Vol. 7, No. 1, 1991.

About the Authors

Walter Allegretto received a B.A.Sc. degree in electrical engineering in 1965 and a Ph.D. degree in mathematics in 1969 from the University of British Columbia, Vancouver, Canada. He joined the mathematics department of the University of Alberta, Edmonton, Canada, in 1970. He is presently a professor of mathematics and an adjunct professor of electrical engineering. In the past, he has also served as associate chair for graduate studies and research. His research deals with the theoretical and practical problems associated with partial differential equations. He is the author or coauthor of approximately 100 publications and is a member of several scientific societies, including the IEEE, the American Mathematical Society, and the Society for Industrial and Applied Mathematics. His current interests include the analysis and simulation of semiconductor devices with special emphasis on microsensors.

Frank Secco d'Aragona received a Ph.D. in geological sciences from the University of Milan, Italy. From 1963 to 1967, he was with the Solid-State Physics Department of CEN-SCK, Mol, Belgium, where he conducted transmission electron microscopy studies of lattice defects in crystalline materials. Before joining Motorola, he was with the Dow Corning Corporation in Midland, Michigan, where he worked on defect characterization in silicon crystals with particular emphasis on etching techniques. In 1974 he joined Motorola, where he has been working on various aspects of silicon semiconductor technology, including epitaxy, crystal growth, silicon purification, photovoltaics, material characterization, and materials/device correlation studies. Dr. Secco d'Aragona has published 20 papers, holds six patents in silicon technology, and is an active member of the Electrochemical Society. He is currently a principal staff engineer in the Materials Research Organization, working on the developmental aspects of silicon direct wafer bonding.

William Dunn graduated from the University of London in 1956 with a B. Sc. (Hons.) degree in physics. He has been with Motorola, the Semiconductor Products Sector for seven years as a circuit design engineer in the Advanced Custom

Technologies Group, working on power devices and sensor signal conditioning circuits. Before joining Motorola, he worked on circuit design and system development in the United States, Canada, and the U.K. He has over 20 patents and has published over 20 papers.

Dr. I. M. Filanovsky received his M.Sc. and Ph.D. degrees in electrical engineering from V.I. Ulianov (Lenin) University of Electrical Engineering, St. Petersburg, Russia (previously USSR). He worked as a senior research scientist in the Research Institute of High-Frequency Currents and in the Girioond Research Institute, St. Petersburg, Russia. He joined the University of Alberta, Canada, in 1976, where he is currently a professor. Dr. Filanovsky is a coauthor of one book on impedance converters (in Russian), and the author and a coauthor of about 150 journal and conference publications on circuit theory (theory of approximation, theory and technical applications of oscillations, and strongly nonlinear oscillations) and applied microelectronics (analog electronic circuits, oscillator and multivibrator circuits, and circuits for sensors). He has three patents on electronic circuits. Dr. Filanovsky is a Senior Member of IEEE, a member of New York Academy of Sciences and a member of Canadian Society for Electrical and Computer Engineering. In 1993 he was a guest editor of the sensors and circuits issue of the *International Journal of Integrated Circuits and Signal Processing*.

Randy Frank is a technical marketing manager for Motorola's Semiconductor Products Sector in Phoenix, Arizona. He has a BSEE, MSEE, and MBA from Wayne State University in Detroit, Michigan, and over 25 years experience in automotive and control systems engineering. For the past ten years he has been involved with semiconductor sensors, power transistors, and smart power ICs. Prior to joining Motorola, he worked for Chrysler Corporation and American Motors Corporation, where he was concerned with various engine electrical and electronic control systems. He is currently a member of SAE, the SAE Sensor Standards and Dual Voltage Committee, and a member of IEEE and the IEEE Sensor Terminology Taskforce. He has taught advanced instrumentation and control at the University of Michigan, has written more than 100 papers and presented numerous papers at technical conferences, and has several patents issued and pending on various aspects of control systems and semiconductor technology.

Jon Geist received a BS in theoretical and applied science from George Washington University and a Ph.D. in electrical engineering from the University of Alberta. He worked in various areas of radiometric, photodiode, and semiconductor physics at the National Bureau Standard from 1964 to 1991, and is the author of over 90 technical papers in these areas, including the photometry and radiometry articles in the 1987 edition of the *McGraw-Hill Encyclopedia of Science and Technology*. His current interests include microsystems and photodiode physics.

Henry G. Hughes holds a B.S. degree in chemistry and a Ph.D. in organic chemistry. He has been employed in the development centers of the Semiconductor Products Sector of Motorola for 23 years. He has held the positions of senior scientist, project leader, section manager, and member of the technical staff, and is currently a senior member of the technical staff. His experience includes photoresist and photopolymer technology, plasma photoresist development, spin-on glass chemistries and manufacturing, and high-voltage passivation techniques. In his present assignment, he heads the Micromachining group in the Sensor organization of the Advanced Custom Technologies center. He has taught classes in chemistry and chemical wafer processing at Motorola. Dr. Hughes has over 24 publications, is coeditor of a proceedings, and has chaired symposia for the Electrochemical Society. He holds several patents. He is a member of the Electrochemical Society (treasurer of the Dielectric Science and Technology Division), the American Chemical Society, and Motorola's Scientific & Technical Society, and holds an honorary doctorate of science from Quincy College.

Dr. Margaret L. Kniffin received her S.B. in materials science from the Massachussetts Institute of Technology in 1983 and her Ph.D. degree in materials science from Stanford University in 1991. She joined Motorola after completing postdoctoral studies in the Department of Electrical Engineering at Stanford. Dr. Kniffin is currently a principal staff engineer with Motorola's Advanced Custom Technologies Group. Her research interests include process development for surface-micromachining, materials for sensor applications, and sensor package development.

James R. Janesick received his masters degree in electronic engineering from the University of California at Irvine, majoring in lasers and masers. Since January, 1973, he has been working at the Jet Propulsion Laboratory—California Institute of Technology as group leader developing charge-coupled devices for the wide field/planetary camera for the Hubble space telescope, the Galileo solid-state imaging camera, and other NASA space imaging systems. He is author of 40 papers on the subject of CCDs, was guest editor for *Optical Engineering* for three special issues on CCDs, has contributed to several NASA technical briefs, and holds three patents on various CCD innovations. He received the Exceptional Engineering Achievement medal from NASA in 1982 and a NASA Achievement Award in 1986. He is currently involved in the design and development of ultra-low-noise CCDs used in the near IR, visible, UV, EUV, and soft X-ray as well as pursing narrow-bandgap monolithic CCD sensors for the far infrared.

Wolfgang Rasmussen is a professor at the College de Sherbrooke in Quebec and a member of the Center for Research in Solid State Physics and the Group for Component and Microstructure Research at the Université de Sherbrooke. Dr. Rasmussen earned his Ph.D. in elementary particle physics at the University of Colorado, a master's degree in electrical engineering at the Université de Sherbrooke and a B.S. degree at New Mexico State University. He has postdoctoral experience at the Johns Hopkins University and held an A. von Humboldt research fellowship at the University of Cologne. He has published in elementary particle physics, quantum optics, mathematical physics, and CMOS thermal sensors. He has been involved in the design, layout, verification, and testing of numerous sensor-related CMOS chips. His present research interests are focussed on GaAs and related materials research, nanostructures, and applications.

Ljubisa Ristic earned his B.Sc., M.Sc., and Ph.D. degrees, all in electrical engineering, from Nis University, Yugoslavia. Currently, Dr. Ristic is a manager and a senior member of the Technical Staff at Motorola Inc. in Phoenix, where he conducts R&D activities in development of sensor technology and devices. He joined Motorola in 1990. From 1985 to 1990, he was a professor in the department of electrical engineering at the University of Alberta, Canada, where he taught undergraduate and graduate courses on VLSI processing, the physics of semiconductor devices, switching electronics, IC circuits, and microsensor devices and technology. He also served as an R&D engineer, a chief engineer, and an R&D manager in the Semiconductor Device Division of the Electronics Industry Corporation, Yugoslavia, from 1975 to 1985. His research interests have included theoretical and experimental study of Zener diodes, power and Schottky rectifiers, Si/SiO_2 interfaces, and MIS structures. The phenomena related to planar transistors including passivation technology of planar structures, power transistors, and both bipolar and CMOS IC circuits have been among the subjects of his research. His research activities have also included the failure mechanisms and reliability investigation of semiconductor devices. In the last several years he has been involved in the research and development of magnetic field sensors, and micromachining technology and related devices. He is the author or coauthor of more than 70 scientific papers and 11 issued or pending patents. Dr. Ristic has given invited lectures in the United States, Canada, Europe, and Japan, and chaired several conferences on microelectronics and related subjects.

Raymond M. Roop received his B.S. in physics from Ohio State University and his M.S. and Ph.D. from the University of Illinois. He joined Motorola in 1973 and has worked in research and development in power transistors, linear integrated circuits, smart power, and sensor devices. He is currently manager of sensor development in Motorola's Advanced Custom Technologies Center in Mesa, Arizona. Dr. Roop is a member of the American Physical Society, IEEE, Sigma XI, and Phi Beta Kappa.

A. Russell Schaefer has 25 years of experience in atomic/molecular physics and astronomy, radiometric physics, photometry, silicon photodiode physics, electro-optics, dye lasers and dye laser spectrophotometry, electron storage rings, adaptive optics, and CCD physics and development. Since July 1987, Dr. Schaefer has been working for Science Applications International Corporation (SAIC), San Diego, CA, as a senior staff scientist responsible for optical characterization and testing procedures, sensor and sensor system characterization, and electro-optical system design and integration. He is involved with designing new CCDs and new approaches to CCD technology. From 1986 to 1987, Dr. Schaefer worked at Western Research Corporation (now ThermoElectron Technologies), San Diego, CA, as chief scientist. He was responsible for optical measurement procedures, sensor characterization, and optical systems integration. From 1970 to 1986, Dr. Schaefer worked at the National Bureau of Standards (now the National Institutes of Standards and Technology), Washington, D.C. His technical contributions while at NBS included the inception and development of a joint program with NASA Ames Research Center to measure relative star brightnesses to previously unattained levels of precision using solid-state detectors. From 1974 to 1986, he was an adjunct professor of physics and astronomy at Montgomery College in Rockville, Maryland, teaching various night courses in physics and astronomy. He has given invited lecture tours in Japan, China, and Europe, and authored 40 papers. Dr. Schaefer is a member of ASA, SPIE, Sigma Xi Research Society and Phi Beta Kappa.

Frank Shemansky received his B.S. degree in chemical engineering from Pennsylvania State University in 1983 and his M.S. degree in chemical engineering from Arizona State University in 1988. His masters research emphasized transport phenomena in biological systems, specifically oxygen delivery to the ischemic myocardium. In 1991, he received his Ph.D. in chemical engineering from Arizona State University. His doctoral dissertation focused on theoretical and experimental studies in low-pressure chemical vapor deposition. Upon graduating in 1991, Dr. Shemansky accepted a position at Motorola in the Semiconductor Research and Development Laboratories and is currently a senior staff engineer in the Advanced Custom Technologies microelectronic sensors research and development group.

Index